Handbook of Practical Electronic Circuits

John D. Lenk
Consulting Technical Writer

Handbook of Practical Electronic Circuits

Prentice-Hall, Inc., Englewood Cliffs, NJ 07632

Library of Congress Cataloging in Publication Data

Lenk, John D.
Handbook of practical electronic circuits.

Includes index.
1. Electronic circuits. 2. Transistor circuits.
I. Title.
TK7867.L46 621.3815'3 81-2227
ISBN 0-13-380741-X AACR2

Editorial/production supervision
and interior design: Barbara Bernstein
Manufacturing buyer: Gorden Osbourne

Printed in the United States of America

10 9 8 7 6 5 4 3 2 1

Prentice-Hall International, Inc., *London*
Prentice-Hall of Australia Pty. Limited, *Sydney*
Prentice-Hall of Canada, Ltd., *Toronto*
Prentice-Hall of India Private Limited, *New Delhi*
Prentice-Hall of Japan, Inc., *Tokyo*
Prentice-Hall of Southeast Asia Pte. Ltd., *Singapore*
Whitehall Books Limited, *Wellington, New Zealand*

To Irene, the "Sandpiper Lady"
and
Mr. Lamb, the "Magic Lamb"

Thank you for making me the number one best-selling technical author!

To Irene, the "Sandpiper Lady"
and
Mr. Lamb, the "Magic Lamb"

Thank you for making me the number one best-selling technical author!

Contents

Chapter Two

RADIO-FREQUENCY CIRCUITS **55**

Chapter Three

POWER-SUPPLY CIRCUITS **81**

Chapter Four

OSCILLATOR CIRCUITS **104**

Chapter Five

FILTER, ATTENUATOR, AND WAVEFORMING CIRCUITS **127**

Chapter Six

SWITCHING CIRCUITS **162**

Chapter Seven

ELECTRONIC CONTROL CIRCUITS **181**

Chapter Eight

UNIJUNCTION AND PROGRAMMABLE UNIJUNCTION TRANSISTOR CIRCUITS **216**

Chapter Nine

OPERATIONAL-AMPLIFIER CIRCUITS **248**

Preface

By the time you have finished reading this handbook, you should be able to recognize over 270 circuits commonly used in all phases of electronics. Not only will you be able to recognize the circuits very quickly, but you will understand how the circuits operate and where they fit into electronic equipment and systems. These two bits of information alone make this handbook an excellent one-stop source or reference for anyone (student, experimenter, technician, or designer) involved with electronic circuits.

However, this handbook goes much further than a mere collection of circuits and theoretical descriptions. Although not intended as a design work, the book also describes the details of circuit design in very practical terms. In some instances, this is done by describing the design problems associated with the particular circuit. In other instances, there is a simple, yet complete, explanation of the "why" and "how" of circuit performance. In any event, using the information contained in this handbook, you can put the circuits to work "as is" or you can change circuit values (thus altering circuit characteristics) to suit your own requirements.

In some cases, actual circuits using proven component values are given in full detail. These are the circuits that can be put to immediate use. In other cases, basic circuits are given, with details as to how circuit values are to be selected to meet a certain circuit performance goal. This information is particularly useful to the student and experimenter.

For example, Chapter 7 includes many circuits involving the use of the currently very popular control rectifiers, such as SCRs, SCSs, triacs, diacs, SUSs, SBSs, LASCRs, LASCSs, and opto-couplers. Chapter 7 also includes full details of practical circuits for *interfacing* between digital electronic equipment and various control systems.

This handbook assumes that the reader has a basic knowledge of transistors and schematics at a level found in the author's best-selling *Understanding Electronic Schematics* and is an excellent follow-up to that work. However, no reference to any of the author's books is required to make full use of this handbook.

Since the handbook does not require advanced math or theoretical study, it is ideal for the experimenter and the engineering designer. On the other hand, the book is suited to schools where the basic teaching approach is circuit analysis with an emphasis on practical circuit design.

Many professionals have contributed their talent and knowledge to the preparation of this book. The author gratefully acknowledges that the tremendous amount of effort needed to make this book so comprehensive would not be possible for one person to undertake, and he wishes to thank all who have contributed directly and indirectly.

The author wishes to give special thanks to the following: Lothar Stern of Motorola Semiconductor Products, Inc.; Walter Dennen of RCA Corporation, Solid State Division; Ross Snyder of Hewlett-Packard; Texas Instruments Incorporated; and Joseph A. Labok of Los Angeles Valley College.

John D. Lenk

Handbook of Practical Electronic Circuits

Chapter 1

Audio-Frequency Circuits

This chapter is devoted entirely to audio-frequency (AF) circuits. These circuits include basic bias, basic two-junction transistor amplifiers, multistage transistor amplifiers, amplifiers with transformer coupling, amplifiers using plastic transistors, and AF operating/adjustment controls. We also discuss the effect of audio-amplifier components on frequency, amplifier coupling circuits, and amplifier classifications.

1-1 BASIC BIAS CIRCUITS

All transistors (and other solid-state devices) require a form of bias. In a typical transistor, the collector-base circuit must be reverse-biased. That is, current should not flow between collector and base. Any collector-base current that does flow is the result of leakage or breakdown. Breakdown must be avoided by proper design. Leakage is an undesirable (but almost always present) condition that must be reckoned with in design.

The emitter-base circuit of a typical transistor must be forward-biased. That is, current should flow between base and emitter during normal operation. Some current flows all the time in certain circuits (class A amplifiers and most oscillators). In other circuits (class C amplifiers, switches, etc.) current flows only in the presence of an operating signal or trigger. Either way, the emitter-base circuit must be biased so that current can flow.

The desired bias is accomplished by applying voltages to the corresponding transistor elements, usually through resistances. The following sections describe several basic bias circuits. These circuits (or variations of them) represent most of the bias methods used in transistor circuits.

Keep one point in mind when studying the following bias circuits: The purpose of the basic bias circuit is to establish collector-base-emitter voltage and current relationships at the *operating point* of the circuit. (The operating point is also known as the quiescent point, *Q*-point, no-signal point, idle point, or static point.) Since transistors rarely operate at this static point, the basic bias circuits are used as a reference or starting point for design. The actual circuit configuration and (especially) the bias circuit values must be selected on the basis of dynamic circuit conditions (desired output voltage swing, expected input signal level, etc.).

1-1-1 Basic-Bias-Design Considerations

The first step in bias circuit design is to determine the characteristics of both the circuit and the transistor to be used. For example, will the circuit be used as an amplifier, oscillator, or switch? What class of operation (A, AB, B, or C) is required? How much gain (if any) is required? What power supply voltages are available? What transistor is to be used? Must input and or output impedances be set at an arbitrary value?

1-1-2 Operating Load Current and Resistance

Once the basic design questions have been answered, the next step is to calculate the operating load current and resistance. In some circuits, an arbitrary load resistance must be used (to achieve a given circuit output impedance, for example). In other circuits, an arbitrary load current is required (to obtain a given power output). In the following examples, it is assumed that both load current and load resistance are chosen solely to obtain a given operating point. Usually, the no-signal collector voltage is one-half the supply voltage at the operating point.

Two major factors determine operating load current for a transistor: leakage current and maximum rated current. Obviously, the load current cannot exceed the maximum rated current for the transistor, Similarly, the load current must not be less than the leakage current, or no current will flow through the load. If the circuit is battery-operated, an operating load current near the low end is usually selected to minimize battery drain. Often, the operating load current is midway between the two extremes (leakage current and maximum rated current).

Once the operating load current has been established, the next step is to calculate the load resistance value. When operating at frequencies up to about 100 kHz, the load resistance value can be calculated on a d-c basis. At higher frequencies, it may be necessary to select a load resistance on the basis of impedance. This is discussed in later chapters, particularly in Chapter 2.

When the operating load current is flowing at the selected no-signal point, the collector load resistance should drop the collector voltage to one-half the supply voltage. This is typical for all class A circuits. (Operating classes of amplifier circuits are discussed further in Sec. 1-4.)

1-1-3 Bias with Emitter Feedback

Once the desired (or arbitrarily selected) load current and load resistance have been determined, the final step in the basic-bias-circuit design is to select the resistances for the emitter and base that will produce the correct operating point. Although there are many bias networks, each with its own advantages and disadvantages, one major factor must be considered for any network. The basic network must maintain the desired base current in the presence of temperature (and frequency, in some cases) changes. This is often referred to as *bias stability*.

There are several methods to provide termperature and frequency compensation for bias circuits. One of the most effective is with *emitter feedback*. Note that all the bias circuits discussed in this section use some form of emitter resistance to produce feedback.

The use of an emitter-feedback resistance in any bias circuit can be summed up as follows. Base current (and, consequently, collector current) depends upon the differential in voltage between base and emitter. If the differential voltage is lowered, less base current (and, consequently, less collector current) will flow. The opposite is true when the differential is increased. All current flowing through the collector (ignoring collector-base leakage) also flows through the emitter resistor. The voltage drop across the emitter resistor is therefore dependent (in part) on the collector current.

Should the collector current increase for any reason, both the emitter current and the voltage drop across the emitter resistor will also increase. This *negative feedback* tends to decrease the differential between base and emitter, thus lowering the base current. In turn, the lower base current tends to decrease the collector current and to offset the initial collector-current increase.

1-1-4 Determining Base Current

If the exact base current required to produce a given collector current is known, bias-circuit design is a simple matter. Unfortunately, the exact relationship between base and collector currents (or gain) is never known. Nor can data sheets be trusted for exact gain information. Gain is temperature- and frequency-dependent as well as dependent upon circuit values.

There are two basic ways to find the approximate base current that will produce a given operating point. The first method involves a *load line* drawn on static collector characteristic curves, as shown in Fig. 1-1. As indicated for the transistor shown in Fig. 1-1, a base current of about 0.2 mA produces a collector current of 2.4 mA. This indicates a beta of about 12. Note that the load line is drawn between the source or supply voltage (20 V) and the maximum permitted current (5 mA). (Note that the values shown in Fig. 1-1 apply only to a 2N332 NPN transistor.)

No matter what method is used, keep in mind that the values are approximate. Therefore, the selected bias resistance values are just as approximate. In addition, all resistors have some tolerance for their values (usually 5 or 10 percent).

In practical applications, when a bias circuit is being designed "from scratch," the bias resistance values are calculated on paper. Then the circuit is connected in experimental form using the calculated resistance values. Power is applied and the voltages (collector, base, and emitter) are measured. Then the resistance values are trimmed to produce the desired results. In the case of a bias circuit, "desired results" are a given collector voltage and current for a given load. When the basic bias circuit is used in another circuit, "desired results" can be overall circuit functions (such as gain for an amplifier, output voltage for a oscillator, etc.). This aspect of bias adjustment is discussed throughout the book.

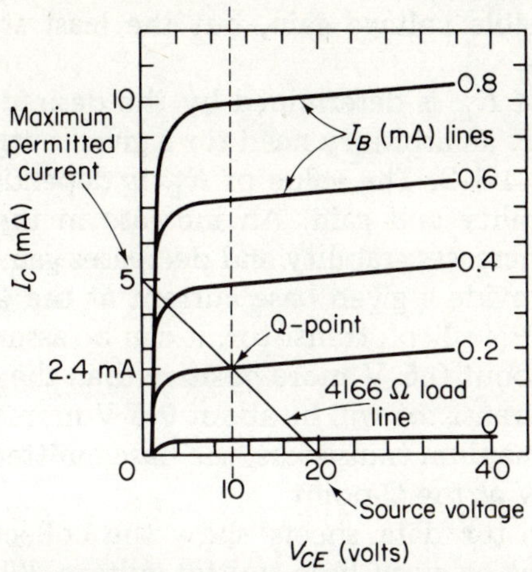

Figure 1-1. Load line drawn on static characteristic curves to find approximate base current and Q-point.

1-1-5 Bias Circuit A

The basic characteristics for bias circuit A are shown on Fig. 1-2 and summarized as follows. The output impedance is approximately equal to R_L (at frequencies up to about 100 kHz). The input impedance is approximately equal to beta times R_E (at frequencies up to about 100 kHz). Since input impedance is dependent on beta, the input impedance is subject to wide variation. The circuit current gain is approximately equal to the a-c beta of the transistor. The voltage gain is approximately equal to the ratio R_L/R_E. Bias circuit A offers

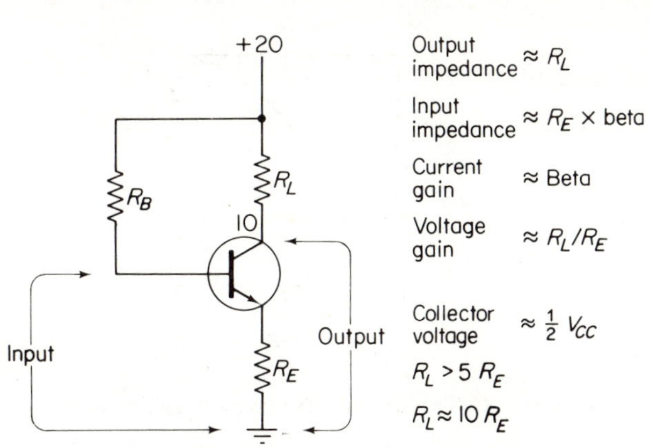

Figure 1-2. Bias circuit A.

the greatest possible voltage gain, but the least stability, of all the basic bias circuits.

The value of R_L is determined by the desired collector voltage and current, or by an arbitrary need for a given output impedance, as discussed in Sec. 1-1-2. The value of R_E is dependent upon a trade-off between stability and gain. An increase in the value of R_E, in relation to R_L, increases stability and decreases gain. The value of R_B is selected to provide a given base current at the Q-point (no-signal point). For an NPN silicon transistor, it can be assumed that the base voltage will be about 0.5 V more positive than the emitter. The base of a PNP silicon transistor will be about 0.5 V more negative than the emitter. For germanium transistors, the base-emitter voltage differential is about 0.2 V at the Q-point.

Some transistor data sheets show the collector current for a given base current or given base-emitter voltage. When such information is available on data sheets, it is shown in the form of curves such as in Fig. 1-3. These curves, usually called *transfer characteristic curves*, make it possible to calculate a more accurate value of R_B. However, in practice, the calculated value of R_B is only a trial value.

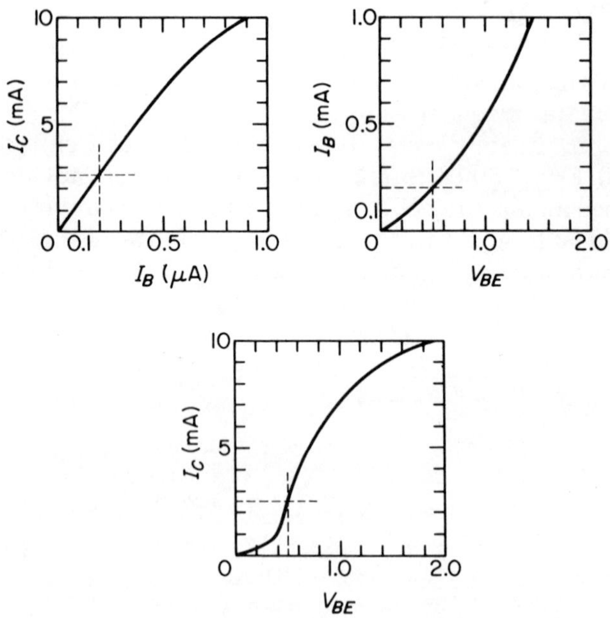

Figure 1-3. Typical transfer characteristic curves.

1-1-6 Bias Circuit B

The basic characteristics for bias circuit B are shown in Fig. 1-4 and are essentially the same as for bias circuit A, except that stability is increased. The increase in stability is brought about by connecting base resistance R_B to the collector rather than to the source. If the collector current increases for any reason, the drop across R_L increases, lowering the voltage at the collector. This lowers the base voltage and current, thus reducing the collector current. The feedback effect is combined with that produced by the emitter resistor R_E (Sec. 1-1-3) to offset any variations in collector current. However, gain for bias circuit B is only slightly less than for circuit A.

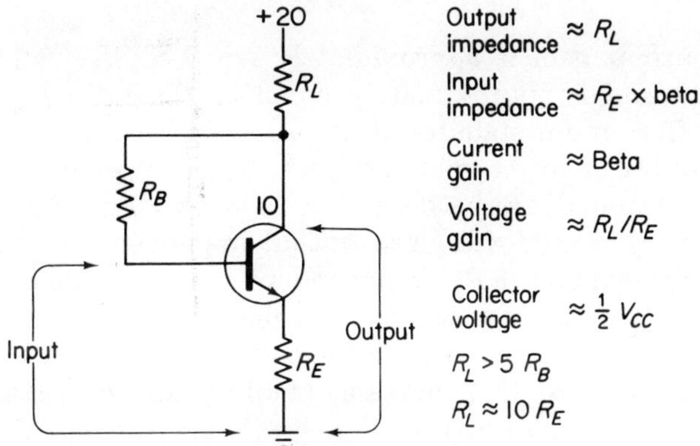

Figure 1-4. Bias circuit B.

1-1-7 Bias Circuit C

The basic characteristics for bias circuit C are shown in Fig. 1-5 and summarized as follows. The output impedance is approximately equal to R_L (at frequencies up to about 100 kHz). The input impedance is approximately equal to R_B (at frequencies up to about 100 kHz). Actually, the input impedance is equal to R_B in parallel with R_E times (beta + 1). However, unless beta is very low, the R_E × (beta + 1) factor is much greater than R_B. Therefore, the value of R_B (or slightly less) can be considered as the stage or circuit input impedance and will be so considered when the bias circuit C or any of its variations appear in examples throughout this book.

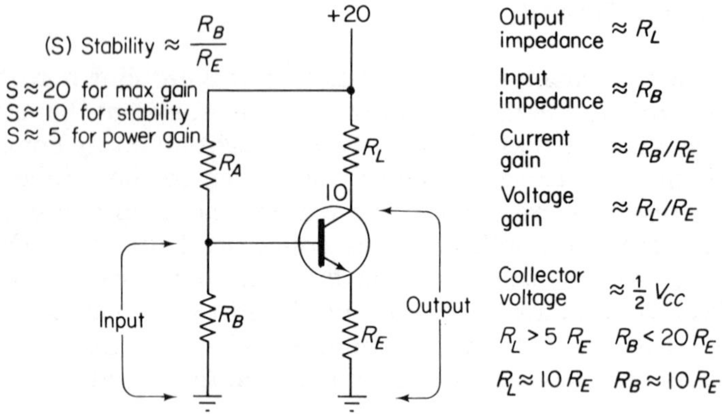

Figure 1-5. Bias circuit C.

The current gain is approximately equal to the ratio R_B/R_E. The voltage gain is approximately equal to the ratio R_L/R_E. Bias circuit C offers more stability than either bias circuit A or B, but with a trade-off of lower gain and lower input impedance. The value of R_L is determined by the desired collector voltage and current, or by an arbitrary need for a given output impedance, as discussed in Sec. 1-1-2. The value of R_E is dependent upon a trade-off between stability and gain. An increase in the value of R_E, in relation to R_L, increases stability and decreases voltage gain. An increase in the value of R_E, in relation to R_B, increases stability and decreases current gain.

The value of R_B is dependent upon trade-offs among the value of R_E, current gain, stability, and the desired input impedance. If the input impedance is of special importance in the circuit, the value of R_B is selected on that basis. This may require a different value of R_E to maintain stability and current-gain relationships. Of course, any change in R_E will result in a change of the voltage gain (assuming that R_L remains the same). In addition to stability, the major advantage of bias circuit C is that the input and output impedances, as well as voltage and current gain, are not dependent upon transistor beta. Instead, circuit characteristics are dependent upon circuit values.

1-1-8 Bias Circuit D

Figure 1-6 is the diagram for bias circuit D. The basic characteristics for bias circuit D are the same as for circuit C except that temperature stability is increased. The increase in temperature stability is brought about by connecting diode *D* between the base and R_B. Diode *D* (forward-biased) is of the same material (silicon or germanium) as the base-emitter junction and is maintained at the same temperature. Thus, the voltage drops across diode *D* and the base-emitter junction are the same and remain the same with changes in temperature.

The values for R_L, R_E, and R_B in circuit D are the same as for circuit C. The value for R_A is slightly different. Since the drop across diode D is the same as across the base-emitter junction, the drop across R_B is the same as across R_E. As in the case of bias circuit C, bias circuit D's characteristics are not dependent upon transistor beta. In practice, diode D is mounted near the transistor so that the base-emitter junction and diode D are at the same temperature.

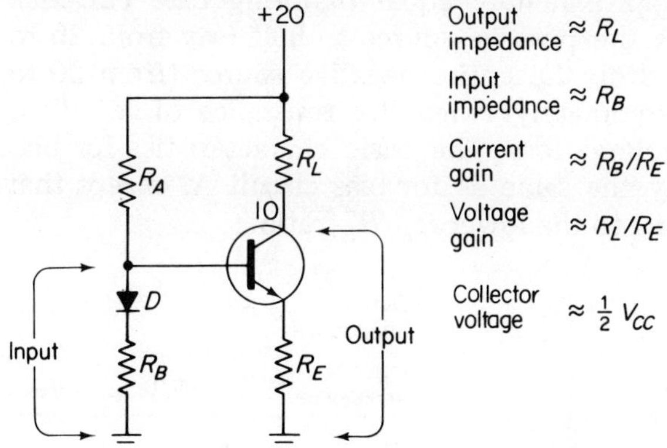

Figure 1-6. Bias circuit D.

1-1-9 Bias Circuit E

Figure 1-7 is the diagram for bias circuit E. The basic characteristics for bias circuit E are the same as for circuit C. However, bias circuit E is used in those special applications that require a negative and a positive voltage, each with respect to ground, to control base current. The values for R_L, R_E, and R_B in circuit E are the same as for circuit C.

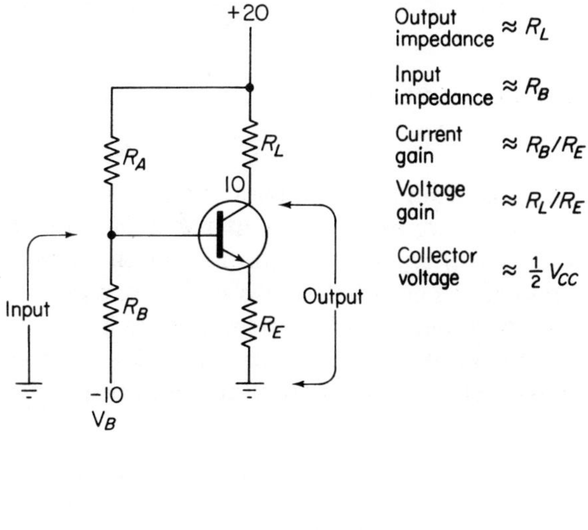

Figure 1-7. Bias circuit E.

1-1-10 Bias Circuit F

Figure 1-8 is the diagram for bias circuit F. This circuit is used in those special applications where it is necessary to supply collector-emitter current from both a positive and a negative source. Since the transistor is NPN, the collector is connected to the positive source through R_L, while the emitter is connected to the negative source through R_E. If both sources are approximately equal, it is difficult to design a circuit that will produce any voltage gain. The collector and emitter currents are approximately equal (ignoring base current). Therefore, if R_L drops the positive source to half (say from 20 to 10 V), then R_E must drop the entire negative source (from 20 to 0 V), and R_E is approximately twice the resistance of R_L. This generally produces a voltage loss. The basic characteristics for bias circuit F are essentially the same as for bias circuit A, except that voltage gain is low, owing to the lower R_L/R_E ratio.

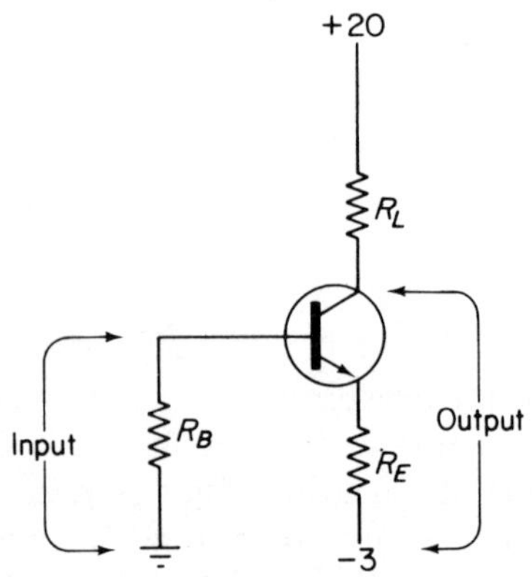

Output impedance	$\approx R_L$
Input impedance	$\approx R_E \times$ beta
Current gain	$\approx$ Beta
Voltage gain	$\approx R_L/R_E$

Figure 1-8. Bias circuit F.

1-2 EFFECT OF AMPLIFIER COMPONENTS ON FREQUENCY

Amplifier components do not attenuate (or pass) signals of all frequencies equally. That is, every circuit component has some *impedance* and is thus *frequency-sensitive*. Even a simple length of wire has some impedance. Wire, being a conductor, has some resistance. If alternating current is passed through the wire, there is some inductive reactance. If the wire is near another conductor (or metal chassis), there is some capacitance between the two conductors and thus some capacitive reactance. The reactance and resistance combine to produce impedance, which, in turn, varies with frequency.

In theory, transistors are capable of operating at any frequency from direct current on up. The top frequency limit is set only by the transit time of electrons across the transistor junctions. Primarily because of the reactance, there are limitations placed on the operating frequency of any amplifier by the transistor characteristics. Also, other components in the amplifier (resistors, capacitors, inductances, etc.) limit the operating frequency of any circuit. In a practical circuit, many of the impedances and reactances presented by components are of little concern. In other cases, certain impedances and reactances have a very pronounced effect on amplifier circuit operation. In addition to transistors, other major components used in AF amplifier circuits include capacitors, resistors, and inductances (both coils and transformers). Let us see how the impedances and reactances of the components affect AF amplifier circuit operation.

1-2-1 Transistor Frequency Limitations

All transistors have some capacitance between the elements (emitter-base junction, collector-base junction, gate-drain, etc.). If any of these elements is at common or ground, the remaining elements have some capacitance to ground. This is shown in Fig. 1-9. Capacitive reactance decreases with an increase in frequency, and vice versa. A capacitance in series with a conductor presents less attenuation to the signal as frequency increases. A capacitor across a conductor (for example, in parallel or shunt from the conductor to ground) acts as a short to signals of increasing frequency.

Consider the circuit of Fig. 1-9 where transistor junction capacitances are across the input and output. As frequency increases, the capacitive reactance drops, producing a short across the input and output and increased attenuation of the signal. At some frequency, the attenuation equals the transistor amplification, so there is no gain. At higher frequencies, the attenuation exceeds the amplification and there is a loss, even though the transistor may still continue to operate.

From a circuit-operating standpoint, the input and output capacitances of transistors have little effect at audio frequencies. Most modern transistors will operate well beyond the AF range and will generally produce equal (or flat) frequency response. That is, all signals up to about 20 kHz (or possibly higher) are amplified by the same amount. However, with most transistors, as signal frequencies increase into the RF range, amplification begins to drop. (RF circuit operation is discussed in Chapter 2.)

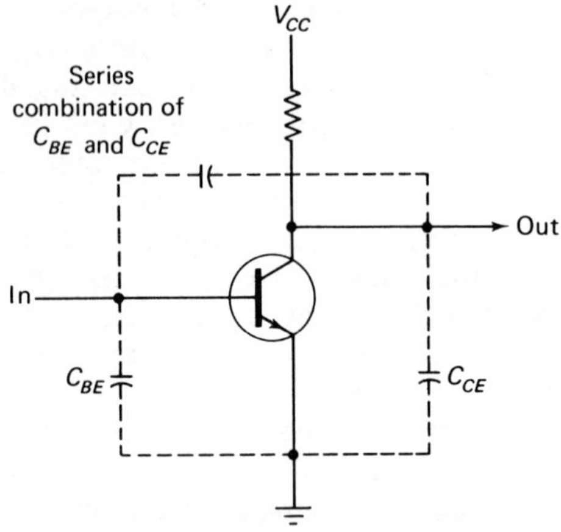

Figure 1-9. Capacitances associated with transistor elements.

All transistors have some inductance in their leads. This produces inductive reactance in series with the transistor elements. Inductive reactance increases with frequency. In the AF range, the inductive reactance is of little concern. However, at radio frequencies, the inductive reactance can produce considerable attenuation.

1-2-2 Resistor Frequency Limitations

At audio frequencies, resistors offer relatively few problems, since resistors attenuate signals equally. Only at very high frequencies, where the resistor leads and body can produce some kind of reactance, is there any particular concern about frequency limits imposed by resistors. However, resistors do produce voltage drops. These voltage drops can be a problem when considering interstage coupling methods (described in Sec. 1-3). Also, resistors offer some problems when used in conjunction with coupling capacitors in amplifier circuits.

1-2-3 Capacitance Frequency Limitations

Capacitors have three major uses in transistor AF amplifier circuits: bypass, decoupling, and coupling.

Bypass capacitors are used to provide a signal path around high resistances, as shown in Fig. 1-10. For example, if the power supply (Chapter 3) of an AF amplifier does not have a filter capacitor, or a battery is used, the collector-emitter current must pass through a high resistance. This can impede the a-c component of the signal.

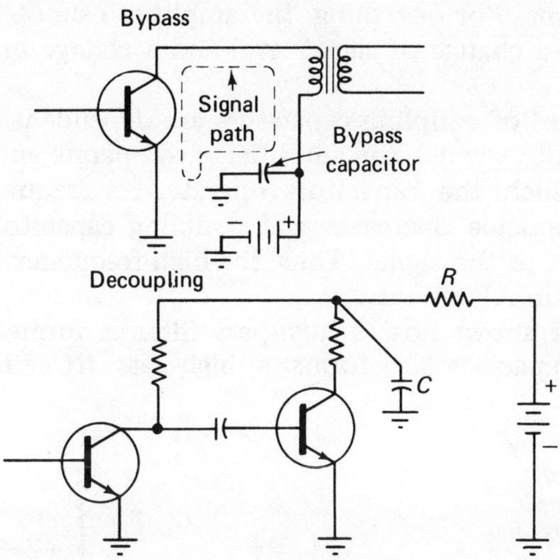

Figure 1-10. Examples of bypass and decoupling capacitors.

When several stages of amplification are connected, they all join at one point, the common power supply. In multistage amplifiers there is the possibility of one stage feeding back through the power supply to a previous stage, thus causing interference with the signal. To avoid this feedback, one or more of these stages may be *decoupled* from the power supply. Fig. 1-10 also shows a typical decoupling network. Resistor R is placed in series between the load resistors of the stages and the power supply. Hence, resistor R offers a high-resistance path from the a-c signal component to the power supply. Capacitor C, conversely, offers a low shunt reactance to this component and thus decouples (bypasses) the component to ground.

Actually, the functions of bypass and decoupling capacitors are the same and the terms are interchangeable. In either case, the main concern is that the reactance be low at the lowest frequency involved. For example, assume that the lowest frequency is 100 Hz and the minimum required reactance is 100 Ω. This requires a reactance of about 16 μF ($C = 1/(6.28\ FX_C)$). If the required frequency is decreased to 10 Hz, the capacitance value must be raised to about 160 μF to keep the reactance below 100 Ω.

Coupling capacitors are used at the input and output of each stage to block direct current. For example, if a coupling capacitor is not used between the transistor stages of Fig. 1-10, the collector of the first stage is connected directly to the base of the following stage, and both transistor elements are at the same bias voltage. Although it is possible to operate transistors in this way, direct coupling does increase problems. For one thing, the amplifier cannot tell the difference between a change of signal level and a change in power-supply voltage.

The values of coupling capacitors are dependent upon the low-frequency limit at which the amplifier is to operate and on the resistance with which the capacitors operate. As frequency increases, capacitive reactance decreases and coupling capacitors become (in effect) a short to the signal. Thus, the high-frequency limit need not be considered in AF circuits.

Fig. 1-11 shows how a high-pass filter is formed by coupling capacitors. Capacitor C_1 forms a high-pass RC filter, with R_B.

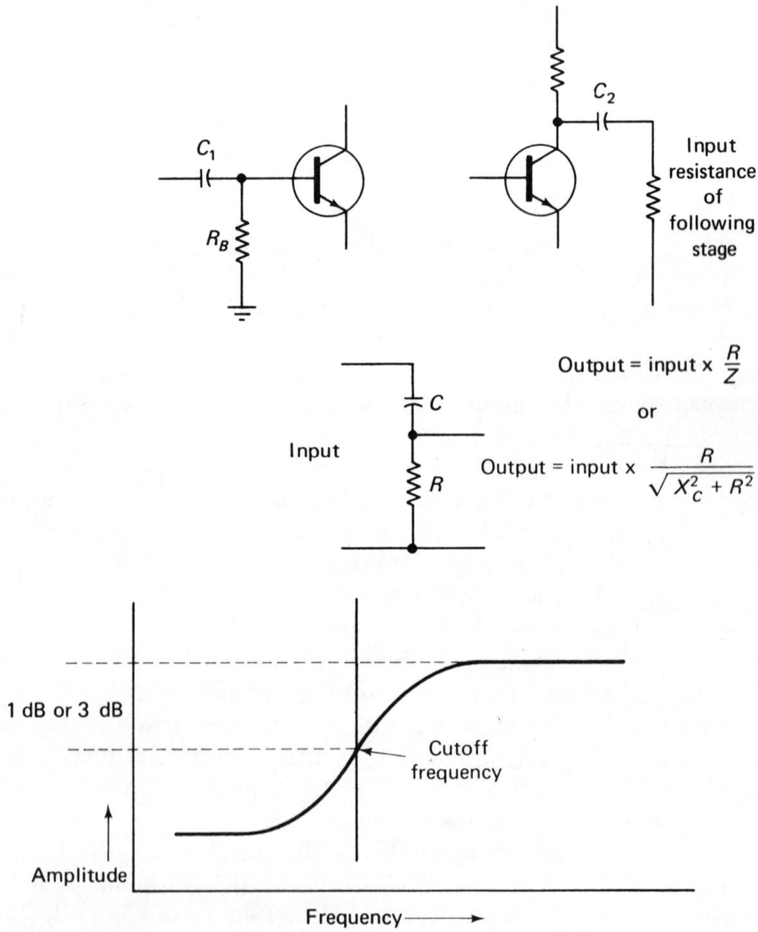

Figure 1-11. Formation of high-pass (low-cut) RC filter by coupling capacitors and related resistances.

Capacitor C_2 forms another high-pass filter with the input resistance of the following stage (or the load). The input voltage is applied across the capacitor and resistor in series. The output voltage is taken across the resistance. The relationship of input voltage to output voltage is

$$\text{output voltage} = \text{input voltage} \times R/Z$$

where R is the d-c resistance value and Z the impedance obtained by the vector combination of series capacitive reactance and d-c resistance. As a guideline, when reactance drops to about one-half the resistance, the output drops to about 90 percent of the input.

1-2-4 Inductance Frequency Limits

Both coils and transformers are used in audio amplifiers. As discussed in Sec. 1-3, coils are sometimes used instead of the collector resistor as a load. This permits the collector to be operated at the full voltage. Similarly, transformers are used for coupling between stages. This provides impedance matching, as discussed in Sec. 1-3.

The inductive reactance of coils and transformers increases with frequency. At the high end of the audio range, the attenuation produced by this increased reactance is usually sufficient to impair operation of the amplifier. At the low end of the AF range, the reactance of a typical transformer drops to a few ohms. This low impedance acts as a short across the line and attenuates the signal. Thus, coils and transformers tend to attenuate signals at both the high end and the low end of the AF range.

1-2-5 Stray Impedances

As discussed, any conductor (wiring, terminals, etc.) can have resistance, reactance, and impedance. Thus, care must be used in the routing of wires and placement of terminals to minimize the effects of this stray impedance. Similarly, the effects of stray impedance can alter the characteristics of components. A classic example of this is stray capacitance, which is added to the input and output capacitances of transistors. The effects of stray impedances are usually not critical at audio frequencies. However, the effects of stray impedance on amplifiers operating at radio frequencies (Chapter 2) can be of considerable importance.

1-3 AMPLIFIER COUPLING CIRCUITS

The operation of coupling networks must be considered in any amplifier circuit. Even a single-stage audio amplifier must be coupled to the input and output devices. If the circuit is multistage, there must be interstage coupling. Amplifier circuits are often classified according to coupling design. For example, the four basic coupling techniques are capacitor (or capacitance), inductive, direct, and transformer, as shown in Fig. 1-12.

The term *resistance-coupled* could be applied to any of the four coupling techniques, since resistors are involved. However, the term resistance-coupled is generally used to indicate that the circuit does not have inductances or transformers between stages, and that the input and/or output impedance is formed by a resistance. Capacitor coupling is often called resistance-capacitance (or *RC*) coupling.

With *direct coupling* (Fig. 1-12a), the collector of one transistor is connected directly to the base of the following transistor. The

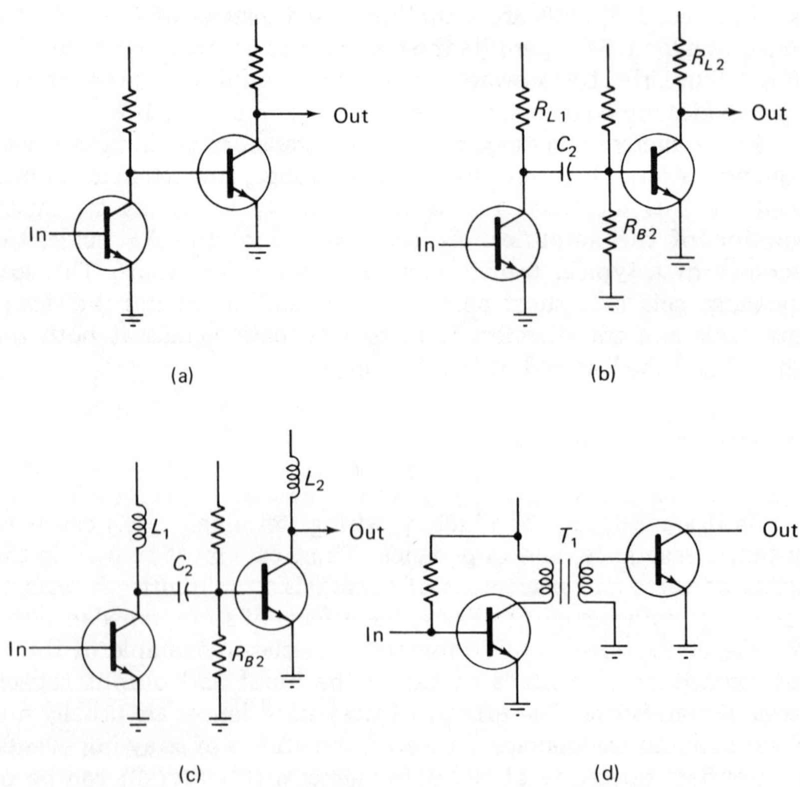

Figure 1-12. Four basic types of coupling used in audio amplifiers.

outstanding characteristic of a direct-coupled amplifier is the ability to amplify direct current and low-frequency signals.

With *capacitor coupling* or *RC coupling* (Fig. 1-12b), the coupling is accomplished by means of the load resistor R_{L1} of stage 1, the base resistor R_{B2} of stage 2, and the coupling capacitor C_2. The original signal is acted upon by stage 1 and appears in amplified form as the voltage drop across R_{L1}. The d-c component of the amplified signal is blocked by C_2, which passes the a-c component to the input section of stage 2 for further amplification. If necessary, more stages may be coupled to the output of stage 2 for still further amplification.

The main advantage of capacitor or RC coupling is that the amplifier produces uniform amplification over nearly the entire audio range, since resistor values are independent of frequency changes. However, as discussed in Sec. 1-2, RC coupling amplifiers do have a low-frequency limit imposed by reactance of the capacitor (which increases as frequency decreases). RC coupling is also small, light, inexpensive, and produces no magnetic field to interfere with the signal. One disadvantage of the RC coupling method is that the supply voltage is dropped (usually to one-half) by the load resistance. Thus, the transistors must operate at reduced voltages.

With *inductive or impedance coupling* (Fig. 1-12c), the load resistors R_{L1} and R_{L2} are replaced by inductors L_1 and L_2. The advantage of impedance coupling over resistance coupling is that the ohmic resistance of the load inductor is less than that of a load resistor. Thus, for a power supply of a given voltage, there is higher collector voltage.

Impedance coupling also suffers from some disadvantages. Impedance coupling is larger, heavier, and more costly than resistance coupling. To prevent the magnetic field of the inductor from affecting the signal, the inductor turns are wound on a closed, iron core and usually shielded extensively. The main disadvantage of impedance coupling is frequency discrimination.

With impedance coupling at very low frequencies, the gain is low owing to the capacitive reactance of the coupling capacitor, just as in the RC-coupled amplifier. The gain increases with frequency, leveling off at the middle frequencies of the audio range. (However, the frequency spread of this level portion is not as great as for the RC amplifier.) With impedance coupling at very high frequencies, the gain drops because of the increased reactance. Impedance coupling is rarely, if ever, used at frequencies well above the audio range.

With *transformer coupling* (Fig. 1-12d), the transformer T_1 serves several purposes. As the fluctuating collector current of the first stage flows through the primary winding of T_1, the current induces an alternating voltage with similar waveform in the secondary of T_1. This voltage forms the input signal to the second stage. Since the secondary of T_1 conveys the a-c component of the signal directly to the base of the second stage, there is no need for a coupling capacitor. Also, since the secondary winding furnishes a base return path, there is no need for a base resistance.

Compared to the *RC*-coupled amplifier, the transformer-coupled amplifier has essentially the same advantages and disadvantages as the impedance-coupled amplifier. The transistor collectors can be operated at higher voltages. The impedances are set by the transformer primary and secondary windings. Transformers are frequency-sensitive (impedance changes with frequency). Thus, the frequency range of the transformer-coupled amplifier is limited.

The inductances and transformers used in AF work are generally of the iron-core type. If air-core transformers are used at audio frequencies, the inductive reactance (and the impedance) will be so small as to be ineffective. At frequencies above the audio range (or at this high end), the reactance of iron-core inductances and transformers is so large that signals cannot pass (or are greatly attenuated). Thus, air-core transformers and inductances are used for higher-frequency amplifiers, as discussed in Chapter 2.

Coupling transformers can also provide for impedance matching between stages. Because the transistor is a current-operated device, impedance matching between the output of one stage to the input of the next is desirable for maximum transfer of power. This can be accomplished by making the primary and secondary transformer windings of different impedances.

Typically, the input impedance of a transformer stage is less than the output impedances. Thus, the secondary impedance of an interstage transformer is typically lower than the primary impedance. When two common-emitter stages are impedance matched, the overall gain is greater than when identical stages are resistance-coupled. Transformer coupling is also effective when the final amplifier output must be fed to a low-impedance load. For example, the impedance of a typical loudspeaker is in the order of 4 to 16 Ω, whereas the output impedance of a transistor stage is several hundred (or thousand) ohms. A transformer at the output of an audio amplifier can offset the obviously undesired effects of such a mismatch.

1-3-1 Effects of Coupling on Audio-Amplifier-Circuit Frequency Response

A simplified *frequency-response graph* or curve is shown in Fig. 1-13, which illustrates the effects of coupling methods on amplifier frequency response. The response is measured by the gain of the amplifier at various frequencies in the audio range.

Note that the gain falls off at the very low frequencies. In an *RC*-coupled amplifier, this drop in gain (generally referred to as *rolloff*) at the low end is due to the capacitive reactance of the coupling capacitor. Since the coupling capacitor is between the output of the first stage and the input to the second stage, the signal is attenuated by the voltage drop across the capacitor. Hence, the lower the frequency, the larger the capacitive reactance and the smaller the signal input is to the second stage. In impedance-coupled or transformer-coupled amplifier circuits, the low-frequency rolloff is caused by the very low inductive reactance, which acts as a short across the signal path. In effect, the low reactance bypasses some of the signal to ground.

As shown in Fig. 1-13, the gain also falls off at the higher frequencies. In *RC*-coupled amplifiers, this high-frequency rolloff is due to the output capacitance of the first stage, the input capacitance of the second stage, and the stray capacitance of the coupling network. These capacitances act to bypass some of the signal to ground. The higher the frequency, the smaller the capacitive reactance becomes, and the greater the amount of signal so bypassed. Thus, the overall gain is reduced. At frequencies between these two extremes, the gain remains fairly constant (in a well-designed amplifier circuit). In impedance-coupled or transformer-coupled amplifiers, the high-frequency rolloff is due to the large inductive reactance that attenuates the signal.

In summary, resistance coupling produces the lowest gain, transformer coupling the highest. As a guideline, three stages of *RC*-coupled amplification produces approximately the same gain as two stages of comparable transformer-coupled amplification. *RC* coupling

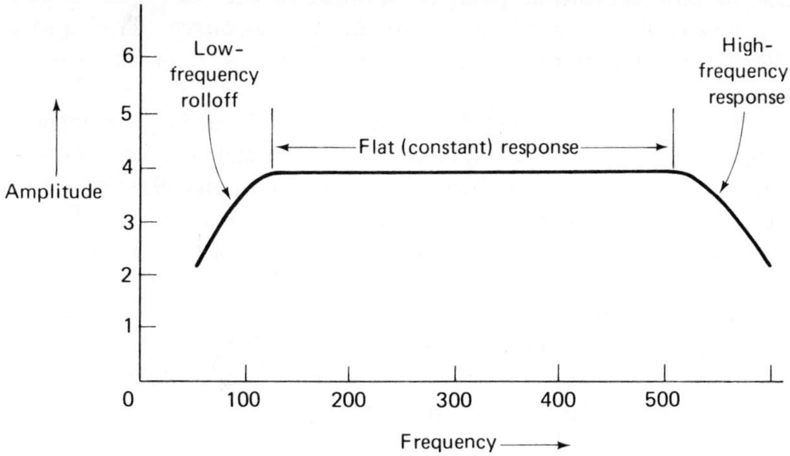

Figure 1-13. Simplified frequency-response graph.

produces the least frequency distortion of the signal. Transformer coupling has the added advantage of providing an impedance match at the amplifier input and output.

1-4 AMPLIFIER-CIRCUIT CLASSIFICATIONS

There are many ways to classify amplifier circuits. One of the most common ways is to use the *operating point* of the transistors. That is, the particular circuit is classified by the amount of current flowing in the amplifier transistors under no-signal conditions. The following is a brief summary of the four basic operating-point classifications.

In all four classifications, the base-collector (or gate-drain in the case of a FET) junction is always reverse-biased at the operating point, as well as under all signal conditions. Thus, no base-collector current flows (with the possible exception of reverse leakage current, called I_{CBO} on most transistor data sheets). The base-emitter (or gate-source) junction is biased such that base-emitter current flows under certain conditions, and possibly under all conditions. When base-emitter current flows, emitter-collector current also flows.

1-4-1 Class A Amplifier Circuits

In the class A amplifier, the base-emitter bias and the input voltages are such that the transistor operates only over the *linear portion* of the *characteristic curve*. Such a curve, representing the relationship between base voltage (input) and collector current (output), is shown in Fig. 1-14. At no point of the input signal cycle does the base become so positive or negative as to cause the transistor to operate at the nonlinear portion of the curve. The transistor collector current is never cut off, nor does the transistor ever reach saturation.

The main advantage of the class A amplifier is the relative lack of distortion. The output waveform follows that of the input waveform except in amplified form. However, with any class of amplifier there is some distortion. The disadvantages of class A amplifiers are their relative inefficiency (low power output for a high power input dissipated by the transistor), and their inability to handle large signal voltage swings. Rarely is a class A amplifier over about 35 percent efficient. Thus, if the power input to a class A amplifier is 1 W (generally the maximum power dissipation capability of a single transistor), the output will be less than 0.3 W.

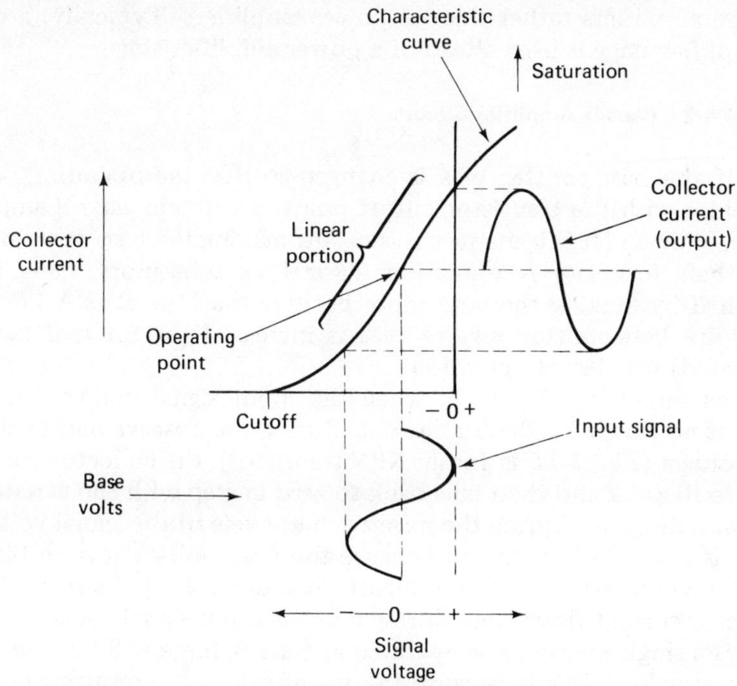

Figure 1-14. Typical class A amplifier characteristic curves.

The peak-to-peak output signal voltage swing of a class A amplifier is limited to something less than the total supply voltage. Since the output voltage must swing both positive and negative, the peak output is less than one-half the supply voltage. For example, assume that the supply voltage is 20 V and the amplifier is biased so that the *Q*-point collector voltage is one-half the supply, or 10 V. (Such a *Q*-point is generally typical for a class A amplifier.) Under these conditions, the output voltage swing cannot exceed ±10 V. If distortion must be kept to a minimum, the output will usually be on the order of ±5 V, so as to keep the transistor on the linear portion of the characteristic curve. (In most cases, the curve becomes nonlinear near the cutoff and saturation points. However, this can be determined only from an actual test of the amplifier circuit.)

The input voltage swing of a class A amplifier is limited by the output voltage swing capability and the voltage amplification factor. For example, if the output is limited to ±10 V and the voltage amplification factor is 100, the input is limited to ±0.1 V (100 mV). Because of these limitations, class A amplifiers are generally used as voltage amplifiers rather than as power amplifiers. Typically, a class A amplifier stage is used ahead of a power amplifier stage.

1-4-2 Class B Amplifier Circuits

If the base-emitter bias is changed so that the operating point coincides with the transistor cutoff point, we obtain class B amplification. For an NPN transistor this means making the base more negative than for class A operation. (For PNP transistors, class B is obtained by making the base more positive than the class A.) Either way, the base-emitter reverse bias is increased (or forward bias is decreased) for class B operation.

As shown in Fig. 1-15, when the input signal voltage is zero, there is no flow of collector current. During the positive half-cycle of the voltage (Fig. 1-15 is for an NPN transistor), the collector current rises to its peak and then falls back to zero in step with the variations of that half-cycle. During the negative half-cycle of the signal voltage, there is no collector current because the base-emitter reverse bias is at all times greater than the cutoff voltage of the transistor. Thus, collector current flows only during half the input signal cycle.

If a single transistor is operated as class B, there will be considerable distortion. This is because the waveform of the resulting collector current resembles that of the positive half-cycle of the input signal and, consequently, does not resemble the complete waveform of the input. It is possible to use two transistors in *push-pull*, to reconstruct an output waveform that resembles the full waveform at the input. (Push-pull class B is discussed further in Sec. 1-8.)

The peak output voltage swing of a class B amplifier is slightly less than the supply voltage. Since the output appears only on half-cycles, it is possible to operate class B amplifiers at a higher current (or power) rating than class A, all other factors being equal. For example, if a transistor is capable of 0.3-W dissipation (without damage) as class A, the same transistor can be operated at 0.6 W, class B, since the transistor is conducting collector current only half the time. (This is a theoretical example. In practice, there are factors that limit class B power dissipation to something less than twice that of class A.)

The peak output of a class B amplifier is equivalent to the peak-to-peak output of a class A amplifier. Thus, if two transistors are connected in push-pull and operated as class B, the output voltage can be twice that of class A. Because of these voltage and power factors, class B amplifiers are generally used as power amplifiers rather than as voltage amplifiers. Typically, two push-pull transistors are operated in class B, preceded by a single class A amplifier circuit.

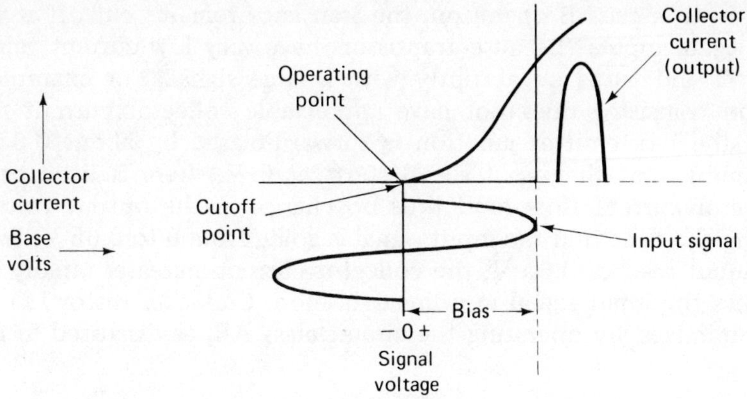

Figure 1-15. Typical class B amplifier characteristic curves.

The class A circuit provides voltage amplification, whereas the class B circuit produces the necessary power amplification.

1-4-3 Class AB Amplifier Circuits

Class B is the most efficient operating mode for audio amplifiers since it draws the least amount of current. That is, the transformers are cut off at the *Q*-point and draw collector current only in the presence of an input signal. Class B operation can, however, result in a form of distortion known as *crossover distortion* (discussed further in Sec. 1-10). The effects of crossover distortion can be seen by comparing the input and output waveforms in Fig. 1-16.

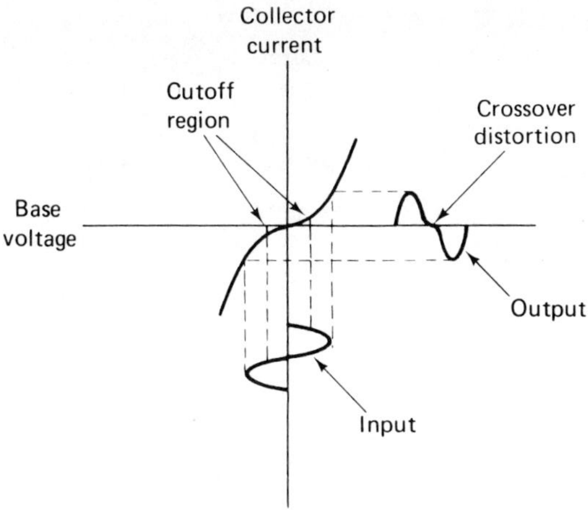

Figure 1-16. Example of crossover distortion.

In true class B operation, the transistor remains cut off at very low signal inputs (because transistors have very low current gain at cutoff) and turns on abruptly with a large signal. For example, a silicon transistor does not have appreciable collector current flow until the base-emitter junction is forward-biased by about 0.65 V. Assuming that the input signal starts at 0 V, there is little or no collector current flow (and thus no change in the output voltage) during the time that the input signal is going from 0 to 0.65 V. When the input reaches 0.65 V, the collector current increases rapidly and follows the input signal in a linear fashion. Crossover distortion can be minimized by operating the circuit class AB, as discussed in Sec. 1-10.

1-4-4 Class C Amplifier Circuits

If the transistor is reverse-biased considerably below the cutoff point, we obtain a class C amplifier. As shown in Fig. 1-17, during the positive half-cycle of the input signal, the signal voltage starts from zero, rises to the positive peak value and falls back to zero. (Figure 1-17 is for an NPN transistor.) A portion of the input signal causes the base-emitter junction to be forward-biased. As a result, there is a flow of collector current for a portion of one-half the input cycle, the negative half-cycle of the input signal between the cutoff point and the peak. The resulting collector current is a pulse, the duration of which is considerably less than a half-cycle of the input signal.

The class C waveform output signal cannot resemble that of the input signal, even with push-pull operation. For this reason, class C

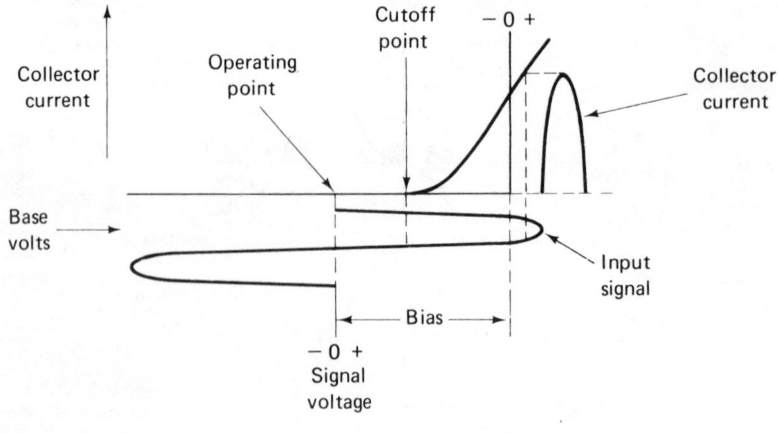

Figure 1-17. Typical class C amplifier characteristic curve.

is limited to those applications where distortion is of no concern. Generally, class C operation is limited to use in RF amplifiers (Chapter 2) and is not found in AF amplifiers.

1-5 BASIC TRANSISTOR AMPLIFIER CIRCUITS

Figure 1-18 is the working schematic of a basic, single-stage audio amplifier using a two-junction transistor. Input and output coupling capacitors are added to prevent d-c flow to and from external circuits. A bypass capacitor C_3 is shown connected across the emitter resistor R_E. Capacitor C_3 is required only under certain conditions. Circuit characteristics are shown in Fig. 1-18. As shown, and as discussed in Sec. 1-1, the purpose of emitter resistor R_E is to stabilize gain. Maximum voltage gain occurs when R_L is made larger in relation to R_E. Similarly, current gain increases when R_B is made larger in relation to R_E. However, circuit stability is greatest when R_B and R_L are made smaller in relation to R_E. That is, the circuit gain will remain more constant in the presence of temperature, supply voltage, or input signal changes when R_E is made larger in relation to R_B and R_L. Thus, there is a trade-off between gain and stability. This form of emitter feedback (current feedback) is known as *stage feedback* or *local feedback*, since only one stage is involved. *Overall feedback* or *loop feedback* is sometimes used where several stages are involved.

The input and/or output impedances are set by the values of R_B and R_L, as shown by the equations of Fig. 1-18. Maximum power transfer occurs when R_B and R_L match the impedances of the previous stage and the following stage, respectively. The maximum peak-to-peak output voltage is set by the source voltage. For class A operation, the collector is operated at approximately one-half the source voltage. This permits the maximum positive and negative swing of output voltage.

The values of C_1 and C_2 are dependent upon the low-frequency limit at which the amplifier is to operate. C_1 forms a high-pass (or low-cut) *RC* filter with R_B (as discussed in Sec. 1-2). Capacitor C_2 forms another filter with the input resistance of the following stage (or the load). For a given resistance value, a lower frequency requires a larger capacitor value. Of course, if the resistance can be made larger (with the same desired frequency), the capacitor value can be reduced. Since two-junction transistors are low-impedance, current-operated devices, the coupling capacitors in two-junction amplifiers are generally large in relation to those of FET (and vacuum-tube) amplifier circuits.

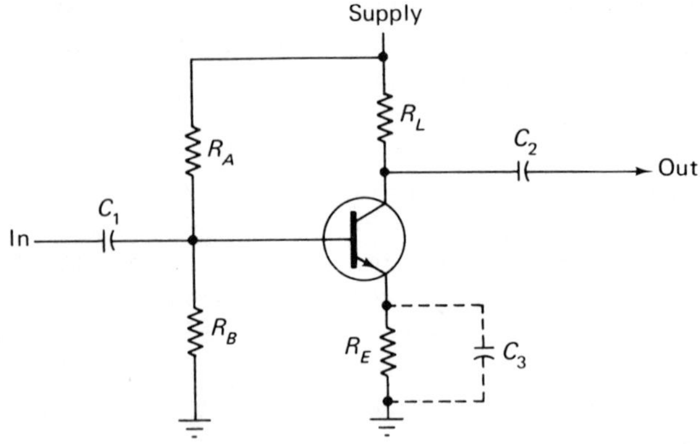

$Z_{in} \approx R_B$

$Z_{out} \approx R_L$

$I_{gain} \approx \dfrac{R_B}{R_E}$

$V_{gain} \approx \dfrac{R_L}{R_E}$

$R_L > 5R_E$

$R_L \approx 10R_E$

$R_B \approx 10R_E$

$R_B < 20R_E$

$R_E \approx 100 - 1000\ \Omega$

$V_{collector} \approx 0.5 \times$ supply (set by R_A)

$I_{gain} \approx \text{stability} \approx \dfrac{R_B}{R_E}$

$S \approx 20$ for high gain

$S \approx 10$ for stability

$S \approx 5$ for power gain

$C_1 \approx \dfrac{1}{3.2\,FR}$ for 1 dB

$C_2 \approx \dfrac{1}{6.2\,FR}$ for 3 dB

Figure 1-18. Basic single-stage audio amplifier using a two-junction transistor.

1-5-1 Emitter Bypass for Transistor Amplifier Circuits

Figure 1-18 shows (in phantom) a bypass capacitor C_3 across emitter resistor R_E. This arrangement permits R_E to be removed from the circuit as far as the signal is concerned but leaves R_E in the circuit (in regard to direct current). With R_E removed from the signal path, the voltage gain is approximately R_L divided by the dynamic resistance of the transistor, and the current gain is approximately equal to the a-c beta of the transistor. Thus, the use of an emitter-bypass capacitor permits the highly temperature stable d-c circuit to remain intact while providing a high signal gain.

The emitter bypass is generally used where maximum gain must be obtained from a single stage of amplification and a stable gain is of little concern. The value of the emitter-bypass capacitor should be such that the reactance is less than the transistor input impedance at the lowest frequency of operation. This will effectively short the emitter (signal path) around R_E. The value of C_3 can be found by

$$\text{capacitance} = \frac{1}{6.2FR}$$

where capacitance is in farads, F is the low-frequency limit in hertz, and R is the maximum input impedance of the transistor in ohms.

1-5-2 Basic Audio Amplifier with Partially Bypassed Emitter

Figure 1-19 shows a basic single-stage two-junction transistor used as an audio amplifier with a partially bypassed emitter resistor. This circuit is a compromise between the basic amplifier without bypass and the fully bypassed emitter. The d-c characteristics of both the unbypassed and partially bypassed circuits are essentially the

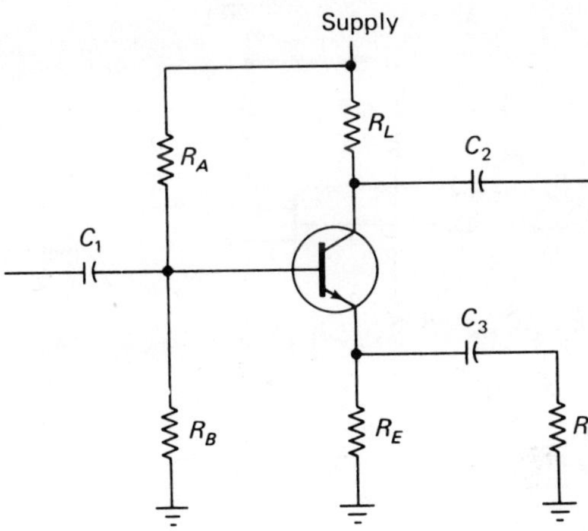

$$I_{gain} \approx \frac{R_B}{R_C} \qquad R_L < 50R_C$$

$$V_{gain} \approx \frac{R_L}{R_C} \qquad R_B < 50R_C$$

$$C_3 \approx \frac{1}{6.2FR_C}$$

Figure 1-19. Basic single-stage two-junction transistor used as an audio amplifier with partially bypassed emitter resistor.

same. The circuit values (except C_3 and R_C) are calculated in the same way for both circuits. However, the voltage and current gains for a partially bypassed amplifier are greater than for an unbypassed circuit but less than for the fully bypassed circuit.

The value of R_C is chosen on the basis of voltage gain, even though current gain is increased when voltage gain increases. R_C should be substantially smaller than R_E. Otherwise, there will be no advantage to the partially bypassed circuit. However, a smaller value for RC requires a larger value for C_3, since the C_3 value is dependent upon the R_C value and the desired low-frequency cutoff point.

With the circuit of Fig. 1-19, current gain and voltage gain are dependent upon the ratios of R_B/R_C and R_L/R_C, respectively. Thus, the value of R_E has little or no effect on circuit gain.

1-5-3 Basic FET Amplifier Circuits

Figure 1-20 shows a basic single-stage FET amplifier circuit. Input and output coupling capacitors C_1 and C_2 are added to prevent d-c flow to and from external circuits. Bypass capacitor C_3, con-

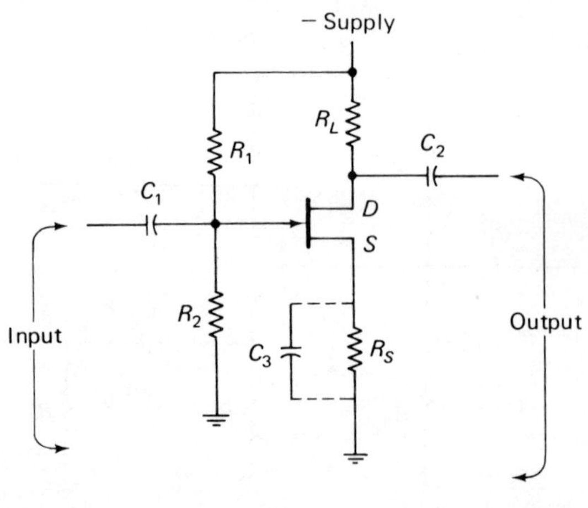

$$A_V \approx \frac{R_L}{\frac{1}{Y_{fs}} + R_S} \approx \frac{R_L}{R_S}$$

Drain voltage = 0.5 x supply

$$Z_{in} \approx R_1 \,||\, R_2 \approx R_2 \qquad C_1 \approx \frac{1}{3.2\,FR_2} \text{ for 1 dB}$$

$$Z_{out} \approx R_L \qquad C_1 \approx \frac{1}{6.2\,FR_2} \text{ for 3 dB}$$

Figure 1-20. Basic common-source FET amplifier stage.

nected across R_S, is required only under certain conditions. Also note that the circuit of Fig. 1-20 is the FET version of the Fig. 1-19 circuit. Thus, many of the circuit characteristics are similar.

One major difference is that voltage gain in the FET circuit (Fig. 1-20) is set by factors of R_L, R_S, and y_{fs}. This y_{fs} factor, also known as *forward transconductance* on FET data sheets, is a measure of the change in output current for a given change in input voltage (and is essentially the same as transconductance for a vacuum tube). Another difference is that the values of coupling capacitors for FET circuits are much smaller than for two-junction transistors. This is because FETs are high-impedance, voltage-operated devices (whereas two-junction transistors are low-impedance, current-operated devices).

Figure 1-20 also shows (in phantom) a bypass capacitor C_3 across source resistor R_S. This arrangement permits R_S to be removed from the circuit as far as signal is concerned but leaves R_S in the circuit (in regard to direct current). With R_S removed from the signal path, the voltage gain is approximately equal to $y_{fs} \times R_L$. Thus, the use of a bypass capacitor permits a temperature-stable d-c circuit to remain intact while providing a high signal gain.

1-6 MULTISTAGE TRANSISTOR AMPLIFIERS

When stable voltage gains greater than about 20 are required and it is not practical to bypass the emitter (or source) resistor of a single stage, two or more transistor amplifier stages can be used in *cascade* (where the output of one transistor is fed to the input of a second transistor). In theory, any number of two-junction or FET amplifiers can be connected in cascade to increase voltage gain. In practice, the number of stages is usually limited to three. The overall gain of the amplifier is equal (approximately) to the cumulative gain of each stage, multiplied by the gain of the adjacent stage.

As an example, if each stage of a three-stage amplifier has a gain of 10, the overall gain is (approximately) 1000 ($10 \times 10 \times 10$). Since it is possible to design a very stable single stage with a gain of 10 and adequately stable circuits with gains of 15 to 20, a three-stage amplifier can provide gains in the range 1000 to 8000. Generally, this is more than enough voltage gain for most practical applications. Using the 8000 figure, a 1-μV signal (say from a low-voltage transducer or delicate electronic device) can be raised to the 8-mV range while maintaining stability in the presence of temperature and power-supply variations.

1-6-1 Basic Considerations for Multistage Amplifier Circuits

Any of the single-stage amplifier circuits described thus far can be connected to form a two-stage or three-stage *voltage amplifier.* For example, the basic stage (without emitter or source bypass) can be connected to two like stages in cascade. The result is a highly temperature stable voltage amplifier circuit. Since each stage has its own feedback, the gain is precisely controlled and very stable.

It is also possible to mix stages to achieve some given design goal. For example, a three-stage amplifier can be designed using a highly stable unbypassed amplifier for the first stage and two bypassed amplifiers for the remaining stages. Assuming a gain of 10 for the unbypassed stage and gains of 30 for the bypassed stages, this results in an overall gain of 9000. Of course, with the bypassed stages the gain is dependent upon the transistor characteristics and is therefore unpredictable. However, once the gain is established for a given amplifier, the gain should remain fairly stable.

Distortion and clipping. As in the case with any high-gain amplifier, the possibility of overdriving a multistage transistor amplifier is always present. As an example, assume an overall gain of 1000 and a supply voltage of 20 V. Typically, this implies a 10-V *Q*-point (for the output collector or drain) and a 20-V (peak-to-peak) output swing (from 0 to 20 V). In practice, a swing from about 0 to 19 V is more realistic. Either way, a 20-mV (peak-to-peak) input signal, multiplied by a gain of 1000, will drive the final output to its limit and possibly into distortion.

Feedback. When each stage of a multistage amplifier has its own feedback (*local* or *stage feedback*), the most precise control of gain is obtained. However, such feedback is often unnecessary. Instead, overall feedback (or *loop feedback*) is used, where a part of the output from one stage is fed back to the input of a previous stage. Usually, such feedback is through a resistance (to set the amount of feedback), and the feedback is from the final stage to the first stage. However, it is possible to use feedback from one stage to the next (second stage to first stage, third stage to second stage, etc.).

Feedback phase inversion. There is a problem of phase inversion when using loop or overall feedback. In a common-emitter or common-source amplifier, the phase is inverted from input to output. That is, if the input signal is positive, the output is negative, and vice versa. If feedback is between two stages, the phase is inverted twice, resulting in *positive feedback.* This usually produces oscillation, or very unstable gain. In any event, positive feedback will not stabilize gain. The phase-inversion problem can be overcome, when multiple stages are involved, by connecting the output collector (or drain) or the second stage back to the emitter (or source) of the first stage. This produces the desired *negative feedback*.

As an example, if the base (or gate) of the first stage is swinging positive, the collector (or drain) of that stage will swing negative, as will the base (or gate) of the second stage. The collector (or drain) of the second stage will swing positive, and this positive swing can be fed back to the emitter (or source) of the first stage. A positive input at the emitter (or source) has the same effect as a negative at the base (or gate). Thus, negative feedback is obtained.

Low-frequency cutoff. Unless direct coupling is used, as described in Sec. 1-7, coupling capacitors must be used between stages, as well as at the input and output. Such capacitors form a low-pass *RC* filter with the base-to-ground (or gate-to-ground) resistance. Thus, each stage has its own low-pass filter. In multistage amplifiers, the effects of these filters are cumulative. As an example, if each filter causes a 1-dB drop at some given cutoff frequency and there are three filters (one at the input and two between stages), the result is a 3-dB drop at that frequency in the final output.

1-7 DIRECT-COUPLED TRANSISTOR AMPLIFIERS

The amplifiers discussed thus far cannot amplify direct currents (or direct voltages) since such currents are not passed by coupling capacitors or transformers. Also, amplifiers using transformers and/or coupling capacitors are not well suited for amplification of very low frequencies. For reasons discussed in Sec. 1-3, direct-coupled amplifiers are used if direct currents and/or very low frequency signals must be amplified. Direct-coupled amplifiers, also known as d-c amplifiers, permit a signal to be fed directly to the transistor without the use of any coupling device. Direct-coupled amplifiers may be single-stage or multistage. However, direct coupling is generally limited to three stages.

1-7-1 Practical Direct-coupled Two-junction Transistor Amplifiers

Figure 1-21 is the working schematic of a two-stage direct-coupled *complementary amplifier*. That is, the output of an NPN is fed into a PNP transistor to increase stabilization. The increased stabilization for the complementary direct-coupled amplifier arises

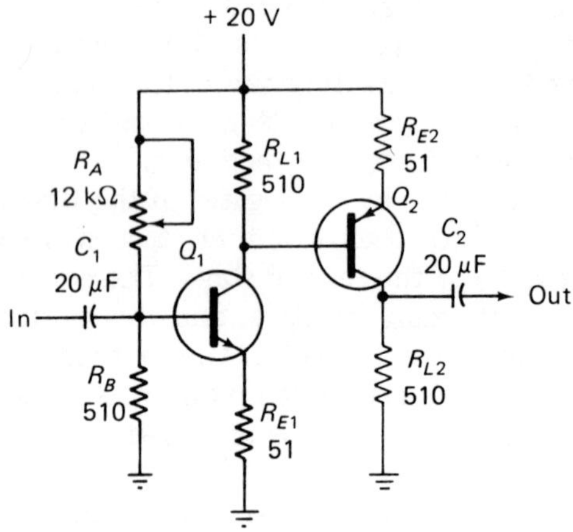

$Z_{in} \approx R_B \approx 510$ $\qquad$ $R_L > 5R_E$ $\qquad$ $R_B < 20R_E$

$Z_{out} \approx R_{L2} \approx 510$ $\qquad$ $R_L \approx 10R_E$

$Q_1\ V_{gain} \approx \dfrac{R_{L1}}{R_{E1}} \approx 10$ $\qquad$ $R_B \approx 10R_E$

$Q_2\ V_{gain} \approx \dfrac{R_{L2}}{R_{E2}} \approx 10$

$Q_2\ V_{collector} \approx 0.5 \times$ supply (as adjusted by R_A)

Figure 1-21. Basic direct-coupled complementary amplifier circuit.

from the fact that a change in collector current of Q_1 (due to temperature, power-supply variation, etc.) is opposed by an equal change in collector current of Q_2 but in the opposite direction. If more stages are added, the complementary system is continued. That is, NPN and PNP are used alternately.

When two stages are direct-coupled in the stabilized circuit of Fig. 1-21, the overall gain is about 70 percent (possibly higher) of the combined gains of each stage. The gain of each stage is approximately 10, as set by the 10:1 ratio of collector emitter resistances. The combined gain of the circuit is theoretically 100, and practically about 70. Thus, an input of 100 mV is increased to about 7 V at the output. Note that the circuit of Fig. 1-21 is highly stable and has a very wide band-frequency response (typically from direct current on up to whatever maximum is set by the transistor high-frequency limitations).

1-7-2 Practical Direct-coupled FET Amplifiers

Both MOSFETs and JFETs can be used in direct-coupled amplifier circuits. However, MOSFETs are especially well suited to direct-coupled applications. Since the gate of a MOSFET acts essentially as a capacitor rather than a diode junction, no coupling capacitor is needed between stages. For a-c signals, this means that there are no low-frequency cutoff problems, *in theory.* In practical circuits, the input capacitance can form an *RC* filter with the source resistance and result in some low-frequency attenuation.

Figure 1-22 is the working schematic of an all-MOSFET three-stage amplifier. Note that all three MOSFETs are of the same type, and all three drain resistors (R_1, R_2, and R_3) are the same value. This arrangement simplifies design. At first glance, it may appear that all three stages are operating at zero bias. However, when drain current flows, there is some drop across the corresponding drain resistor, producing a voltage at the drain of the stage and an identical voltage at the gate of the next stage. The gate of the first stage is at essentially the same voltage as the drain of the last stage, because of feedback resistor R_F. There is no current drain through R_F, with the possible exception of reverse gate current (which can be ignored for practical purposes).

The circuit of Fig. 1-22 requires one coupling capacitor (C_1) at the input. This is necessary to isolate the input gate from any d-c voltage that may appear at the input. This makes the circuit of Fig.

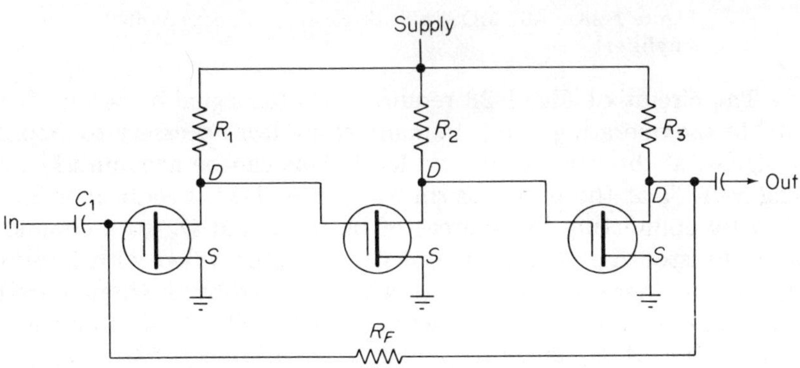

Gain without $R_F \approx (R_1 \times y_{fs})(R_2 \times y_{fs})(R_3 \times y_{fs})$

$$R_{IN} \approx \frac{R_F}{\text{Gain}}$$

Figure 1-22. All MOSFET three-stage amplifier.

1-22 unsuitable for use as a d-c amplifier. The circuit can be converted to a d-c amplifier when the coupling capacitor is replaced by a series resistor R_{in}, as shown in Fig. 1-23. The considerations concerning operating point are the same for both circuits. However, the series resistance must be terminated at a d-c level equivalent to the operating point. For example, if the operating point is −7 V, point *A* must be at −7 V. If point *A* is at some other d-c level, the operating point is shifted.

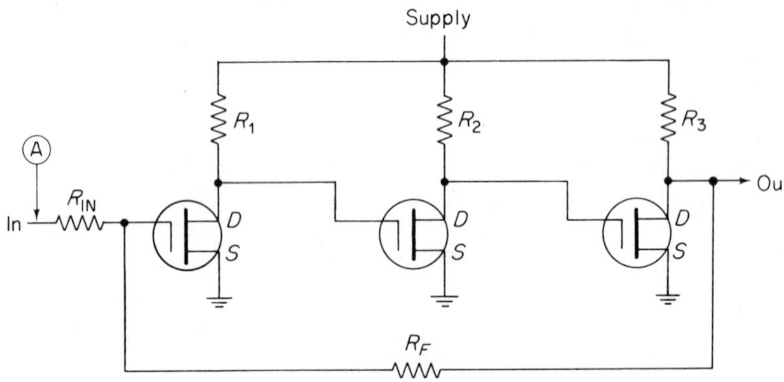

$$\text{Gain} \approx \frac{R_F}{R_{IN}}$$

$$Z_{IN} \approx R_{IN}$$

$$Z_{OUT} \approx R_3$$

Figure 1-23. All MOSFET three-stage direct-coupled amplifier.

The circuit of Fig. 1-23 requires that the signal be at a d-c level equal to the operating point. In many cases, it is necessary to amplify d-c signals at the zero or ground level. This can be accomplished by using a JFET at the input, as shown in Fig. 1-24. Feedback is introduced by connecting the sources of both Q_1 and Q_3 to a common-source resistor R_2. The source of Q_2 is not provided with a source resistor, but there is some bias on Q_2 produced by the drain current drop across R_3. The input impedance of the Fig. 1-24 circuit is set by the value of R_1 and the gate-drain capacitance of Q_1. Gain can be sacrificed for stability by increasing the value of R_2. With the values shown and typical FETs, the gain should be on the order of 3000 to 5000. The bias and operating point for Q_2 and Q_3 is set by R_3, shown as 33 KΩ. In practice, the value of R_3 is approximated by calculation and then adjusted for a desired operating point at the output (drain) of Q_3.

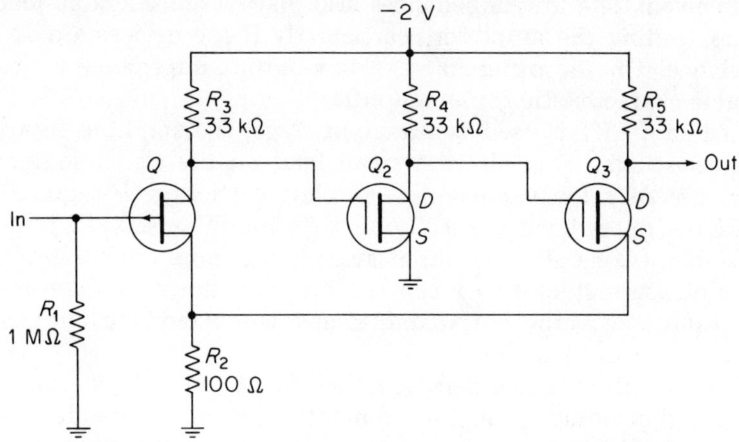

Figure 1-24. Direct-coupled amplifier with JFET input, followed by two MOSFET stages.

1-7-3 Combination FET and Two-junction Amplifiers

In certain applications, FET stages and two-junction transistor stages can be combined. The classic example is where a single FET stage is used at the input, followed by two two-junction circuits. Such an arrangement takes advantage of both the FET and two-junction transistor characteristics.

FETs are voltage-operated devices. This permits large voltage swings with low currents, making it possible to use high-resistance values (resulting in high impedances) at the input and between stages. In turn, these high-resistance values permit the use of low-value coupling capacitors and eliminate the need for bulky, expensive electrolytic capacitors. If operated at the *zero-temperature coefficient* point (OTC point) with the gate-source biased at a specific voltage and held constant at the voltage, the drain current will not vary with changes in temperature. The OTC point is different for each type of FET.

Two-junction transistors are essentially current-operated devices, permitting large currents at about the same voltage levels as the FET. Thus, with equal supply voltage and signal current swings, the two-junction transistor can supply much more current gain (and power gain) than the FET. Since currents are high, the impedances (input, interstage, and output) must be low in two-junction transistor circuits. This requires large-value coupling capacitors if low frequencies are involved. The low impedances also place a considerable load on devices feeding the amplifier, particularly if the devices are of high impedance. On the other hand, a low-output impedance is often a desirable characteristic for an amplifier.

When a FET is used as the input stage, the amplifier input impedance is high. This places a small load on the signal source and allows use of a low-value input coupling capacitor (if required). If the FET is operated at the 0TC point, the amplifier input is temperature-stable. (Generally, the input stage is the most critical in regards to temperature stability.) When two-junction transistors are used as the output stages, the output impedance is low, and current gain (as well as power gain) is high.

Combination amplifiers can be direct-coupled or capacitor-coupled, depending upon requirements. The direct-coupled design offers the best low-frequency response, permits d-c amplification and is generally simpler (uses less parts). The capacitor-coupled combination amplifier permits a more stable design and eliminates the voltage regulation problem. That is, a d-c amplifier cannot tell the difference between signal-level changes and power-supply changes, so the power supply must be very stable for a d-c amplifier.

The FET can be combined with any of the classic two-stage two-junction transistor amplifier combinations. The two most common combinations are the Darlington pair described in Sec. 1-8 (for no voltage gain, but high-current gain and low-output impedance), and the NPN-PNP complementary pair amplifier of Sec. 1-7-1 (for both current gain and voltage gain).

Figure 1-25 is the working schematic of a direct-coupled amplifier using a FET input stage and a two-junction transistor pair as the output. Note that local feedback is used in the FET stage (provided by R_S) as well as overall feedback (provided by R_4). As a design example, assume that the circuit of Fig. 1-25 is to provide an input impedance of 1 MΩ, an output impedance of 500 Ω, and maximum gain. Temperature stability is not critical.

Under these conditions, the values of R_2 and R_4 are set at 1 MΩ and 500 Ω, respectively, as shown by the equations of Fig. 1-25. For practical purposes, with the voltage drop across R_S ignored, Q_1 operates at zero gate-source voltage. Assume that the drain current for Q_1 is 0.2 mA under these conditions. (The actual drain current depends on the type of FET used for Q_1, and the gate-source voltage, as found on the FET data sheet.) The voltage drop across R_3 is the normal base-emitter drop of a two-junction transistor (about 0.5 to 0.7 V for silicon and 0.2 to 0.3 V for germanium). The drop across R_L is twice this value (about 1 to 1.5 V for silicon and 0.4 to 0.6 V for germanium). With the required drop of 1.5 V and 0.2-mA drain

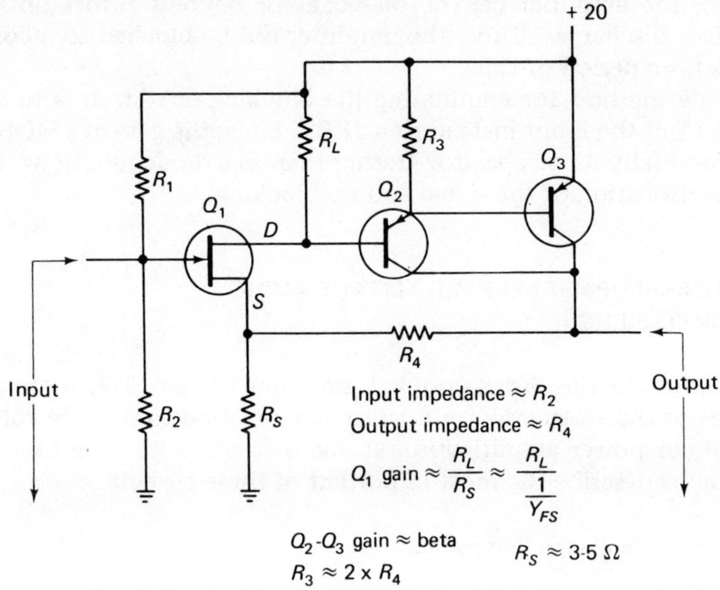

Figure 1-25. Direct-coupled hybrid amplifier.

current, the value of R_L is approximately 7.5 kΩ. Since R_4 is 500 Ω, R_3 should be 1 kΩ (as indicated by the equations). If temperature stability is not critical, the FET can be operated at zero bias by omitting R_1. If temperature stability is critical, the value of R_1 is selected to produce the necessary bias to operate Q_1 at the 0TC point. The key component in setting up this circuit is R_L. With the circuit operating in experimental form, R_L is adjusted for the desired Q-point voltage at the output (collectors of Q_2 and Q_3).

One problem with the circuit of Fig. 1-25 is that the input to the JFET Q_1 may result in a *blocking condition*. Generally, a direct-coupled amplifier does not require any coupling capacitors. One exception is a coupling capacitor at the input to isolate the amplifier from direct current (when the signal is composed of direct current and alternating current). When a coupling capacitor is used at the gate of a JFET (or at the base of some two-junction transistors), a condition known as blocking can occur.

Blocking is produced by the fact that the gate junction of a JFET is similar to that of a diode (as is the base junction of a two-junction transistor). The "diode" acts to rectify the incoming signals. If a capacitor is connected in series with the "diode," large signals can charge the capacitor. On one half-cycle, the diode is forward-biased and charges rapidly. On the opposite half-cycle, the diode is reverse-biased and discharges slowly. If the signal and charge are large enough, the amplifier can be biased at or beyond cutoff until the capacitor discharges. Thus, the amplifier can be blocked to incoming signals for a period of time.

One method for eliminating the blocking condition is to use a MOSFET at the input instead of a JFET. Since the gate of a MOSFET acts essentially as a capacitor (rather than as a diode junction), there is no rectification of the signal and no blocking.

1-8 TRANSFORMERLESS MULTISTAGE AUDIO AMPLIFIERS

In addition to the direct-coupled amplifier of Sec. 1-7, there are a number of transformerless multistage circuits used to provide voltage, current, or power amplification at audio frequencies. The following paragraphs describe the most important of these circuits.

1-8-1 Darlington Compounds

Figure 1-26 shows the basic Darlington circuit (known as the *Darlington compound*), together with two practical versions of the circuit. As shown, the Darlington compound is an emitter follower (or common collector) driving a second emitter follower. An emitter follower (where the output is taken from the emitter resistor) provides no voltage gain but can provide considerable power gain. The main reason for using a Darlington compound (especially in audio work) is to produce high current (and power) gain. For example, Darlington compounds are often used as audio drivers to raise the power of a signal from a voltage amplifier to a level suitable to drive a final power amplifier. Darlingtons are also used as a substitute for a driver section (or to eliminate the need for a separate driver).

Darlingtons as basic common collectors (emitter followers). When the Darlington is used as a common collector, as shown in Fig. 1-26, the output impedance is approximately equal to the load resistance R_L. The input impedance is approximately equal to beta^2 $\times$ R_L. The current gain is approximately equal to the average beta of the two transistors, squared. However, in most common collectors, power gain is of primary concern. That is, the designer is interested in how much signal power can be increased across a give output load.

As an example, assume that the value of R_L (in Fig. 1-26c) is 1000 Ω and that the average beta is on the order of 15. This results in an input impedance of about 225 kΩ ($15^2 \times 1000$) and an output impedance of 1000 Ω. Now assume that a 2.5-V signal is applied at

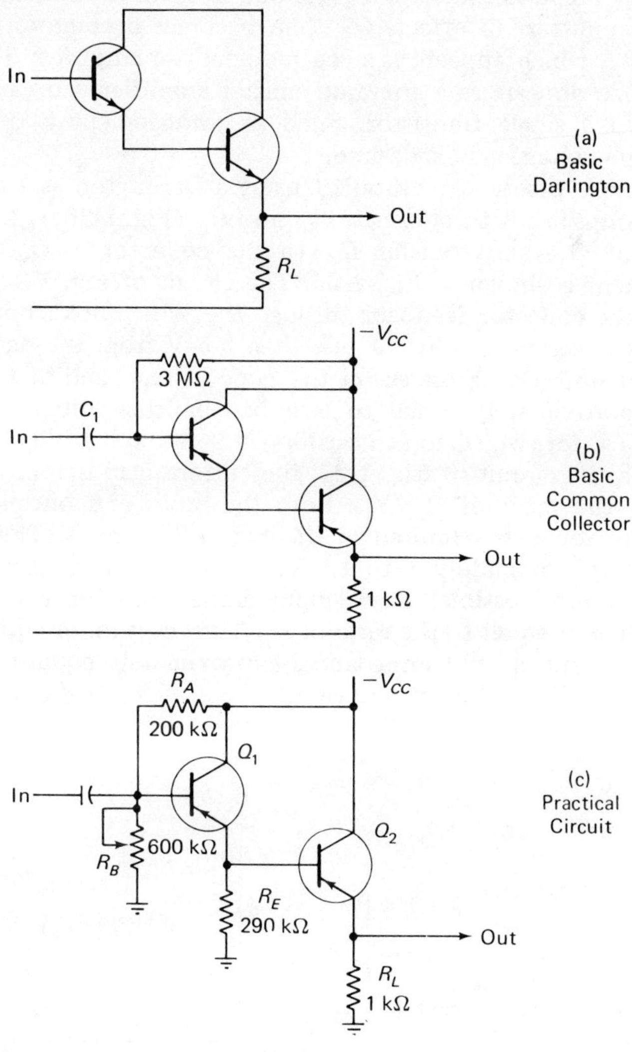

Figure 1-26. Basic and practical Darlington compounds.

the input and an output of 2 V appears across R_1. This results in a voltage loss. However, the input power is $2.5^2/225{,}000$, or approximately 0.028 mW. The output power is $2^2/1000$, or 4 mW. Thus, the power gain is 4/0.028, or about 140.

Darlingtons as basic common-emitter amplifiers. Darlington compounds can be used as common-emitter amplifiers to provide voltage gain. This is done by adding a collector resistor to any of the circuits in Fig. 1-26, and taking the output from the collector rather than the emitter. In effect, Q_1 then becomes a common-collector driving Q_2, which appears as a common-emitter amplifier. The entire circuit then appears as a common-emitter amplifier and can be used to replace a single transistor. Such an arrangement is often used where high-voltage gain is desired.

A more practical method of using a Darlington as a common-emitter amplifier is to eliminate R_B and R_E (Fig. 1-26c), ground the emitter of Q_2, and transfer R_L to the collector of Q_2. Such an arrangement is shown in Fig. 1-27. The circuit of Fig. 1-27 is stabilized by the collector feedback through R_A, which holds both collectors at a potential somewhat less than 0.5 V from the base of Q_1. Note that both collectors are at the same voltage, and that this voltage is approximately equal to two base-emitter voltage drops (or about 1.5 V for two silicon transistors).

With the circuit of Fig. 1-27, the current gain is approximately equal to the ratio of R_A/R_L. Both the input and output voltage swings are somewhat limited in the Fig. 1-27 circuit. The input is biased at approximately 1 to 1.5 V. However, voltage gains of 100 (or more) are possible, since input impedance (or resistance) is approximately equal to the ratio of R_A /current gain, or equal to R_L. (With input and output impedances approximately equal to R_L, the voltage gain follows the current gain.)

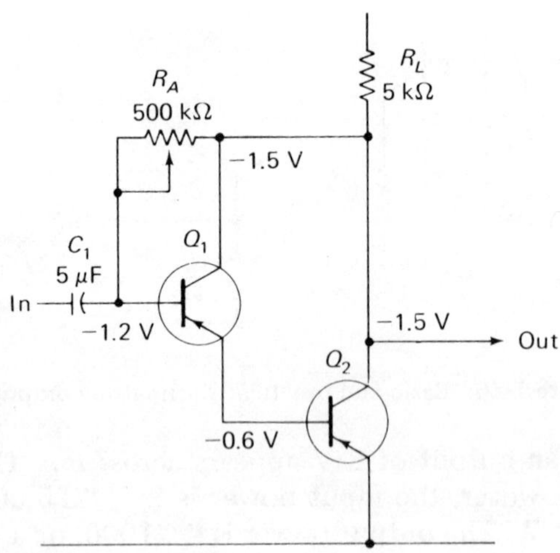

Figure 1-27. Darlington compound with collector feedback and common-emitter output.

Multistage Darlingtons. Darlington compounds need not be limited to two transistors. Three (and even four) transistors can be used in the Darlington circuit. A classic example of this is the familiar General Electric circuit of Fig. 1-28. This circuit is essentially a common-collector and common-emitter Darlington, followed by a common-emitter amplifier.

With R_I out of the circuit, both the input and output impedances are set by R_L. (In practice, the input impedance is slightly higher than R_L, typically on the order of 700 to 800 Ω.) With R_I removed, the voltage gain is about 1000. When R_I is used, the input impedance is approximately equal to R_I, and the voltage gain is reduced accordingly. For example, if R_I is 10 kΩ, the 1000 voltage gain drops to about 50.

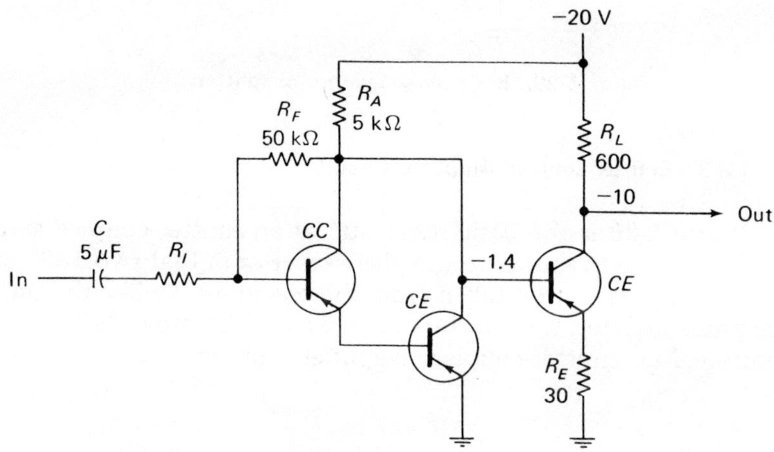

Figure 1-28. General Electric CC-CE-CE multistage Darlington compound.

1-8-2 Phase Inverter or Splitter Circuits

Figure 1-29 is the basic schematic of a phase inverter or splitter. As shown, the circuit is essentially a single-stage amplifier, with an output taken from the emitter as well as the collector. The values of R_L and R_E are usually equal, or nearly equal, so that there is no voltage gain. The collector output is 180° out of phase with the input and, therefore, 180° out of phase with the emitter.

The main reason for using a phase inverter or splitter in audio work is to produce two 180° out-of-phase signals to drive a push-pull output stage. (A phase inverter is the transformerless version of the interstage transformer with an untapped primary and a center-tapped secondary.)

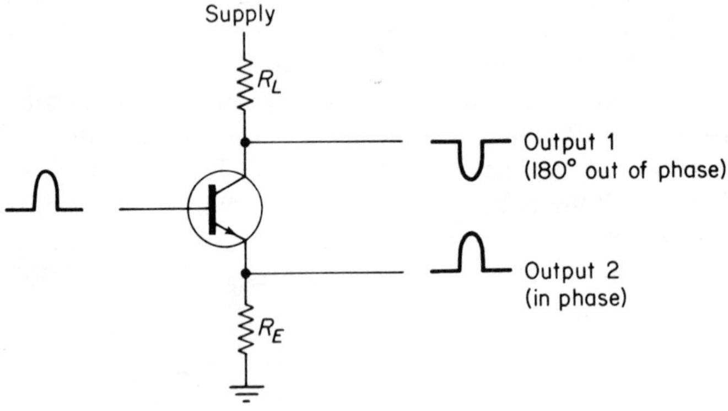

Figure 1-29. Basic phase inverter or splitter.

1-8-3 Emitter-coupled Amplifier Circuits

Figure 1-30 is the basic schematic of an emitter-coupled amplifier. This circuit is similar to the phase inverter in that two 180° out-of-phase outputs can be taken from the one input. Unlike the single-stage phase inverter, the emitter-coupled amplifier provides high gain. Therefore, an emitter-coupled amplifier can be used in a design

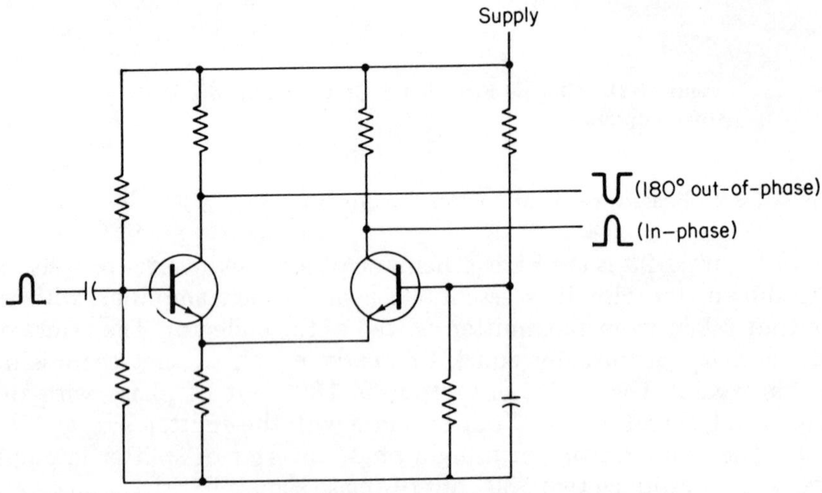

Figure 1-30. Basic emitter-coupled amplifier.

where a low-voltage input must be amplified to drive a push-pull output stage.

1-8-4 Differential Amplifier Circuits

Figure 1-31 is the basic schematic of a differential amplifier. The differential amplifier is similar to the emitter-coupled amplifier except that the two output signals are the result of a *signal difference* between the two inputs. An output is produced only when there is a difference in signals at the input. Differential amplifiers are particularly useful (especially in oscilloscope, electronic meter, and record-

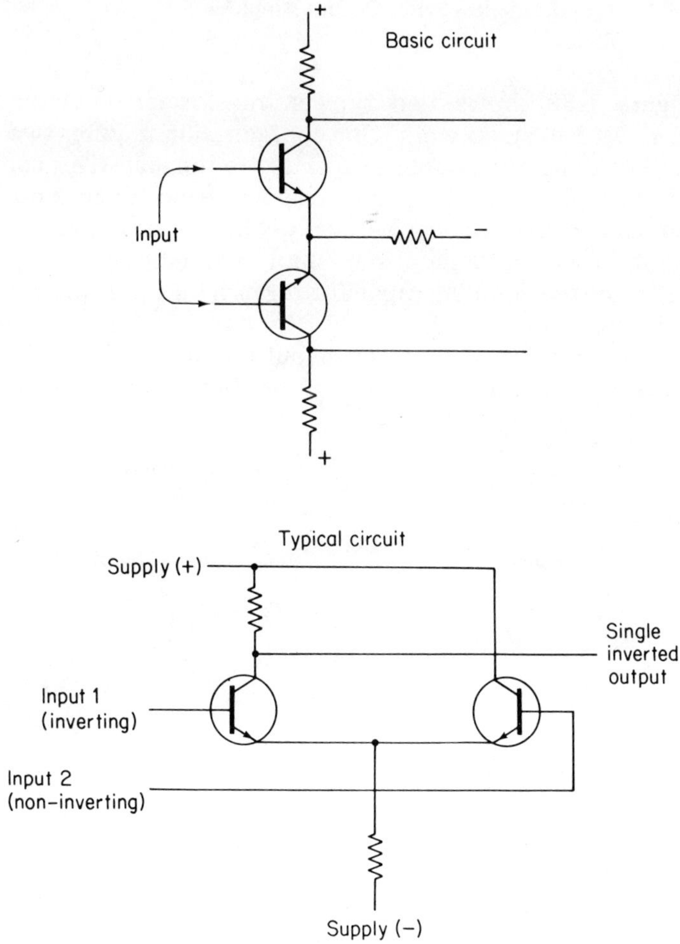

Figure 1-31. Basic differential amplifier.

ing instrument amplifiers) because signals common to both inputs (known as *common-mode signals*) are eliminated or greatly reduced. A common-mode signal (like power-line pickup) drives both bases in phase with equal amplitude a-c voltages, and the amplifier behaves as though the transistors are in parallel. The emitter resistor introduces emitter feedback, which reduces the common-mode signal gain without reducing the differential signal gain.

As described in Chapter 9, a differential amplifier is used as the input circuit of an operational amplifier. The ability of a differential amplifier to prevent conversion of a common-mode signal into a difference signal is expressed by its common-mode rejection ratio.

1-8-5 Transformerless Series-Output Amplifier Circuits

Figure 1-32 shows two typical transformerless series-output amplifiers used in audio work. One configuration requires two power supplies but omits the coupling capacitor to the load. This configuration provides better low-frequency response (since there is no capacitor), but can be inconvenient because of the two power supplies. The configuration with a single power supply has reduced low-frequency response since the coupling capacitor forms a high-pass filter with the load resistance.

Either configuration of series output has two drawbacks. A phase inverter is required to drive the series-output stage, even if gain

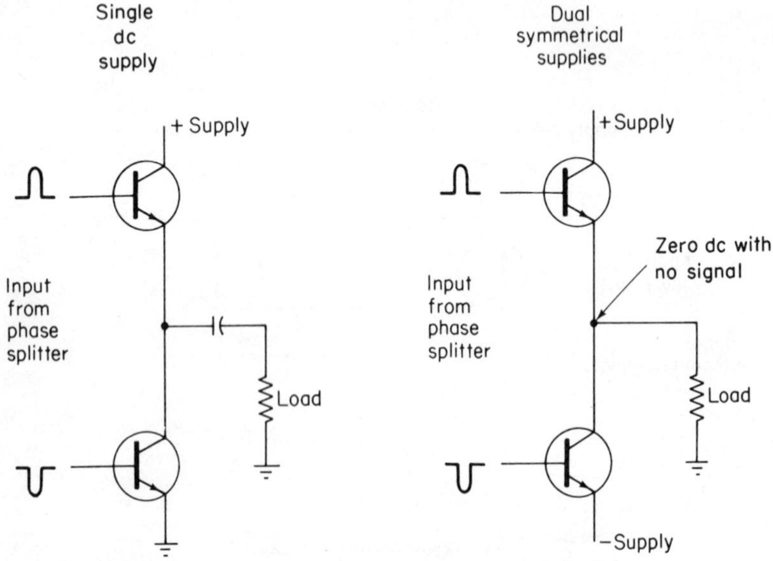

Figure 1-32. Basic transformerless series output amplifiers.

is not required. Also, an additional driver stage may be necessary to bring the power up from the output of a voltage amplifier to a level required by the series-output stage. These problems are overcome by means of a *complementary* circuit, either quasi-complementary or full-complementary.

1-8-6 Quasi-Complementary Amplifier Circuits

Figure 1-33 is the schematic of a quasi-complementary output. This circuit consists of a Darlington compound using NPN transistors and a direct-coupled compound using an input PNP and an output NPN. Both base signals can be in phase (although they are often at different voltage levels), so that a phase inverter is not needed. For example, if the input is positive-going, Q_1 is forward-biased, as Q_2. A positive-going input at Q_3 reverse-biases Q_3 since it is PNP. This produces a negative-going output from Q_3 to Q_4, which is NPN, and reverse-biases Q_4. When the signal at the bases of Q_1 and Q_3 is negative-going, the condition reverses (Q_3 is reverse-biased; Q_4 is forward-biased).

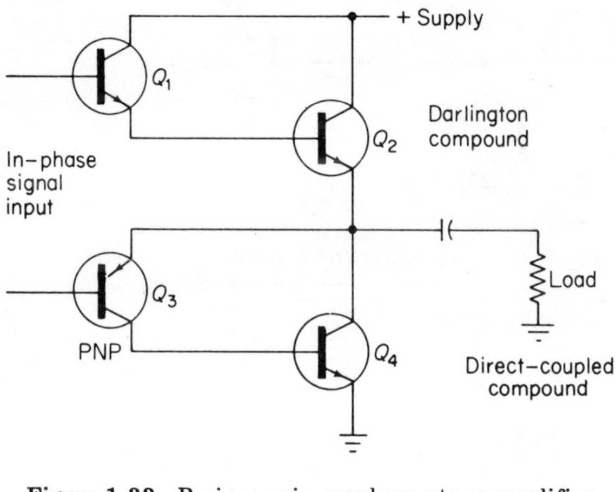

Figure 1-33. Basic quasi-complementary amplifier.

1-8-7 Full-Complementary Amplifier Circuits

Figure 1-34 shows two versions of the full-complementary output. Either version has an advantage over the quasi-complementary in that both halves of the circuit are identical. This makes it easier to match both halves (for positive and negative signals) to minimize distortion that could be caused by uneven amplification of the signal.

One complementary circuit uses two Darlington compounds. This circuit is also known as a *dual-Darlington* output and is used where power gain is needed. The other circuit uses two direct-coupled compounds and is used where voltage gain is most needed. Phase inversion is not required for either circuit. A positive-going input forward-biases Q_1 and Q_2 and reverse-biases Q_3 and Q_4. A negative-going input produces opposite results. As in the case of series-output circuits (Sec. 1-8-5) the load-coupling capacitor can be omitted if two power supplies are used (one positive and one negative). The trade-off between the inconvenience of two power supplies versus improved low-frequency response must be decided by circuit require-

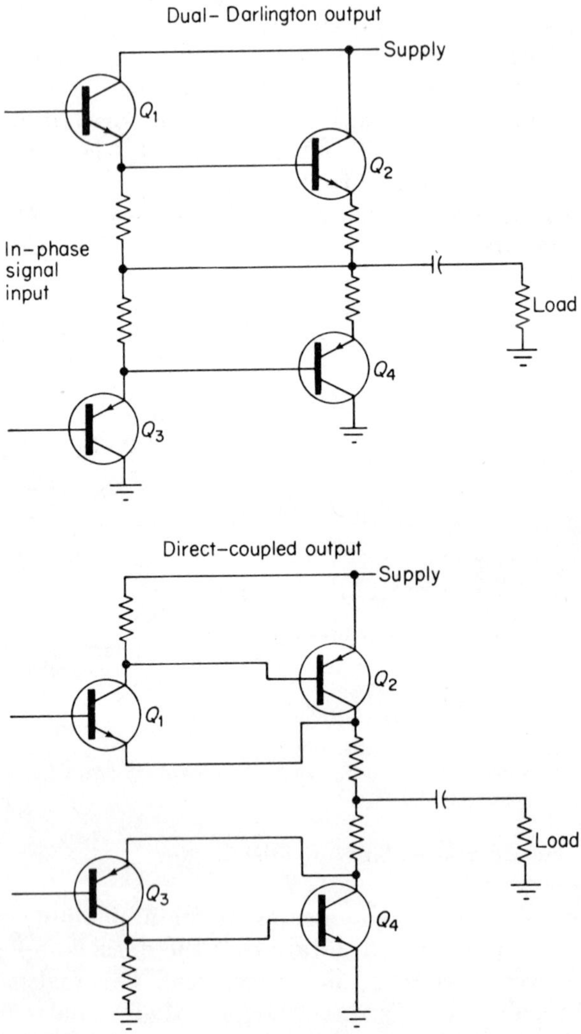

Figure 1-34. Basic full-complementary amplifier.

ments. As a guideline, a 2000 μF capacitor into a 4-Ω load (such as a 4-Ω loudspeaker) produces an approximate 3-dB drop at 20 Hz (a 10-V output will drop to 7 V).

1-8-8 Secondary Breakdown and Short-Circuit Protection

One problem with any of the output circuits is that one (or more) of the collectors will operate without a load resistance. In effect, the output load forms the load for the collector. Although the working load does present a load similar to that of a collector resistor, there is one major difference. The working load in any of the output circuits is not in series with the collector, as is the load resistor in a basic class A amplifier. In the simple class A amplifiers (Secs. 1-4 and 1-5), the collector load resistor reduces the collector voltage when heavy current is flowing and raises the voltage only when a light current is present. If there should be a short in the output, heavy current may flow but the collector voltage will drop.

In the various series- and complementary-output circuits, a short circuit of the load produces heavy current without reducing the collector voltage. This can result in the destruction of the transistors. Therefore, some short-circuit protection should be considered for any output configuration involving circuits without collector resistors. (This includes most class B and AB low-frequency amplifiers, as well as class C RF circuits, discussed in Chapter 2.)

The destructive condition is usually known as *secondary breakdown* or *second breakdown* and results from a sudden channeling of collector current into a localized area of the transistor. Secondary breakdown is usually prevented by limiting the collector current-voltage product.

1-9 EXAMPLES OF TRANSFORMERLESS MULTISTAGE AUDIO-AMPLIFIER CIRCUITS

This section describes the design, as well as operation, of a typical transformerless multistage amplifier circuit using plastic transistors. Such transistors, used in the output circuits of low-power audio amplifiers, permit maximum performance at minimum cost. Motorola Semiconductor Products, Inc., has developed several such transistors, together with complementary-symmetry audio-amplifier circuits that take full advantage of the transistor characteristics.

The circuit described here can be fabricated using the component values and supply voltages shown in the parts list of the related schematic diagram. However, the circuit is quite flexible and can serve as a starting point for design of similar circuits. To make full use of this flexibility, the circuit diagrams and parts list are supplemented with design notes that explain the why and how of circuit performance.

1-9-1 Three-Transistor Audio-Amplifier Circuits

The basic three-transistor shown in Fig. 1-35 can provide 2-W output with suitable component and power-supply values. Table 1 in Fig. 1-35 shows the performance of various versions of the circuit. Table 2 in the figure lists the output power of each version operated at the higher-impedance loads. The following design notes apply to all versions of the circuit in Fig. 1-35.

Supply voltage. The supply voltage (V_{CC}) is chosen to provide the necessary signal swing across the speaker. V_{CC} must also be of sufficient magnitude to offset the losses in the transistors and associated parts.

Center voltage. The center voltage (or the voltage at the junction of emitter resistors R_7 and R_8) must be one-half of the supply voltage to ensure maximum signal swing at the speaker.

Closed-loop gain. The closed-loop a-c gain in the Fig. 1-35 circuit is set by the ratio of R_3 to R_1. The d-c conditions of the circuit determine the value of R_3. The input impedance of the circuit is approximately equal to the resistance of R_1.

Crossover distortion. To offset crossover distortion, which is most evident at low signal levels, the output transistors are forward-biased to allow a small "idling" current through their collectors. (This is similar to class AB operation described in Secs. 1-4-3 and 1-10.) In the Fig. 1-35 circuit, the idling current is produced by the forward voltage of D_1 and the voltage drop across R_5.

Thermal runaway. The heating of the output transistors by signal power and higher ambient temperatures can cause the idling current to increase to a value that might cause thermal runaway. For those not familiar with thermal runaway, a brief description is in order. When current passes through a transistor junction, heat is generated. If all of this heat is not dissipated by the case, the junction temperature increases. In turn, this causes more current to flow through the junction even though the voltage, circuit values, and so on, remain the same. This causes the junction temperature to increase even further, with a corresponding increase in current flow. If the heat is not dissipated by some means, the transistor will burn out and be destroyed.

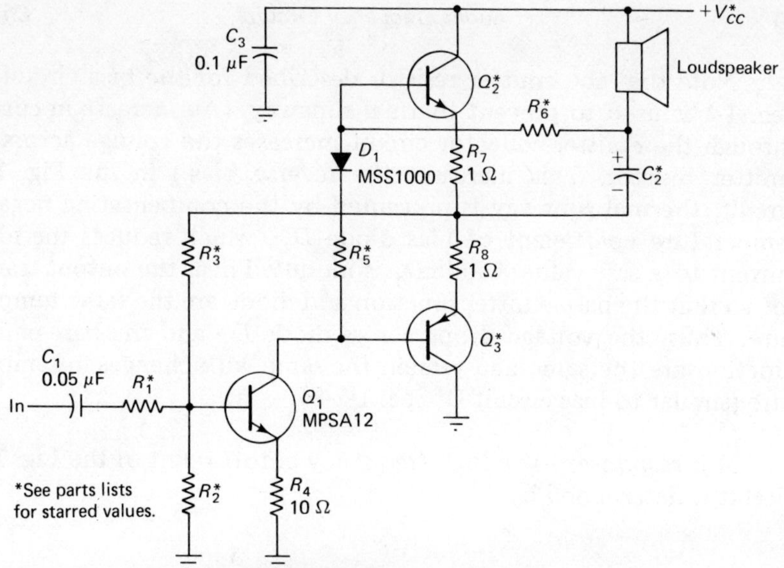

Parts List

Part no.	Loudspeaker characteristics				
	1 W, 8 Ω	1 W, 16 Ω	1 W, 40 Ω	2 W, 16 Ω	2 W, 8 Ω
V_{CC}	12 V	14 V	20 V	18 V	15 V
R_1	560 kΩ	680 kΩ	820 kΩ	560 kΩ	680 kΩ
R_2	560 kΩ	560 kΩ	750 kΩ	470 kΩ	560 kΩ
R_3	2.2 MΩ	2.7 MΩ	5.6 MΩ	3.3 MΩ	2.7 MΩ
R_5	27 Ω	27 Ω	33 Ω	22 Ω	12 Ω
R_6	560 Ω	820 Ω	1 kΩ	220 Ω	200 Ω
C_2	250 μF	150 μF	100 μF	150 μF	250 μF
Q_2	MPS6560	MPS6561	MPSA05	MPS6560	MPSU01 (Uniwatt)
Q_3	MPS6562	MPS6563	MPSA55	MPS6562	MPSU51 (Uniwatt)

Table 1. Performance

Performance feature	1W, 8 Ω	1 W, 16 Ω	1 W, 40 Ω	2 W, 16 Ω	2W, 8 Ω
Sensitivity at rated P_{out}	0.8 V	1.15 V	0.82 V	1.2 V	1.4 V
Input current, rated P_{out}	160 mA	110 mA	70 mA	180 mA	240 mA
Idling current	9 mA	8 mA	9 mA	10 mA	17 mA
Input impedance	560 kΩ	680 kΩ	820 kΩ	560 kΩ	680 kΩ
Approximate high-frequency cutoff	20 kHz	13 kHz	10 kHz	14 kHz	13 kHz

Table 2. Performance with different impedances

Amplifier version	P_{out}	16-Ω load, average d–c current	P_{out}	40-Ω load, average d–c current
1 W, 8 Ω	0.8 W	110 mA	0.35 W	60 mA
1 W, 16 Ω	1 W	110 mA	0.65 W	70 mA
2 W, 8 Ω	1 W	110 mA	0.65 W	70 mA
2 W, 16 Ω	2 W	180 mA	1.1 W	90 mA

Figure 1-35. Three-transistor audio-amplifier circuit using plastic transistors.

Note that the emitter resistor described for the bias circuits of Sec. 1-1 is used to prevent thermal runaway. (An increase in current through the emitter-collector circuit increases the voltage across the emitter resistor. This increases the reverse bias.) In the Fig. 1-35 circuit, thermal runaway is prevented by the compensating negative temperature coefficient of bias diode D_1, which reduces the idling current to a safe value. Diode D_1 is mounted near the output transistor so that the base-emitter junction and diode are the same temperature. Thus, the voltage drops across diode D_1 and the base-emitter junction are the same and remain the same with changes in temperature (similar to bias circuit D, Sec. 1-1-8).

A-c response. The high-frequency cutoff point of the Fig. 1-35 circuit is determined by

$$\text{high-frequency cutoff} = \frac{1}{6.28F \times R_{\text{in}} \times C_{\text{in}}}$$

where R_{in} is the resistance of R_1 and C_{in} is the input capacitance of Q_1. Typically, the high-frequency cutoff is between 10 and 20 kHz, depending on which version of the circuit is used. The low-frequency cutoff point is determined by the input impedance in series with C_1 and the speaker impedance in series with C_2. An approximate low-frequency cutoff point can be found using the equations of Sec. 1-2. Bypass capacitor C_3 is connected from V_{CC} to ground to prevent possible oscillation caused by high-current transients in the power supply leads.

Grounding the loudspeaker. In some systems, it may be desirable to ground the "cold" side of the speaker instead of connecting the speaker to the supply voltage. This is done by using the alternative circuit of Fig. 1-36. The positive half-cycle drive current for Q_2 is furnished through R_6. Assuming that C_4 is absent, it is apparent that the voltage drop across R_6 is going to be reduced as the output is driven positive, thus reducing the drive current to Q_2. This causes clipping at a point much lower than V_{CC}.

In the basic circuit of Fig. 1-35, output capacitor C_2 is charged to one-half of V_{CC}. When the output voltage rises to V_{CC}, the voltage on the load side of C_2 is effectively 1.5 times V_{CC}. This "extra" d-c charge allows full drive current to Q_2 on the positive swing of the output. In the alternative circuit of Fig. 1-36, the bootstrap capacitor C_4 provides the same effect.

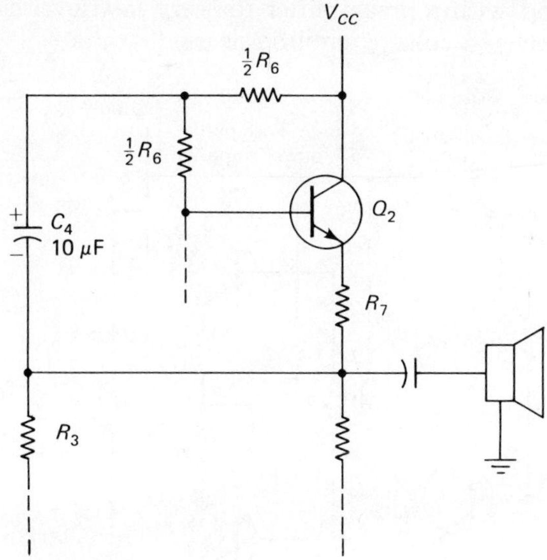

Figure 1-36. Alternate circuit for grounding the loudspeaker.

1-10 MULTISTAGE TWO-JUNCTION TRANSISTOR AMPLIFIERS WITH TRANSFORMER COUPLING

Audio-amplifier circuits can be coupled by means of transformers. Iron-core transformers are used in the AF stage, particularly when power amplification is required. As with any coupling methods, transformers impose certain problems and have certain advantages. These factors can be traded off to meet a specific design need.

1-10-1 Circuit Analysis

Figure 1-37 is the working schematic for a classic transformer-coupled audio-amplifier circuit. The circuit has a class A input or driver stage and a class B push-pull output stage. The class A stage provides both voltage and power gain as needed to raise the low-input signal to a level suitable for the class B power output stage.

The class A stage can be transformer-coupled or *RC*-coupled at the input as needed. Transformer coupling is used at the input where a specific impedance-match problem must be considered in design. The class A input stage can be driven directly by the signal source, or can be used with a preamplifier for very low level signals. When required, a high-gain voltage amplifier is used.

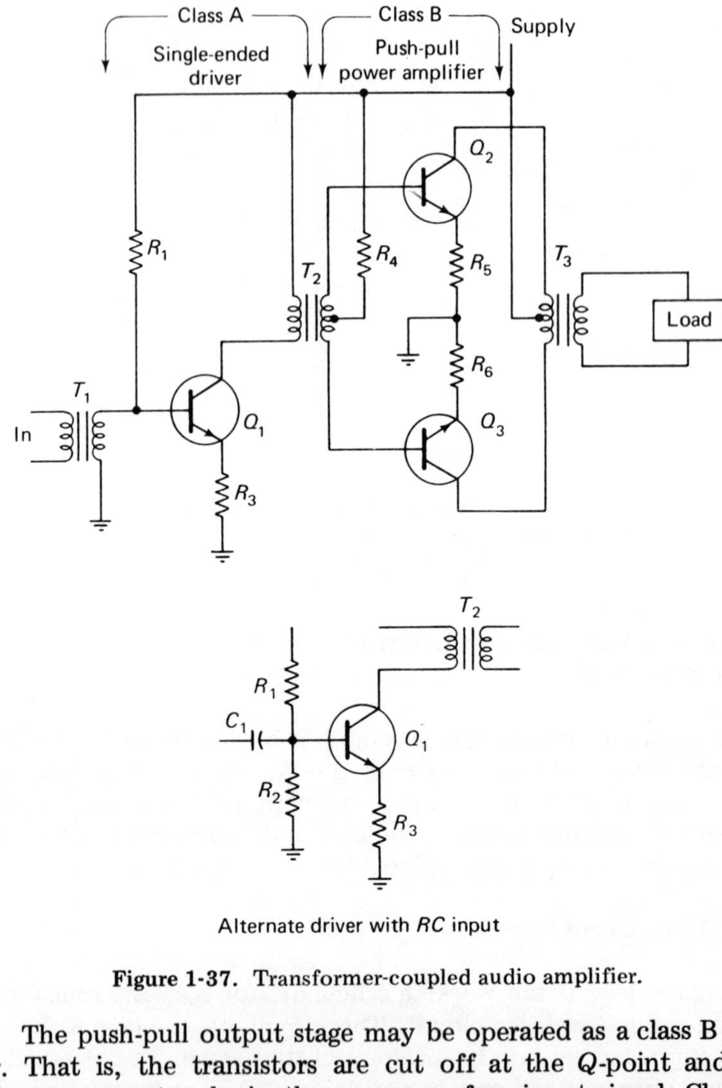

Alternate driver with *RC* input

Figure 1-37. Transformer-coupled audio amplifier.

The push-pull output stage may be operated as a class B amplifier. That is, the transistors are cut off at the Q-point and draw collector current only in the presence of an input signal. Class B is the most efficient operating mode for audio amplifiers, since it draws the least amount of current (and no current where there is no signal). However, true class B operation can result in crossover distortion, as discussed in Sec. 1-4.

The effects of crossover distortion can be analyzed by comparison of the input and output waveforms on Fig. 1-38. In true class B operation, the transistor remains cut off at very low signal inputs (because transistors have low-current gain at cutoff) and turns on abruptly with a large signal. As shown in Fig. 1-38, there is no conduction when the base-emitter voltage V_{BE} is below about 0.65 V (for a silicon transistor). During the instantaneous pause when one transistor stops conducting and the other starts conducting, the output waveform is distorted.

Distortion of the signal is not the only bad effect of crossover distortion. The instantaneous cutoff of collector current can set up large voltage transients equal to several times the size of the supply

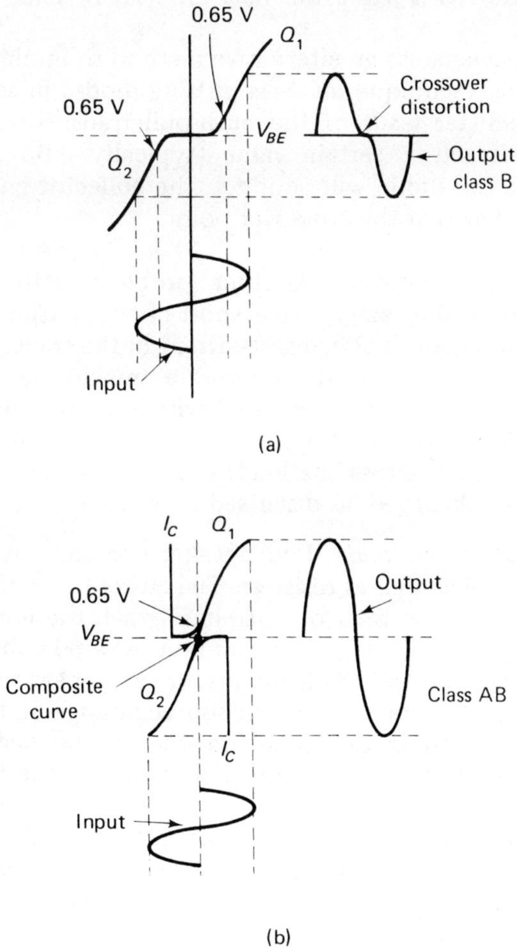

Figure 1-38. Effects of crossover distortion, and how it is eliminated by forward bias of the base-emitter junction.

voltage. This can cause the transistor to break down. Crossover distortion can be minimized by operating the output stage as class AB (or somewhere between B and AB). That is, the transistors are forward-biased just enough for a small amount of collector current to flow at the *Q*-point. Some collector current is flowing at the lowest signal levels, and there is no abrupt change in current gain. The effects of this are shown in Fig. 1-38b. The combined collector currents result in a *composite curve* that is essentially linear at the crossover point. This produces an output that is a faithful reproduction of the input, at least as far as the crossover point is concerned. Of course, class AB is less efficient than class B, since more current must be used.

Some designers use an alternative method to minimize crossover distortion. This technique involves putting diodes in series with the collector or emitter leads of the push-pull transistors. Because the voltage must reach a certain value (typically 0.65 V for silicon diodes) before the diode will conduct, the collector current curve is rounded (not sharp) at the crossover point.

Short-circuit burnout. Another problem with audio transformer circuits is the danger of a short-circuit output, resulting in excessive, simultaneous voltage and current in the transistor collector. There is very little voltage drop across a transformer winding in a transistor collector circuit (compared with the drop across the load in an *RC* or direct-coupled amplifier). If a short circuit in the output (say due to a short across the load) causes heavy current flow, the transistor can be damaged, as discussed in Sec. 1-8-8.

Loudspeaker overload. Loudspeaker systems can appear capacitive or inductive as well as resistive. The current and voltage appearing in the amplifier is thus out of phase when the load appears reactive. A 60° phase shift is not uncommon. At a 60° phase shift, half the supply voltage and all the load current can appear simultaneously at the output transistor, or the full source voltage and one-half the load current can appear, depending on whether the load is capacitive or inductive. This can result in breakdown unless the transistor can dissipate the resulting power. If the loudspeaker is disconnected for any reason, the power is not transferred from the transformer primary to secondary. Under these conditions, the full voltage and most of the current is applied to the output transistors. This often results in breakdown.

Chapter 2

Radio-Frequency Circuits

This chapter is devoted entirely to radio-frequency (RF) circuits. These circuits include resonant circuits, tuning networks, VVC networks, voltage amplifiers, mixers and converters, AVC-AGC circuits, RF detectors, power amplifiers, and multipliers.

2-1 RESONANT CIRCUITS FOR RF AMPLIFIERS

RF amplifier circuits are based on the use of resonant circuits (tank circuits) consisting of a capacitor and a coil (inductance) connected in series or parallel, as shown in Fig. 2-1. At the resonant frequency, the inductive and capacitive reactances are equal, and the circuit acts as a high impedance (if it is a parallel circuit) or a low impedance (if it is a series circuit). In either case, any combination of capacitance and inductance has some resonant frequency.

Either (or both) the capacitance or inductance can be variable to permit tuning of the resonant circuit over a given frequency range. When the inductance is variable, the circuit is usually tuned by means of a metal slug (typically powdered iron) inside the coil. The metal

Capacitive Reactance

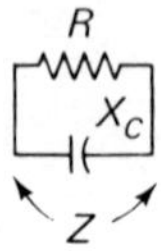

$$Z = \frac{RX_C}{\sqrt{R^2 + X_C^2}} \qquad Q = \frac{R}{X_C}$$

Parallel

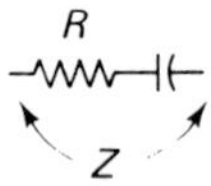

$$Z = \sqrt{R^2 + X_C^2} \qquad Q = \frac{X_C}{R}$$

Series

$$X_C = \frac{159}{F\ (\text{kHz}) \times C\ (\mu\text{F})}$$

$$F\ (\text{kHz}) = \frac{159}{(X_C) \times C\ (\mu\text{F})}$$

$$C\ (\mu\text{F}) = \frac{159}{F\ (\text{kHz}) \times X_C}$$

Inductive Reactance

$$Z = \frac{RX_L}{\sqrt{R^2 + X_L^2}} \qquad Q = \frac{R}{X_L}$$

Parallel

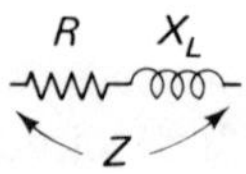

$$Z = \sqrt{R^2 + X_L^2} \qquad Q = \frac{X_L}{R}$$

Series

$$X_L = 6.28 \times F\ (\text{kHz}) \times L\ (\text{mH})$$

$$F = \frac{X_L}{6.28L}$$

$$L = \frac{X_L}{6.28F}$$

Impedance and Resonance

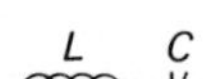

Series (zero impedance)

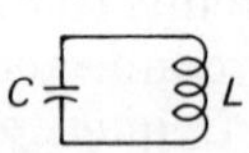

Parallel (infinite impedance*)

$$C\ (\mu\text{F}) = \frac{2.54 \times 10^4}{F\ (\text{kHz})^2 \times L\ (\mu\text{H})}$$

$$F\ (\text{kHz}) = \frac{159}{\sqrt{L\ (\mu\text{H}) \times C\ (\mu\text{F})}}$$

$$L\ (\mu\text{H}) = \frac{2.54 \times 10^4}{F\ (\text{kHz})^2 \times C\ (\mu\text{F})}$$

*When circuit Q is 10 or higher

Figure 2-1. Resonant circuit equations.

slug is screwdriver-adjusted to change the inductance (and thus the inductive reactance) as required. Typical RF circuits used in receivers (AM, FM, communications, etc.) often include two resonant circuits in the form of a transformer (RF or IF transformer, etc.). Either the capacitance or inductance can be variable. In the case of RF transmitter circuits, it is sometimes necessary to design the coil portion of the resonant circuit. This is because coils of a given inductance and physical size may not be available from commercial sources.

2-1-1 Basic Design Considerations for Resonant Circuits

The two most important considerations for RF resonant circuits are *resonant frequency* and the *Q* (or *quality*)-factor.

Resonant frequency. Figure 2-1 contains equations that show the relationship between capacitance, inductance, reactance, and frequency as they relate to resonant circuits. Note that there are three sets of equations. Two sets include reactance (inductive and capacitive) and the third set omits reactance. The reason for three sets of equations is that some design approaches require the reactance to be calculated for resonant networks. Solid-state RF transmitter circuits are a good example of this.

Quality factor and selectivity. A resonant circuit has a *Q*-factor that is directly related to the selectivity of the circuit, and is dependent upon the ratio of reactance to resistance. If a resonant circuit has pure reactance, the *Q* is high (theoretically infinite). However, this is not practical. For example, all coils and the leads of all capacitors have some d-c resistance. Also, as frequency increases, the a-c resistance presented by the leads increases due to skin effect. The sum total of these reactances is usually lumped together and considered as a resistor in series or parallel with the resonant circuit. The total resistance is usually termed the *effective* resistance and is not to be confused with reactance.

The resonant circuit *Q* depends on the individual *Q*-factors of both the inductance and capacitance used in the circuit. For example, if both the inductance and capacitance have a high *Q*, the circuit will have a high *Q*, provided that a minimum of resistance is produced when the inductance and capacitance are connected to form a resonant circuit. Usually, resonant circuit *Q* is measured at points on either side of the resonant frequency, where the signal amplitude is down 0.707 of the peak resonant value, as shown in Fig. 2-2. Notice that the resonant circuit with a high *Q* produces a sharp resonance curve (narrow bandwidth), whereas a low Q produces a broad resonance curve (wide bandwidth). For example, a high-Q resonant circuit provides good harmonic rejection (tends to pass only the fundamental frequency) and efficiency in comparison with a low-Q circuit, all other factors being equal. Thus, the *selectivity* or a resonant circuit is related directly to the Q.

A very high Q (or high selectivity) is not always desired. In some applications, it is necessary to add resistance to a resonant circuit to broaden the response (increase the bandwidth, decrease the selectivity). An example of this is the damping resistor used across peaking coils in a video amplifier. If a given bandwidth must be maintained but the resonant frequency is increased, the Q must also increase. For example, if the resonant frequency is 30 MHz with a bandwidth of 3 MHz, the required Q is 10. If the resonant frequency is increased to 54 MHz with the same 3-MHz bandwidth, the required Q is 18. Also, Q must be decreased for increases in bandwidth if the same resonant frequency is to be maintained.

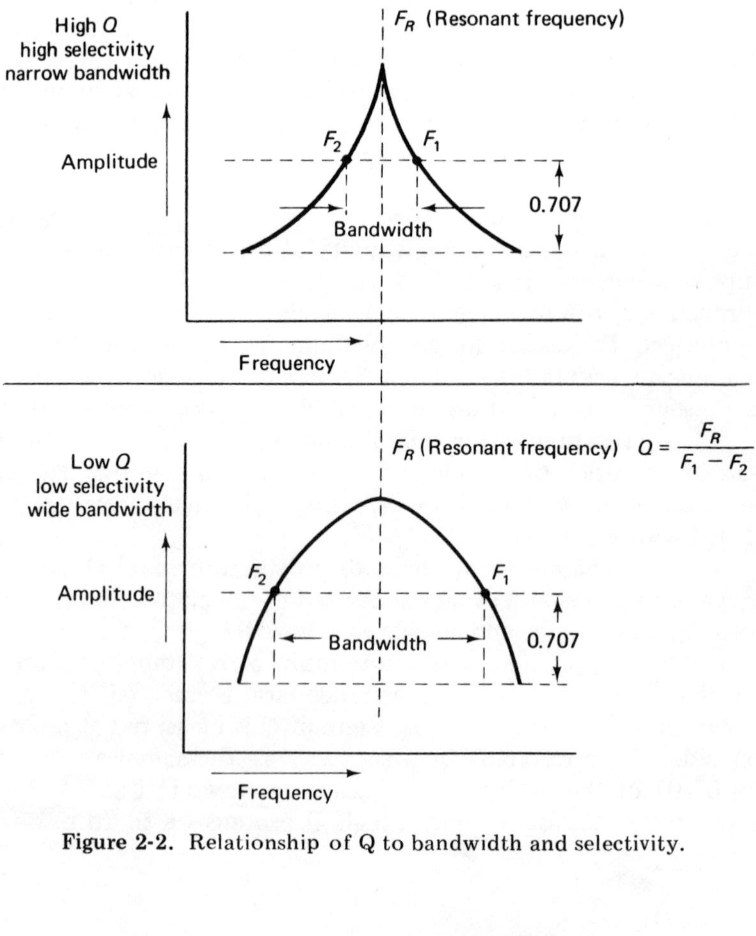

Figure 2-2. Relationship of Q to bandwidth and selectivity.

2-1-2 Basic Design Considerations for RF Coils

The equations necessary to calculate the self-inductance of a single-layer, air-core coil are given in Fig. 2-3. This type of coil is most efficient (maximum inductance for minimum physical size) when the ratio of coil radius to coil length is 1.25, that is, when the length is 0.8 of the radius. The equations of Fig. 2-3 are approximations only and do not take into account such factors as uneven sizes of turns, spacing between the turns, and the like. From a practical standpoint, use the equations to find the nearest number of turns (for a given inductance) and then spread or compress the turns as necessary to obtain a precise value of inductance (as measured on an inductance bridge).

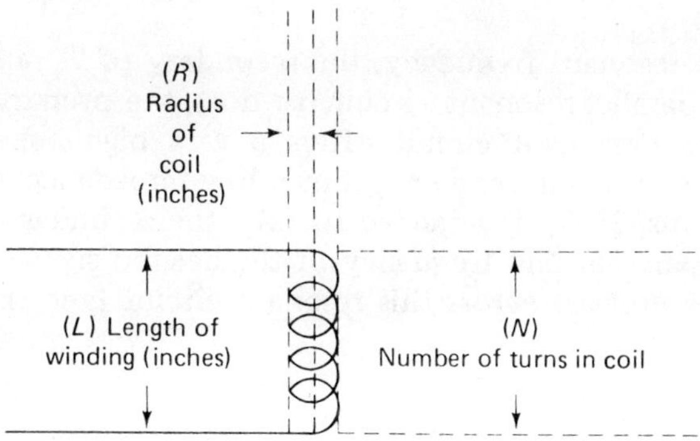

$$\text{Inductance } (\mu H) \approx \frac{(RN)^2}{9R + 10L}$$

When $L = 0.8 \times R$ or $R/L = 1.25$

$$N \approx \sqrt{\frac{17 \times \text{inductance } (\mu H)}{R}}$$

$$\text{inductance } (\mu H) \approx \frac{(RN)^2}{R17}$$

Figure 2-3. Calculations for self-inductance of single-layer air-core coil.

2-2 TYPES OF RF AMPLIFIER TUNING CIRCUITS

As far as the tuning circuits are concerned, there are only three basic types of RF amplifiers: voltage amplifiers, power amplifiers, and multipliers. Going further, there are only two basic methods by which the tuning networks match transistor impedances to source and load impedances. One method uses combinations of coils and capacitors which are tuned to cancel out impedance differences. The other method uses taps on the tank coils to match impedances.

2-2-1 RF Voltage Amplifier Tuning Circuits

The circuit of Fig. 2-4 is a typical narrowband RF amplifier, such as those found in broadcast- and communications-type radio receivers. The circuit is a single stage of tuned RF voltage amplification. Input to the transistor is by means of tuned RF transformer, and output is obtained by a similar device. Transformer T_1 is the input transformer and T_2 is the output transformer. The secondary winding of T_1 is tuned to resonance at the frequency of the incoming signal by means of C_1. The primary of T_2 is tuned to the same frequency by C_2.

At the resonant frequency, the secondary of T_1 and capacitor C_1 form a parallel-resonant circuit, as does the primary of T_2 and C_2. A parallel-resonant circuit offers a very high impedance to a signal at the resonant frequency but a low impedance at other frequencies. Thus, if C_1 is adjusted to tune the secondary winding of T_1 to resonance at the frequency of the desired signal, a relatively large voltage appears across this resonant circuit (and the transistor

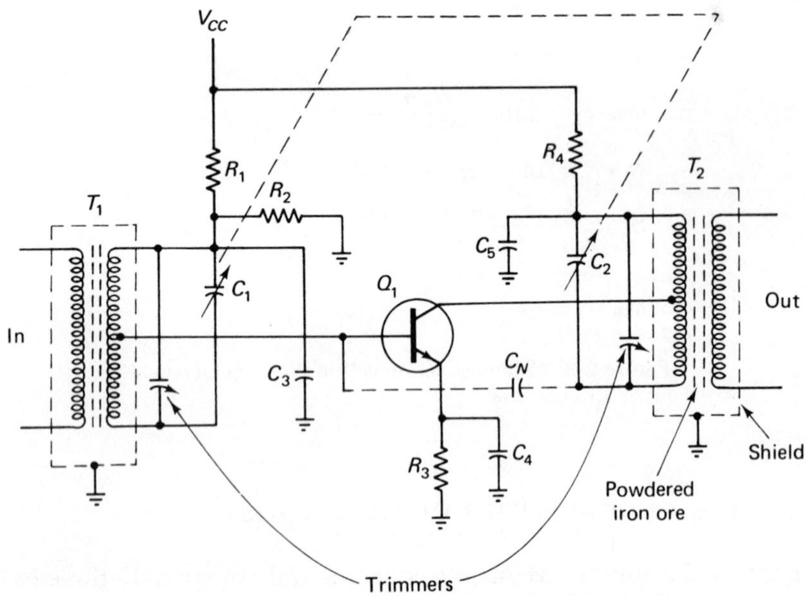

Figure 2-4. Tuned RF voltage amplifier.

base) for signals of this frequency. For signals of other frequencies, the voltage is low.

Similarly, if C_2 is adjusted to tune the primary of T_2 to resonance at the frequency of the desired signal, this resonant circuit will show a large impedance for signals of this frequency and a very low impedance for other signals. The C_2-T_2 resonant circuit also forms the *collector load*. As discussed in Chapter 1, voltage amplification of a stage is set by collector load impedances (all other factors being equal). These conditions provide high-voltage amplification at the desired frequency. At all other frequencies, where the collector load impedance is low, the amplification is also low.

The shunting effect of the transistor input and output capacitances is also minimized by the capacitances of the tuned resonant circuits. For example, the typically small input capacitance of the transistor is in parallel with the relatively large capacitance of variable capacitor C_1 and thus has but a small additive effect.

Bias networks. Resistors R_1 and R_2 form a voltage divider across the power supply (V_{CC}) to bias the emitter-base junction. The bias circuits for RF amplifier circuits are essentially the same as for audio circuits described in Chapter 1. However, the operating point may be different. Typically, audio amplifiers are operated class A (except for push-pull), where collector current flows at all times. RF amplifiers are generally operated as class B or C, where current flows only in the presence of a signal.

Resistor R_2 and capacitor C_3 form a decoupling network (Chapter 1) to prevent the RF signal from entering the power supply (through which the signal may be fed to the output circuit or another stage). Resistor R_3 is the emitter stabilization resistor and C_4 is the bypass capacitor.

Impedance match. Note that the base of the transistor is connected to a tap on the secondary winding of T_1. Since the input impedance of the transistor is relatively small, only a portion of the secondary winding is used to obtain a proper impedance match between the two. However, the entire secondary winding is tuned by C_1 to form the parallel-resonant circuit at the signal frequency. For similar reasons the collector is connected to a tap on the primary of T_2.

Feedback problems. One difficulty often found in RF amplifiers is the prevention of feedback from the output of a stage to its own input or to another stage. There are two types of undesired feedback: *radiated feedback* and *feedback through the transistor.* (Of course, as discussed in Chapter 1, there is feedback that is deliberately introduced to stabilize gain, temperature response, etc.) To eliminate radiated feedback, the amplifier must be properly shielded and designed so as to separate the base and collector leads. The problem of RF amplifier shielding is extensive and will not be covered here because we are primarily concerned with transistors. Fortunately, most modern transistors are so constructed that there is little danger of feedback through the transistor at moderately high frequencies. However, as frequency increases, internal feedback can produce undesired conditions in an RF amplifier.

On feedback problem is known as the *Miller effect*. As shown in Fig. 2-5, there is a capacitance between the base and emitter of a

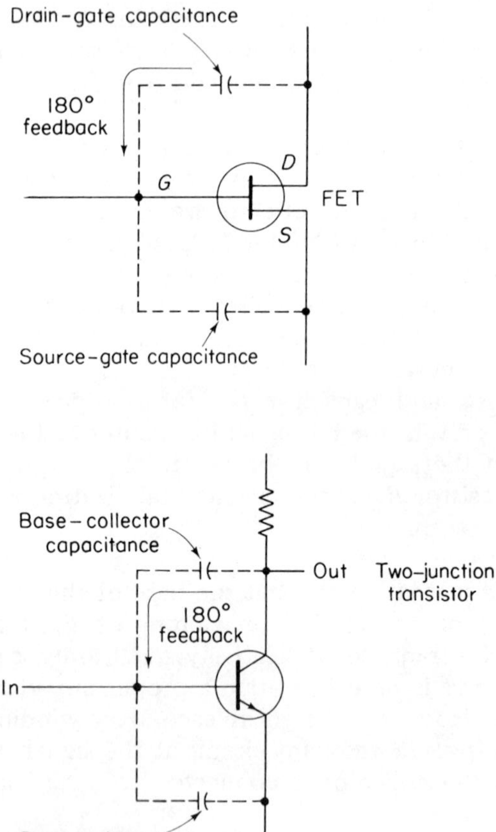

Figure 2-5. Input and feedback capacitances in transistors.

two-junction transistor (or between the gate and source of a FET). This forms the input capacitance of the circuit. There is also a capacitance between the base and collector (or the gate and drain). This capacitance feeds back some of the collector signal to the base. The collector signal is amplified and is 180° out of phase with the base signal (in a common-emitter amplifier). The collector signal feedback opposes the base signal and tends to distort the input signal. Similarly, the collector-base capacitance is, in effect, in series with the base-emitter capacitance and thus changes the input capacitance.

These conditions result in a constantly changing amplitude-modulated relationship of signals in an amplifier. For example, if the input signal amplitude changes, the amount of feedback changes, changing the input capacitance. In turn, the change in input capacitance changes the match between the transistor and the input tuned circuit, changing the amplitude. Similarly, if the input signal frequency changes, the feedback changes (since the collector-base capacitive reactance changes) and there is a corresponding change in amplification.

The Miller effect is not necessarily a problem in all solid-state RF amplifiers. The FET RF amplifier is usually more susceptible to the Miller effect than are two-junction transistors. However, when the Miller effect becomes severe with any RF amplifier, the effect can be eliminated or minimized to a realistic level of *neutralization.*

Neutralization is a method for reducing the amount of unwanted feedback, either from radiation or internal feedback. With neutralization, a portion of the voltage from the output circuit of the stage is fed back to the input circuit in such a way that it cancels the base voltage caused by the unwanted feedback. Neutralization is accomplished by impressing a voltage on the base that is equal in magnitude, but opposite in phase, to the undesired feedback. Thus, the two voltages cancel each other.

The two ends of the primary winding of the output transformer (such as T_2 in Fig. 2-4) are of opposite phase. If this opposite-phase voltage is fed to the base through the neutralizing capacitor (C_N of Fig. 2-4), the two voltages cancel out.

Another method for reducing the unwanted feedback, without neutralization, is to use the common-base (or common-gate) amplifier configuration. A common-base RF amplifier circuit is shown in Fig. 2-6. The base is grounded. The input signal is applied to the emitter. The output is obtained between the collector and base, which is common to the input and output circuits. The grounded base acts as a shield between the input and output circuits, thus reducing feedback.

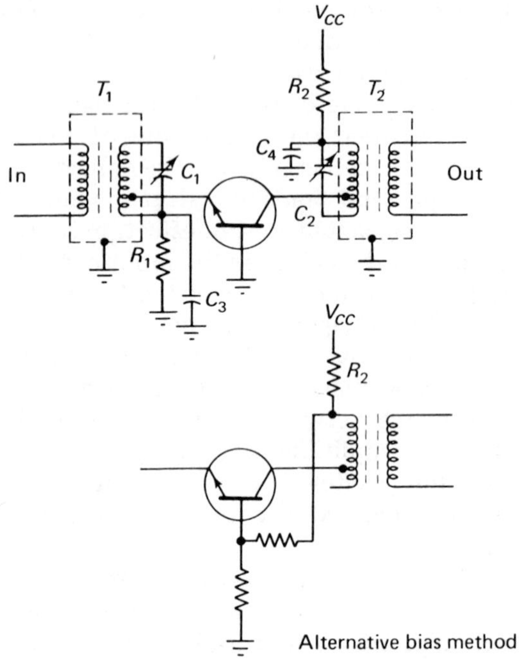

Figure 2-6. Common-base (grounded-base) RF amplifier.

2-2-2 RF Power-Amplifier Tuning Circuits

Most radio transmitters use some form of power amplifier to raise the low-amplitude signal developed by the oscillator to a high-amplitude signal suitable for transmission. For example, most oscillators develop signals of less than 1 W, whereas a solid-state transmitter may require an output of 100 W (or more).

Fig. 2-7 shows two basic RF power-amplifier circuits. In the circuit of Fig. 2-7a, the collector load is a parallel-resonant circuit (called a *tank circuit*) consisting of variable capacitor C_1 and inductor L_1, tuned to resonance at the desired frequency. The output, which is an amplified version of the input voltage, is from L_2, which together with L_1 forms an output transformer.

The circuit of Fig. 2-7b has both advantages and disadvantages. The winding of L_2 can be made to match the impedance of the load (by selecting the proper number of turns and by positioning L_2 in relation to L_1). Although that may prove to be an advantage in some cases, it also makes for an interstage coupling network that is subject to mismatch and detuning by physical movement or shock.

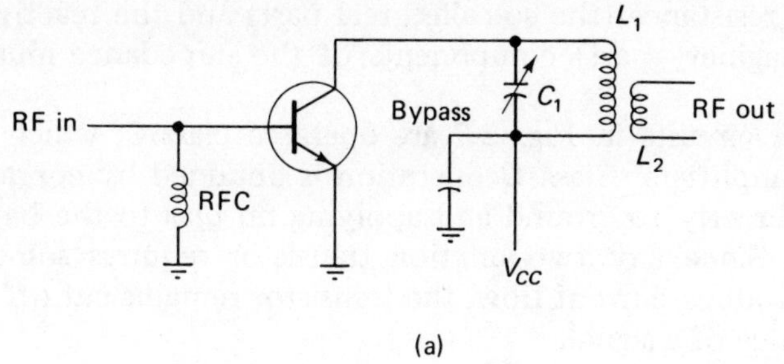

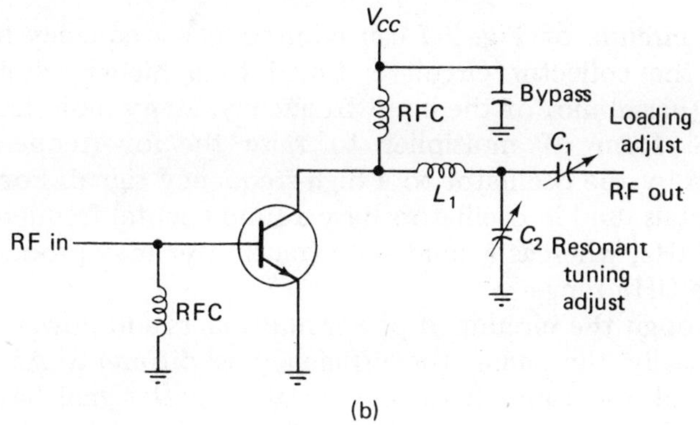

Figure 2-7. Radio-frequency power amplifiers.

Another disadvantage of the Fig. 2-7a circuit is that all of the current must pass through the tank circuit coil. Also, for best transfer of power, the impedance of L_1 should match that of the transistor output. Since two-junction transistor output impedances are generally low, the value of L_1 must be low, often resulting in an impractical size for L_1. The circuit of Fig. 2-7a is a carryover from vacuum-tube circuits and, as such, is not often found in modern two-junction transistor amplifiers. However, the circuit is found in FET RF amplifiers (which are generally of low power and higher impedance).

The circuit of Fig. 2-7b, or one of its many variations, is commonly found in solid-state radio transmitters using two-junction transistors (as discussed in Sec. 2-8). The collector load is a resonant circuit formed by the network of L_1, C_1, and C_2. Note that C_1 is marked "loading adjust," whereas C_2 is marked "resonant tuning adjust." These networks provide the dual function of frequency selection (equivalent to the tank circuit) and impedance matching between transistor and load. For impedances to be properly matched, both the resistance (the so-called real part) and the reactive (the so-called imaginary part) components of the impedance must be considered.

Both circuits in Fig. 2-7 are operated class C, which is typical for RF amplifiers. Class C operation is obtained by connecting the emitter directly to ground and applying no bias to the base-emitter junction. Since any two-junction transistor requires some forward bias to produce current flow, the transistor remains cut off except in the presence of a signal.

2-2-3 RF Multiplier Tuning Circuits

The circuits of Fig. 2-7 can be used as a frequency multiplier. That is, the collector circuit is tuned to a higher whole-number multiple (harmonic) of the input frequency. Many radio transmitters use some form of multiplier to raise the low-frequency signal developed by the oscillator to a high-frequency signal. For example, most crystals used in oscillators have a fundamental frequency of less than 10 MHz, whereas a solid-state transmitter may produce an output in the UHF range.

Although the circuits of power multipliers and power amplifiers are essentially the same, the efficiency is different. An amplifier operating at the same frequency as the input signal has a higher efficiency than does an identical circuit operating at a multiple of the input frequency.

2-2-4 RF Amplifier/Multiplier Tuning Circuit Combinations

The circuits of Fig. 2-7 can be cascaded to provide increased power amplification and/or frequency multiplication. Typically, no more than three stages are so cascaded. The stages can be mixed. That is, one or two stages can provide frequency multiplication, with the remaining one or two stages providing power amplification.

2-3 TUNING RF CIRCUITS WITH VOLTAGE-VARIABLE CAPACITORS

Voltage-variable capacitors (VVC) are used in many applications to tune RF circuits. The capacitance of a VVC is controlled by external voltage. That is, the capacitance can be varied when the external voltage is varied. If a VVC is used in an RF tuning circuit, it is possible to vary the circuit capacitance and thus vary the circuit resonant frequency with a variable external voltage.

A VVC is sometimes called a *voltage-variable diode*, because the device is constructed more like a diode than a capacitor. However, VVC is the accepted term. The reader can obtain a more comprehensive discussion of the theory of VVC operation by consulting the author's best selling *Practical Semiconductor Databook for Electronic Engineers and Technicians* (Prentice-Hall, Inc., Englewood Cliffs, N.J., 1970).

Most VVC resonant circuits are in the form shown in Fig. 2-8 for parallel circuits and in the form shown in Fig. 2-9 for series circuits. In some cases, for biasing purpose, there are additional RF chokes connected in the circuit. The main concern with VVCs in resonant circuits is the *tuning range* of the circuit. All other factors being equal, the tuning range depends on the capacitance range of the VVC.

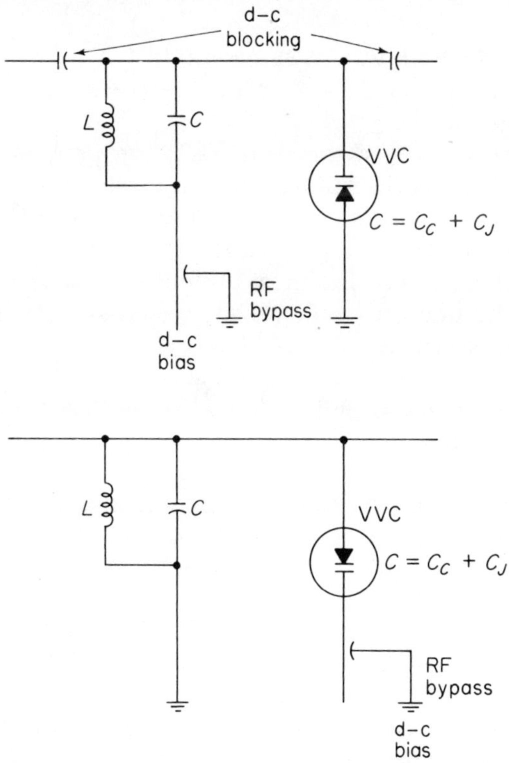

Figure 2-8. Typical parallel circuits for VVC control.

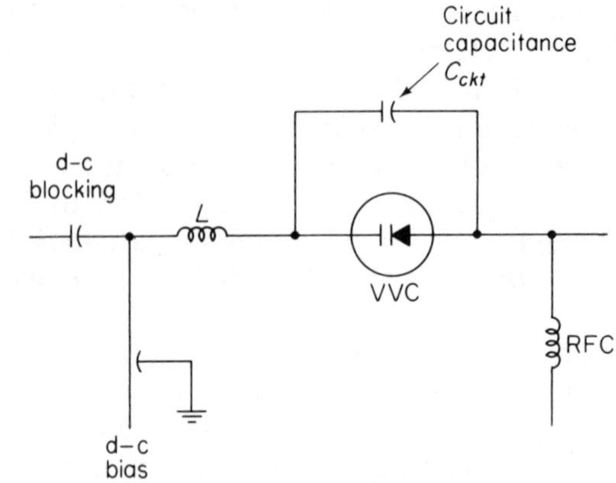

Figure 2-9. Typical series circuit for VVC control.

2-4 EXAMPLES OF RF VOLTAGE-AMPLIFIER CIRCUITS

Both two-junction transistors and FETs are used as RF amplifiers. Generally, FETs are limited to voltage-amplifier applications. Two-junction transistors can be used in either voltage-amplifier or power-amplifier circuits (Sec. 2-8). Since the number of different RF voltage-amplifier circuits is almost unlimited, it is impossible to cover all such circuits here. Instead, we show a cross section of typical RF voltage-amplifier circuits.

2-4-1 RF Voltage Amplifiers Using a Two-junction Transistor

Figure 2-10 shows the working schematic of a typical RF voltage amplifier using a two-junction transistor. Such voltage amplifiers are used primarily in radio receivers and receiver-type circuits. An IF (intermediate-frequency) amplifier or IF limiter amplifier is also an example of an RF voltage amplifier. The input or first stage of a receiver may include a separate RF voltage amplifier (such as with some communications receivers). However, most solid-state receivers combine the RF voltage amplifier function with that of the location oscillator. Such circuits are discussed in Sec. 2-5.

The basic circuit of Fig. 2-10 can be used with few modifications as an IF amplifier, IF limiter, or a separate RF voltage amplifier. Both the input and output are tuned to the desired operating frequency by means of resonant circuits. In this case, the resonant circuits are composed of transformers with a capacitor across the primary. The capacitors can be variable but are often fixed. The resonant circuit is tuned by an adjusting slug between the windings.

Resistor R_E is the emitter resistor used to provide the necessary circuit stability (to prevent thermal runaway and keep gain constant, within limits). Capacitor C_1 is the emitter bypass to provide maximum gain. Resistors R_A and R_B form a voltage divider to provide the necessary fixed bias, and maintain transistor Q_1 at the desired operating point. Capacitor C_3 provides the necessary bypass for the supply voltage line from the power supply.

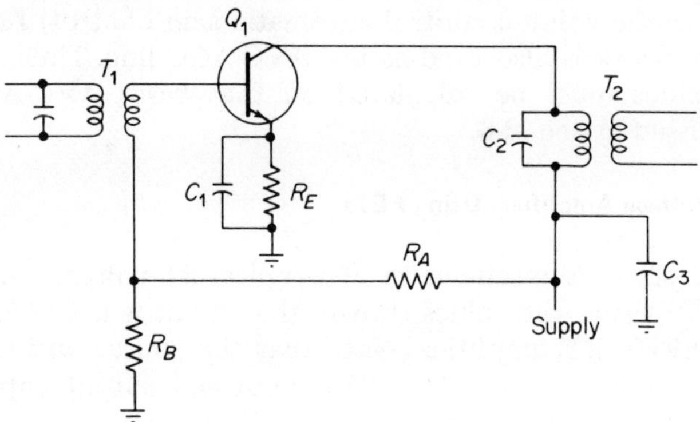

Voltage gain $\approx \beta\left(\frac{1}{N}\right)$

Where N is turns ratio of T_2

$N \approx \sqrt{\frac{Z_P}{Z_S}} \quad N^2 = Z_P/Z_S$

Where $Z_P = T_2$ primary impedence

$Z_S = T_2$ secondary impedence

Supply $\approx$ 3 to 4 times desired output voltage

Voltage drop across $R_A \approx$ supply − drop across R_B; $R_B \approx 10R_E$

Voltage drop across $R_E \approx 0.2$ for germanium, 0.5 for silicon

At operating frequency: $XC_1 \lessapprox Z_{IN}$ of Q_1; $XC_2 \approx$ primary of T_2; $XC_3 \lessapprox 100\ \Omega$

Figure 2-10. Basic RF voltage-amplifier design.

Transformer characteristics. Some commercial transformers (T_1 and T_2) are provided with a built-in fixed capacitor across the primary (and/or secondary in some cases). When a capacitor is used, the transformers are rated as to the resonant frequency range or midpoint (455 kHz for an AM broadcast IF; 500 to 1600 kHz for an RF input transformer, often of the ferrite "loopstick" type; 10.7 MHz for an FM broadcast IF; and so on). In other transformers, a fixed for variable capacitor must be connected across the transformer windings. For example, a loopstick requires a variable capacitor of the given range to provide full tuning across the AM broadcast band.

Stage gain. The required gain of an RF voltage amplifier depends on the circuit application. As guidelines, a communications receiver RF stage requires a gain of between 10 and 20, an AM broadcast IF stage requires a gain of between 30 and 40, and an IF amplifier for FM requires a gain between 40 and 50; a television IF amplifier (broadband) requires a gain between 15 and 20.

Bias resistances. When the circuit of Fig. 2-10 is to incorporate AVC-AGC (automatic volume control/automatic gain control) functions, the bias network is also used as the AVC-AGC line. Thus, the bias network values must be calculated on that basis. AVC-AGC circuits are discussed in Sec. 2-6.

2-4-2 RF Voltage Amplifiers Using FETs

Figure 2-11 shows the schematic of a typical RF voltage amplifier using a FET. Using the values shown, the circuit is a 200-MHz neutralized MOSFET RF amplifier. Note that the source and load impedances are matched to the MOSFET input and output impedances, respectively, to obtain maximum gain. The input impedance of 50 Ω is matched by means of transformation (coupling autotransformer turns ratio). The output of 50 Ω is matched by means of coupling capacitor reactance. The input is tuned by C_1; the output is tuned by C_2. Neutralization capacitor C_5 is made variable so that the exact amount of neutralization can be obtained when the circuit is tuned to different frequencies.

Figure 2-12 shows the circuit of another RF voltage amplifier using a FET. With the values shown, the circuit is a 100-MHz unneutralized MOSFET RF amplifier, used as the front end of a broadcase FM tuner. Capacitors C_4 and C_5, connected to the tuning dial, permit the resonant circuits to be tuned across the FM band from about 88 to 108 MHz. Capacitors C_2 and C_7 act as trimmer capacitors to keep both resonant circuits tracking across the entire FM band. Capacitors C_1 and C_8 provide some fixed capacitance for the resonant circuits. Resistors R_1 and R_2 provide a fixed gate-source bias. Capacitor C_6 is the source bypass. The input impedance at L_1 is 300 Ω, to match a typical FM antenna. The impedance of the input resonant circuit is matched to the FET input by means of a tap on

L_1. The circuit output impedance is matched to a mixer stage by means of a tap on L_2. The circuit output signal is combined with a signal from a local oscillator and is sent to the mixer stage from the junction of C_9 and C_{10}.

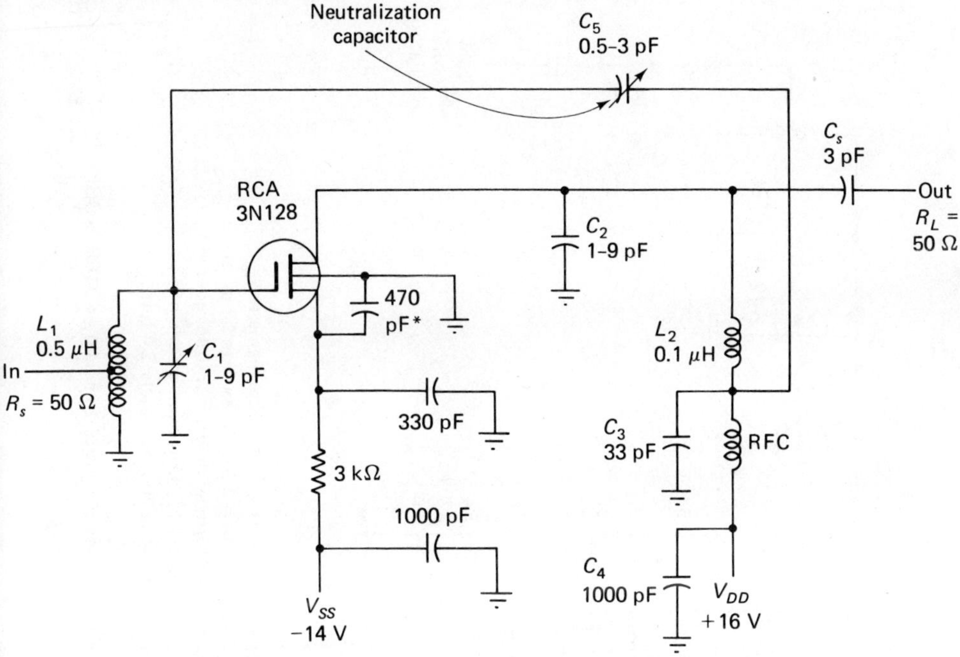

*Leadless disk capacitor.
L_1: $4\frac{1}{2}$ turns of No. 20 wire, $\frac{3}{16}$ in. diameter, $\frac{1}{2}$ in. long, tapped at 1 turn.
L_2: $3\frac{1}{2}$ turns of No. 20 wire, $\frac{3}{8}$ in. diameter, $\frac{1}{2}$ in. long.

Figure 2-11. 200-MHz common-source amplifier circuit with neutralization.

2-5 FREQUENCY MIXER AND CONVERTER CIRCUITS

Figure 2-13 shows the working schematic of a typical frequency mixer and converter such as might be found in an AM broadcast radio receiver. This circuit is actually a combination of an RF voltage amplifier and an RF oscillator. (Oscillators are discussed further in

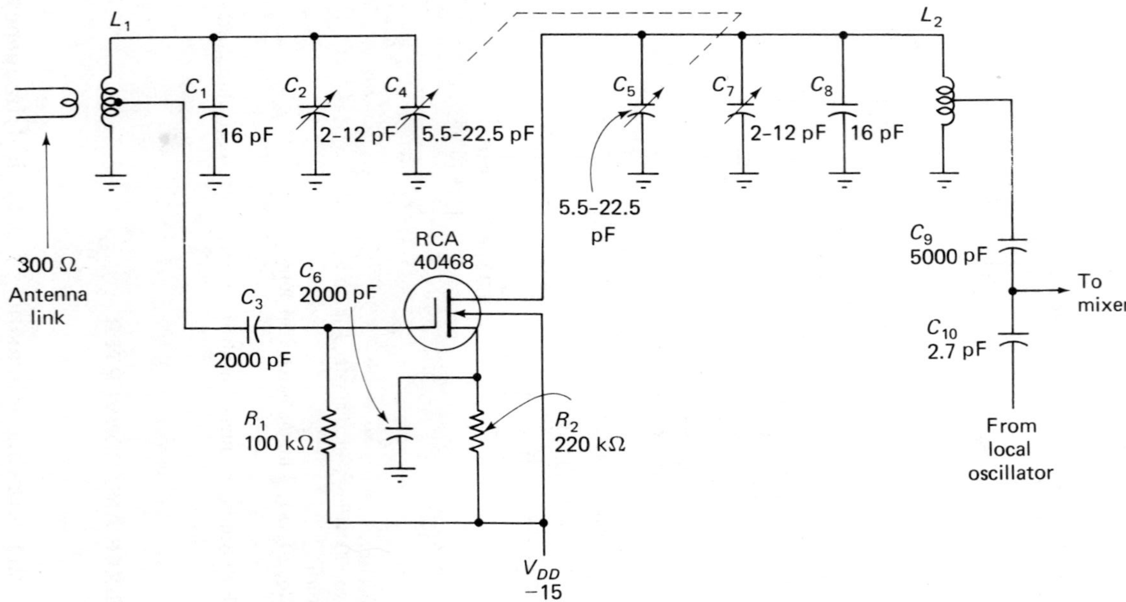

L_1: No. 18 bare copper wire, 4 turns, $\frac{1}{2}$ in. inside diameter, $\frac{7}{16}$ in. winding length, Q at 100 MHz = 120. Tunes with 34 pF at 100 MHz. Antenna link approximately 1 turn from ground end. Gate tap approximately $1\frac{1}{2}$ turns from ground end.

L_2: same as L_1 except mixer tap at approximately 3.4 turns.

Figure 2-12. Typical FM receiver RF amplifier using 40468 MOSFET.

Chapter 4.) The individual outputs of the two sections are combined to produce an intermediate-frequency (IF) output. Usually, the RF oscillator operates at a frequency above the RF amplifier, with the difference in frequency being the intermediate frequency.

The resonant circuit of T_1 is tuned to the incoming RF signal, T_2 is tuned to the oscillator frequency (RF + IF), and T_3 is tuned to the IF. The resonant circuits of T_1 and T_2 are usually tuned by means of variable capacitors, ganged together so that both the oscillator and RF amplifier remain at the same frequency relationship over the entire tuning range. For example, if T_1 tunes from 550

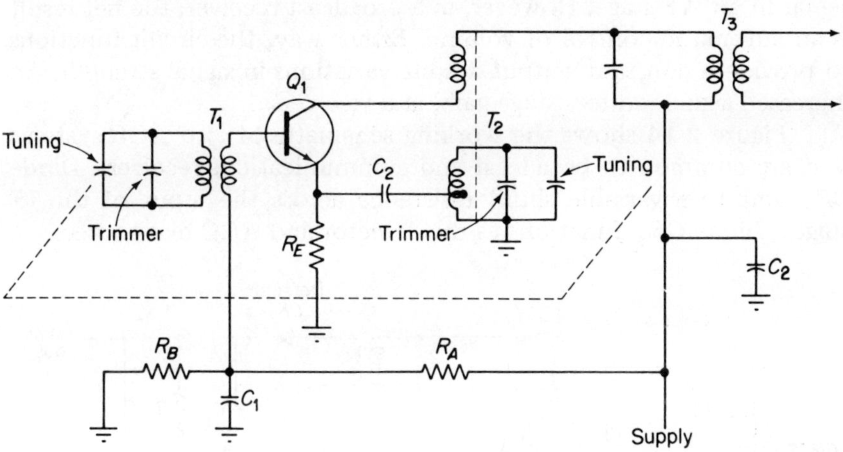

$R_E \approx$ impedance of tap on T_2

Drop across $R_B \approx 0.2 \times$ supply voltage

Drop across $R_A \approx 4 \times R_B$

$R_A + R_B \approx 15\text{–}20 \times R_E$
at operating frequency X_{C1} and $X_{C2} \approx 50\ \Omega$ (or less)

$$\text{Power output of } T_2 \approx 0.5 \times \frac{\text{collector voltage}^2}{\text{Z of } T_2 \text{ collector winding}}$$

$$\text{Power output of } T_3 \approx 0.125 \times \frac{\text{collector voltage}^2}{\text{Z of } T_2 \text{ collector winding}}$$

Figure 2-13. Basic RF mixer and converter (RF amplifier and local oscillator).

to 1600 kHz and T_3 is at an IF of 455 kHz, T_2 tunes from 1005 to 2055 kHz. Usually, trimmer capacitors are placed in parallel with variable capacitors, to permit adjustment over the tuning range.

2-6 AVC-AGC CIRCUITS FOR RF AMPLIFIERS

Most radio receivers have some form of AVC-AGC circuit. The terms AVC and AGC are used interchangeably. AGC is a more accurate term since the circuits involved control the gain of an IF or RF stage (or several stages simultaneously) rather than volume of an audio signal in an AF stage. However, in a broadcast receiver, the net result is an automatic control of volume. Either way, the circuit functions to provide a constant output despite variations in signal strength. An increased signal reduces stage gain, and vice versa.

Figure 2-14 shows the working schematic of two AGC systems that are common to broadcast and communications receivers. Diode CR_1 acts as a variable shunt resistance across the input of the IF stages. Diode CR_2 functions as the detector and AGC bias source.

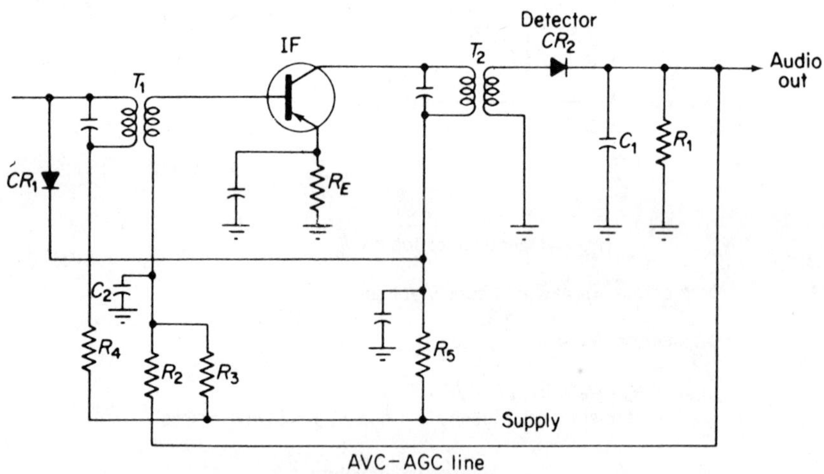

CR_1 is reverse biased with no signal

$C_2 \approx 10\,\mu F$

Drop across $R_1 \approx 0.5-1.0$ V

Drop across $R_1 + R_2 \approx 1.0-2.0$ V

Drop across R_3 = supply − $(R_1 + R_2)$

$R_1 + R_2 \approx 10 \times R_E$

Figure 2-14. Basic AVC-AGC circuit.

Under no-signal conditions, or in the presence of a weak signal, diode CR_1 is reverse-biased and has no effect on the circuit. In the presence of a very strong signal, CR_1 is forward-biased and acts as a shunt resistance to reduce gain.

The output of CR_2 is developed across R_1 and applied to the audio stages. Resistor R_1 also forms part of the bias network for the IF stage transistor. The combined fixed bias (from the network) and variable bias (from the detector) is applied to the IF stage base-emitter circuit. The detector bias varies with signal strength and is of a polarity that opposes variations in signal. That is, if the signal increases, the detector bias then reverse-biases the transistor.

2-7 DETECTORS FOR RF CIRCUITS

AM radio receivers use RF voltage detectors to convert RF signal voltages into a d-c voltage. If there is audio-frequency information present on the RF signal (in the form of amplitude modulation) the detector functions to convert the audio voltages into pulsating d-c voltages. In either case, the detector uses a diode that acts as a rectifier.

FM radio receivers use discriminators or ratio detectors. Both circuits function to remove audio-frequency information present on the RF signal (in the form of frequency modulation) and to convert the audio voltages into pulsating d-c voltages. In either circuit, a diode is used as the basic detector. A ratio detector has an advantage in that it can be directly driven by an IF stage. A discriminator usually requires a limiter stage between the IF amplifier output and the discriminator input.

2-7-1 AM Detectors

Figure 2-15 shows the working schematics for two typical diode detectors such as those used in AM radio receivers. Usually, the detector stage follows one or more IF stages. The output of the detector stage (pulsating d-c voltages at audio frequencies) is applied to the receiver audio section. In the circuit of Fig. 2-15a, the detector output is applied to a volume control in the audio circuits. In Fig. 2-15b, the detector load resistance acts as the volume control, eliminating the need for further controls in the audio circuits. The AVC-AGC circuits of a receiver are associated with the detector circuit, since the AVC-AGC signal is obtained from the detector output, as discussed in Sec. 2-6.

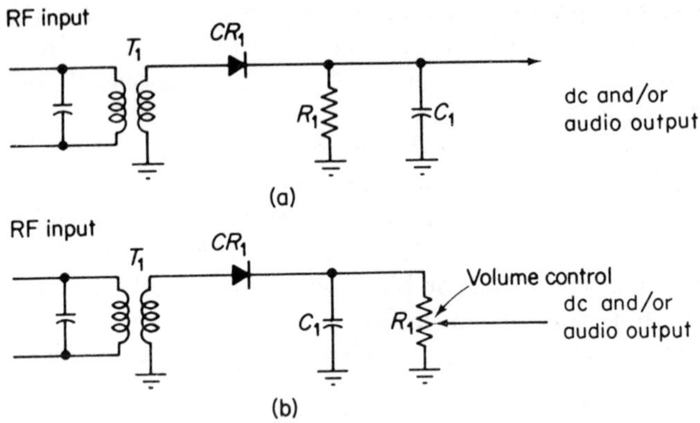

At operating frequency $X_{C1} \lessapprox 50\ \Omega$

R_1C_1 time constant $\lessapprox 250\ \mu$sec

Figure 2-15. Basic AM detector.

2-7-2 FM Discriminator and Ratio Detectors

Figures 2-16 and 2-17 show the working schematics of a discriminator and ratio detector, respectively. In both cases, the input transformer T_1 shows an extra pickup coil or winding in addition to the primary and secondary windings. A portion of the primary signal (from the IF amplifier or limiter stages) is sampled by the pickup coil and injected into the electrical center of the circuit. Not all discriminator and ratio detector transformers are so constructed. An alternative method is to obtain the sampled portion by eliminating the pickup coil and coupling this line to the primary by a capacitor.

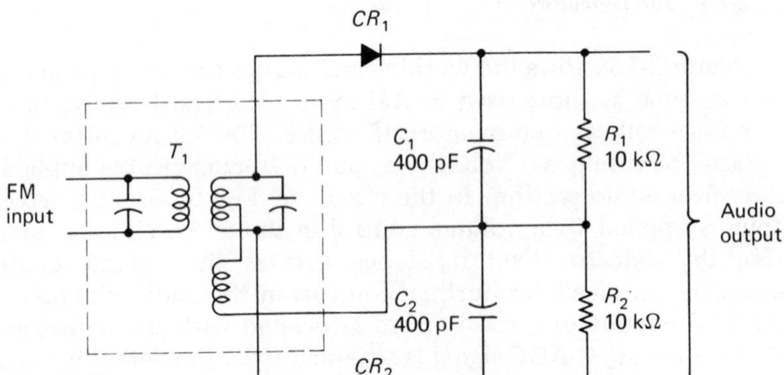

Figure 2-16. Basic FM discriminator.

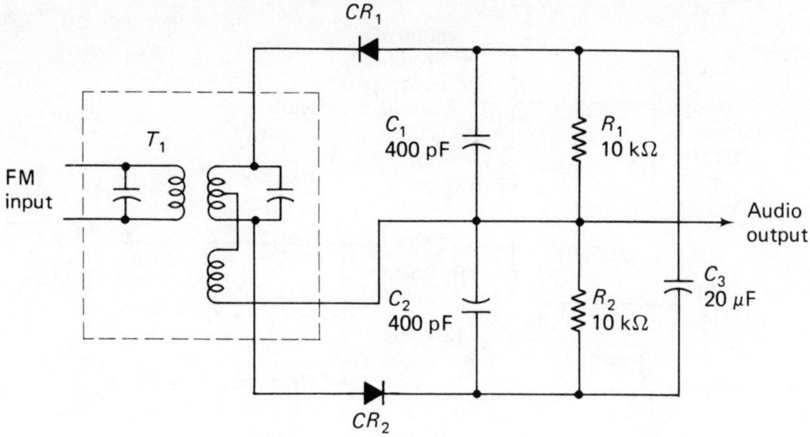

Figure 2-17. Basic FM ratio detector.

2-8 EXAMPLES OF RF POWER-AMPLIFIER CIRCUITS

Figure 2-18 shows the working schematics of typical RF power amplifiers. The same basic circuits can be used as frequency multipliers. However, in a multiplier circuit, the output must be tuned to a multiple of the input. A multiplier may or may not provide amplification. Usually, most of the amplification is supplied by the final amplifier stage, which is not operated as a multiplier. That is, the input and output of the final stage are at the same frequency.

A typical RF transmitter has three stages: an oscillator to provide the basic signal, an intermediate stage that provides amplification and/or frequency multiplication, and a final stage for power amplification. In some cases, there are three stages of amplification following the oscillator stage.

2-8-1 Tuning and Adjustment Controls

Note that the circuit of Fig. 2-18a has two tuning controls (variable capacitors) in the output network. The network of the Fig. 2-18b circuit has only one adjustment control. The circuit of Fig. 2-18a is typical for power amplifiers, where the output must be tuned to the resonant frequency by one control and adjusted for proper impedance match by the other control (often called the loading control). In practice, both controls affect tuning and loading (impedance matching). The circuit of Fig. 2-18b is typical for multipliers or intermediate amplifiers, where the main objective is tuning to the resonant frequency.

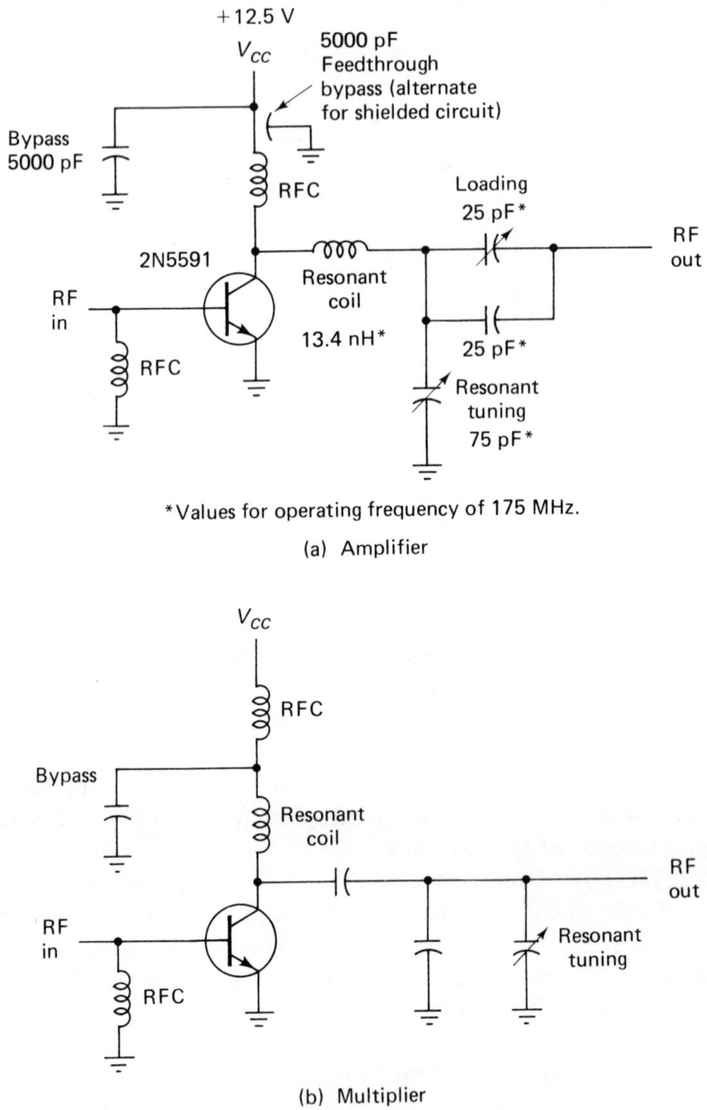

Figure 2-18. Typical RF power amplifier and multiplier circuits.

2-8-2 Minimum Capacitance

Also note that variable capacitors are connected in parallel with fixed capacitors in both networks. This parallel arrangement serves two purposes. First, it provides a minimum fixed capacitance in case the variable capacitor is adjusted to its minimum value. In some cases, if a minimum capacitance is not included in the network, a severe mismatch can occur when the variable capacitor is at its minimum, resulting in damage to the transistor. Another reason for the parallel capacitor is to reduce required capacitance range (and thus the physical size) of the variable capacitor.

2-8-3 Class of Operation

Normally, amplifiers and multipliers using the circuits of Fig. 2-18 are operated class B or C. The transistors remain cut off until a signal is applied and are never conducting for more than 180° (half-cycle) of the 360° input signal cycle. In practice, the transistors conduct about 140° of the input cycle, either on the positive half or negative half, depending on the transistor type (NPN or PNP). No bias, as such, is required for this class of operation.

2-8-4 Emitter Connections

In RF power amplifiers, the emitter is usually connected directly to ground. In those transistors where the emitter is connected to the case (typical in many RF power transistors), the case can be mounted on a chassis that is connected to the ground side of the supply voltage. A direct connection between emitter and ground is of particular importance in high-frequency applications. If the emitter is connected to ground through a resistance (or even a long lead wire), an inductive or capacitive reactance can develop at high frequencies, resulting in undesired changes in the network. Another reason for direct connection between emitter and ground is to produce maximum gain. All other factors being equal, a decrease in emitter resistance (in relation to collector impedance) produces an increase in amplifier gain.

2-8-5 Power-Supply Connections

The transistor base is connected to ground through an RF choke (RFC). This provides a d-c return for the base as well as RF signal isolation between base and emitter or ground. The transistor collector is connected to the supply voltage through an RFC and (in some cases) through the coil portion of the resonant network. The RFC provides d-c return, but RF signal isolation, between collector and power supply. When the collector is connected to the power supply through the resonant network (Fig. 2-18b), the coil must be capable of handling the full collector voltage. For this reason, the circuit of Fig. 2-18a is used for final amplifier networks where high power is involved. The circuit of Fig. 2-18b is used for applications where current is low.

2-8-6 Bypass Capacitance

The power-supply circuits must be bypassed. The feedthrough bypass capacitors shown in Fig. 2-18 are used at higher frequencies where the RF circuits are physically shielded from the power supply and other circuits. The feedthrough capacitor permits direct current to be applied through the shield but prevents radio-frequency signals from passing outside the shield (RF signals are bypassed to the ground return).

2-8-7 Efficiency

A class C RF amplifier has a typical efficiency of about 65 to 70 percent. That is, the RF power output is 65 to 70 percent of the d-c input power. This is when the input and output of a stage is operated as the same frequency. The efficiency of a second harmonic amplifier or multiplier (with the output tuned to twice the input frequency) is about 40 percent. The efficiencies of third, fourth, and fifth harmonic amplifiers are in the approximate range 28, 21, and 18 percent, respectively.

Chapter 3

Power-Supply Circuits

This chapter is devoted entirely to power-supply circuits. These circuits include half-wave, full-wave, full-wave bridge, three-phase, voltage-doubling, voltage-tripling, regulation, overload protection, and power conversion.

3-1 HALF-WAVE POWER-SUPPLY CIRCUITS

A basic half-wave diode rectifier and filter circuit is shown in Fig. 3-1. The diode CR_1 cathode is connected to a filter capacitor C_1 and to filter choke L_1. (Filters are dicussed further in Chapter 5.) Another filter capacitor, C_2, is also used to smooth out the pulsating direct current. A resistance R_1 is placed across the power-supply output. This resistance is not used in all circuits. Instead, the output is connected directly to the load circuit. If resistor R_1 is used, it is known as a *bleeder* resistor, since it places a small current drain on the power supply and helps to stabilize the output.

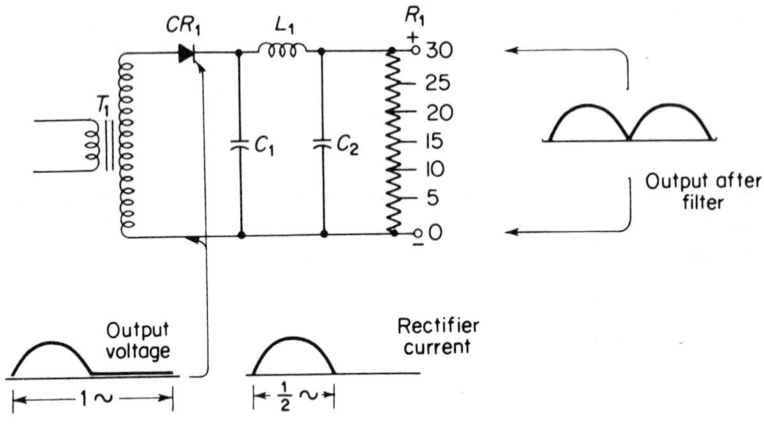

Figure 3-1. Basic single-phase, half-wave rectifier circuit.

If resistance R_1 is made up of a single tapped resistor or a series of several resistors, it is possible to take several different voltages from the bleeder network. For example, if the power-supply output is 30 V and the bleeder is made up of six equal-value resistors, the available voltages are 5, 10, 15, 20, 25, and 30 V, as shown.

The circuit output is a direct current but not necessarily a constant voltage. In practical applications, there is some "ripple" or variation in amplitude. The output voltage decreases slightly between cycles (negative peak of the ripple voltage), and then increases at the peak of each half-cycle (positive peak of the ripple voltage). Usually, the peak-to-peak ripple voltage is expressed as a percentage of the total power-supply output voltage. For example, if the power supply produces 100 V of direct current across the bleeder resistor and the ripple is 3 V peak to peak, there is a 3 percent ripple. The amount of ripple for any given type of power-supply circuit (half-wave, full-wave, etc.) depends on the type of filter.

3-2 FULL-WAVE POWER-SUPPLY CIRCUITS

Both the positive and negative alternations of the a-c cycle are used in a full-wave supply. For this reason, the full-wave circuit requires two diodes and a transformer with a center tap. As shown in Fig. 3-2, one lead of the transformer secondary is connected to diode CR_1, while the opposite end of the secondary is connected to diode CR_2. The center tap is connected to the common or ground circuit. The total voltage across the secondary of the transformer is about twice the voltage that appears at the power-supply output.

The filter capacitors charge at a rate which is twice that for half-wave rectification, since current flows through the bleeder resistor in the same direction on both alternations or half-cycles. Consequently, the ripple frequency is twice that of half-wave rectification ripple frequency. Since the discharge time between the peaks of pulsating d-c is only half that found in half-wave rectification, because there is a shorter time between peaks, the filter capacitors do not have as long to discharge. This makes it easier to maintain a relatively high charge and makes the output smoother than that of a half-wave supply.

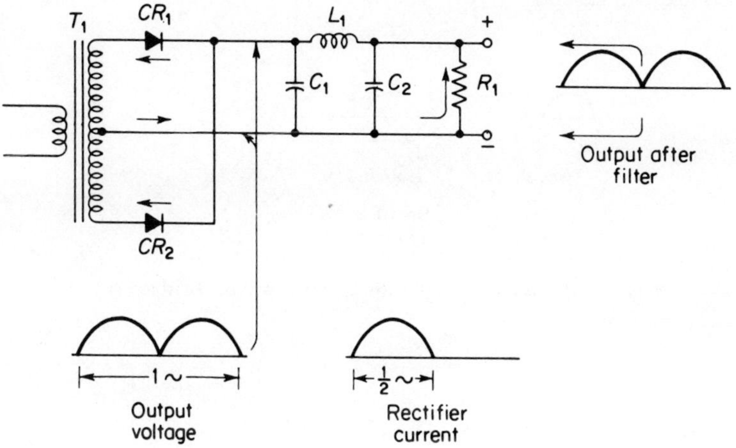

Figure 3-2. Basic single-phase, full-wave rectifier circuit.

3-3 FULL-WAVE BRIDGE POWER-SUPPLY CIRCUITS

A bridge circuit makes it possible to have full-wave rectification using a transformer without a center tap. As shown in Fig. 3-3, four rectifier diodes are required in the basic bridge circuit. When a positive alternation occurs, with the top end of the transformer secondary positive, current flows from the secondary bottom, through diode CR_3, the load or bleeder resistor R_1 and diode CR_2 to the secondary top. On the next alternation, when the bottom end of the transformer secondary is positive, current flows from the secondary top, through diode CR_1, the load or bleeder resistor, and diode CR_4 to the secondary bottom. Full-wave rectification occurs since current flows through the bleeder or load resistance in the same direction on both half-cycles. This produces both a higher ripple frequency and more efficient filtering.

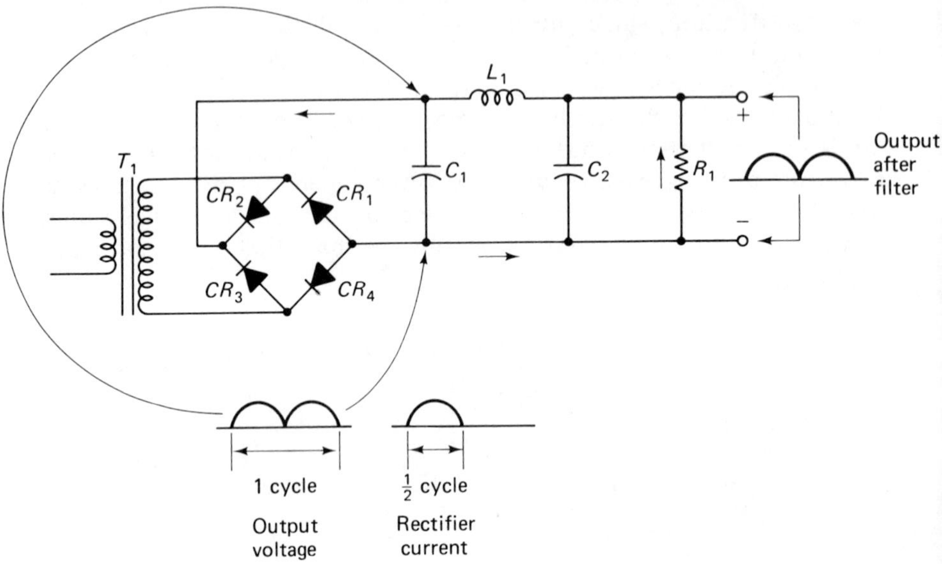

Figure 3-3. Basic single-phase, full-wave, bridge rectifier circuit.

3-4 THREE-PHASE POWER-SUPPLY CIRCUITS

The three-phase power supplies shown in Figs. 3-4 through 3-7 are similar to the basic half-wave and full-wave circuits, except that each phase is provided with its own diode (for half-wave) or diodes (for full-wave). This reduces the amount of current through each diode (making it possible to use smaller diodes) and increases the ripple frequency (making it easier to filter).

The three-phase Y half-wave circuit of Fig. 3-4 uses three rectifiers. One-third of the total output current flows through each rectifier, and the ripple frequency is three times that of a corresponding half-wave. That is, the ripple frequency is three times the line frequency.

The three-phase full-wave circuit of Fig. 3-5 uses six rectifiers. This circuit delivers twice as much voltage output as the circuit of Fig. 3-4 for the same transformer conditions. One-third of the total output current flows through each rectifier, and the ripple frequency is three times that of a corresponding full-wave (or six times the line frequency).

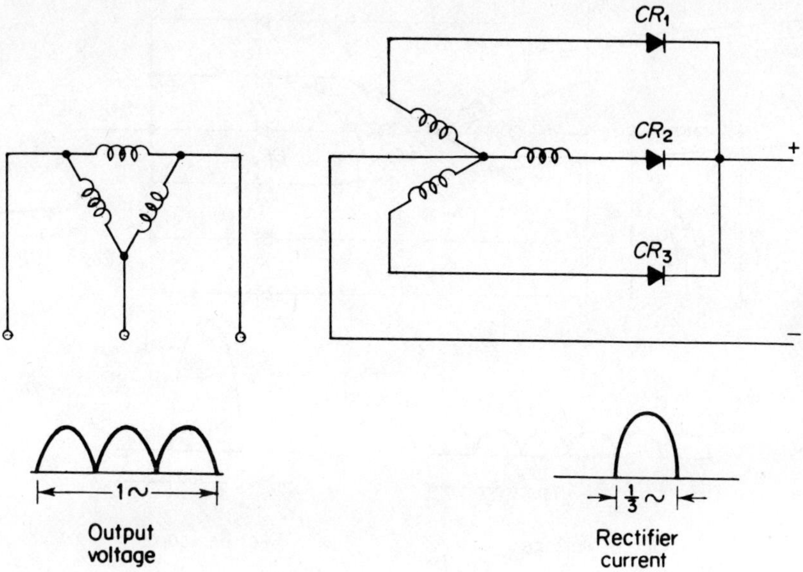

Figure 3-4. Basic three-phase "Y" half-wave rectifier circuit.

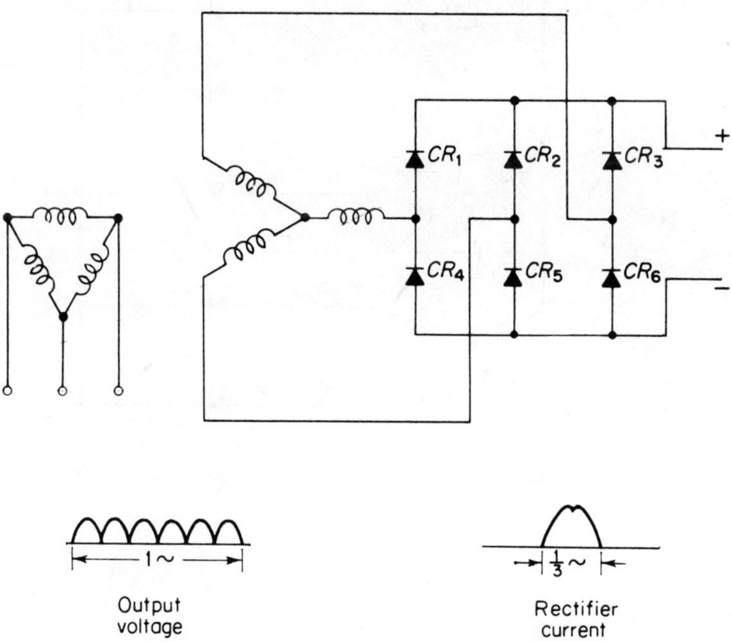

Figure 3-5. Basic three-phase "Y" full-wave rectifier circuit.

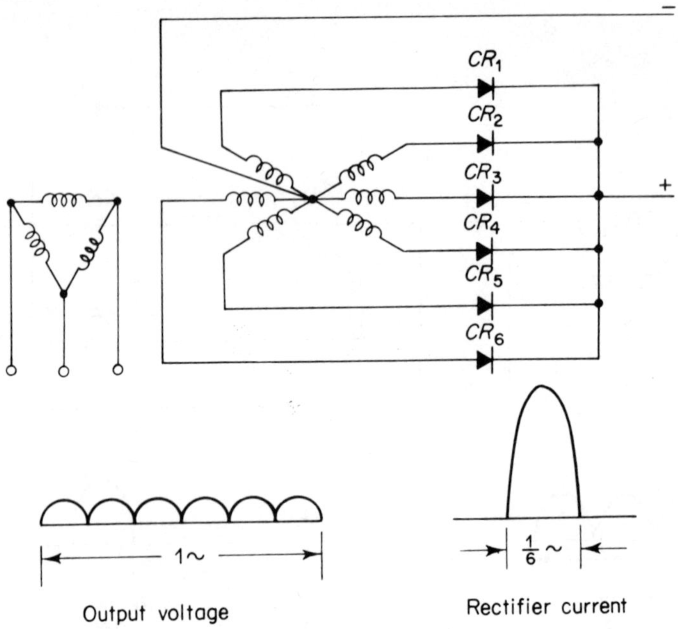

Figure 3-6. Basic six-phase "star" rectifier circuit.

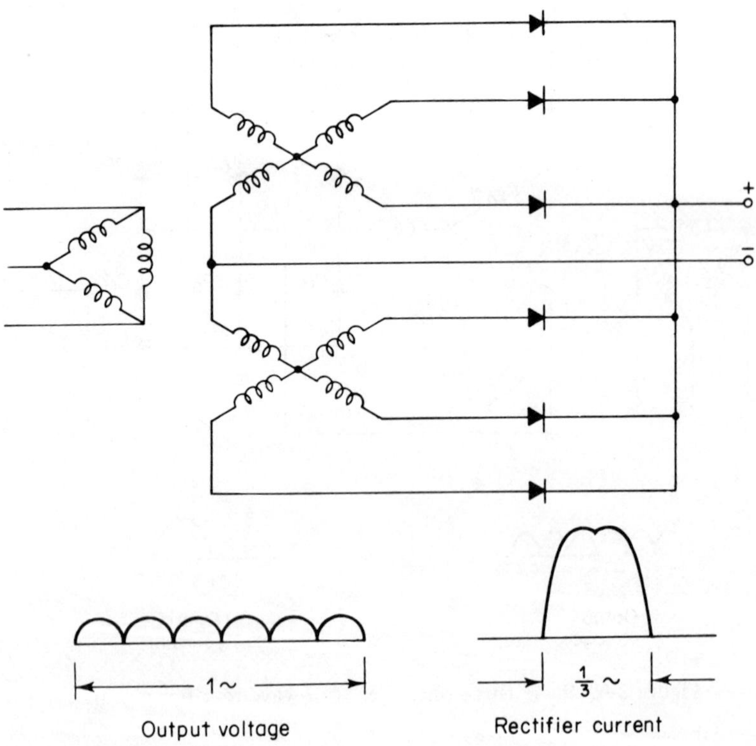

Figure 3-7. Basic three-phase "double-Y" and interphase transformer rectifier circuit.

The six-phase "star" circuit of Fig. 3-6 also uses six rectifiers. However, only one-sixth of the total output current flows through each rectifier.

The three-phase double-Y and interphase transformer circuit of Fig. 3-7 uses six half-wave rectifiers connected in parallel. This provides twice the output current of the half-wave circuit (Fig. 3-4) as well as a ripple frequency six times the line frequency.

3-5 VOLTAGE-DOUBLING AND -TRIPLING CIRCUITS

Voltage-doubling and -tripling are used where a higher voltage is required. Using such circuits, it is possible to increase an available a-c voltage without a transformer.

3-5-1 Voltage-Doubling Circuits

As shown in Fig. 3-8, the basic voltage-doubling circuit requires two rectifier diodes and two capacitors. When the a-c line alternation is such that terminal 1 is negative and terminal 2 is positive, as shown

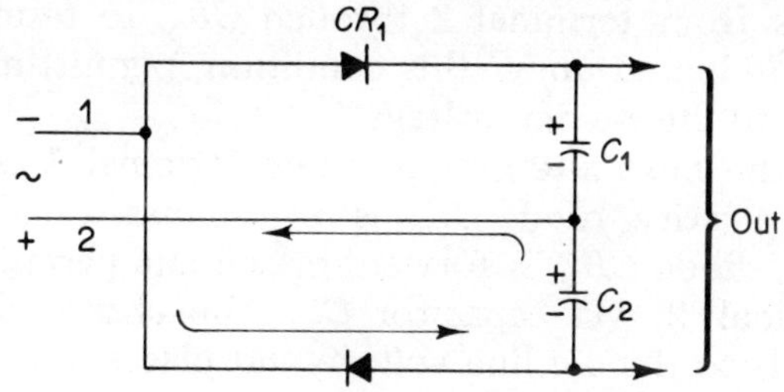

(a)

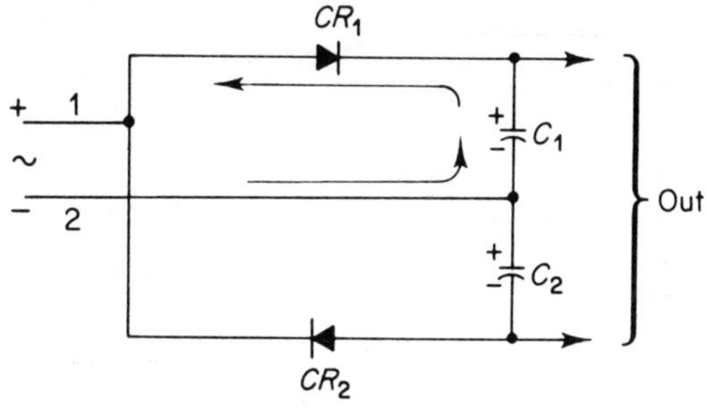

(b)

Figure 3-8. Basic voltage-doubling rectifier circuit.

in Fig. 3-8a, current flows from terminal 1 in the direction shown by the arrows. The path of current from terminal 1 includes diode CR_2 and capacitor C_2, then returns to the positive terminal 2. Capacitor C_2 is charged during this current flow with a polarity as shown.

During the next alternation, when terminal 1 is positive and terminal 2 is negative, current flows as shown in Fig. 3-8b. As a result, current flows from terminal 2 toward C_1 and charges capacitor C_1 to the peak of the a-c voltage. Current continues through diode CR_1 to the positive terminal 1.

Capacitors C_1 and C_2 are charged to the peak values of the a-c line voltage on alternate half-cycles. The d-c output voltage is taken from across the two capacitors. Thus, the d-c voltage is approximately double that of the a-c voltage. The two capacitors are, in effect, in series, with the polarities adding to increase the total voltage.

3-5-2 Voltage-Tripling Circuits

As shown in Fig. 3-9, the basic voltage-tripling circuit requires three rectifier diodes and three capacitors. When the a-c line alternation is such that terminal 1 is positive and terminal 2 is negative, current flows from terminal 2 through CR_1 to terminal 1. Diode CR_1 is forward-biased under this condition, permitting capacitor C_1 to be charged to the a-c line voltage.

During the next alternation, when terminal 1 is negative and terminal 2 is positive, diode CR_1 is reverse-biased and does not conduct. Instead, diode CR_2 is forward-biased and permits current flow toward terminal 2, via capacitor C_1. The charge that is already across C_1 adds to the a-c line voltage and places a charge across C_2. This charge is proportional to the line voltage added to the charge already across C_1.

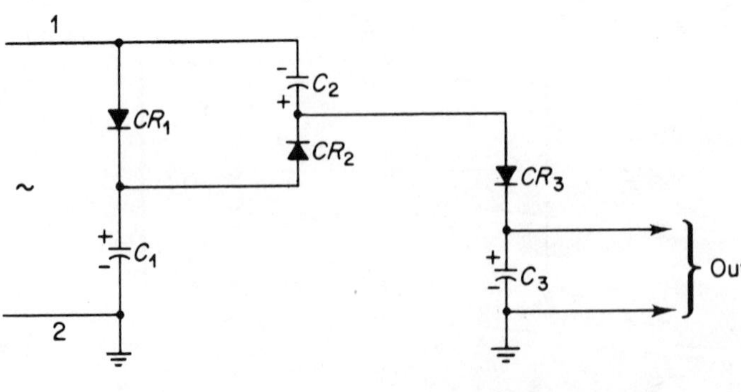

Figure 3-9. Basic voltage-tripling rectifier circuit.

During the next alternation, when terminal 1 is again positive and terminal 2 is negative, capacitor C_1 is again charged as before. But at the same time, diode CR_3 conducts and charges capacitor C_3. When diode CR_3 conducts, the charge across capacitor C_3 is composed of the line voltage plus the existing charge across capacitor C_2. Since the capacitor C_2 charge is already double voltage, this is added to the line voltage and produces a triple voltage across capacitor C_3.

3-6 REGULATION CIRCUITS

The amount of voltage that can be obtained from a power supply is determined by the voltage available at the source (line or transformer), less any voltage drop through the rectifiers and filter. This voltage drop is determined by the amount of current flowing through the power-supply components (rectifiers, chokes, etc.). As more current is drawn from the power supply, the voltage drop across the power-supply components increases and the output voltage is reduced. When less current is drawn from the power supply, the output is increased, since the power-supply components have a minimum voltage drop.

The variation of voltage output with respect to the amount of current drawn from the power supply is known as *voltage regulation*, or simply *regulation*. Usually, regulation is expressed as a percentage. The value is determined by

$$\text{percentage of regulation} = \frac{(\text{no-load voltage}) - (\text{full-load voltage})}{\text{full-load voltage}} \times 100$$

This equation applies to voltage regulation and takes into account the proportions of voltage increase and decrease with a change of load on a power supply. The smaller the difference between the full-load and no-load voltages, the better the regulation.

3-6-1 Filter-Circuit Regulation

The filter components (chokes and capacitors) provide some measure of regulation for the power-supply circuit. Should the output voltage drop, the capacitors discharge and serve to keep the voltage constant. Similarly, a voltage drop causes the magnetic field around the filter chokes to collapse (or decrease). The moving magnetic lines of force produce a current in the coil, but in the opposite direction to the current flow causing the lines of force.

Thus, the filter choke sets up an opposition to any change in voltage amplitude. The regulation provided by the filter network is sufficient for many applications. In other cases, the power-supply output must be maintained at some critical voltage or current. A number of solid-state circuits have been developed to provide both voltage and current regulation. The following paragraphs summarize these circuits.

3-6-2 Zener Diode Regulation

The most common method of providing voltage regulation for solid-state circuits is with a zener diode. (Zener diodes maintain the voltage drop across their terminal at a constant value, no matter what current is being drawn, within limits.) In the simplest form, a zener diode regulator consists of a series resistance R_S and a shunt-connected diode CR_1, as shown in Fig. 3-10. The value of R_S is set by the load requirements, as shown in Fig. 3-10. If R_S is too large, the zener is not able to regulate at large values of load current. If R_S is too small, the zener dissipation rating (power or wattage rating) may be exceeded at low-load current values.

Sometimes it is necessary to regulate a voltage not available in standard zener diodes. This problem can be overcome by various circuit arrangements. For example, several zener diodes can be connected in series, as shown in Fig. 3-11. The total regulated voltage is the sum of the individual zener voltages. If the zener voltages are as shown in Fig. 3-11 (5, 10, and 15 V), the total regulated output is 30 V. The zeners need not have equal voltage ratings since the arrangement is self-regulating. However, the power-handling ability of each diode should be the same. Similarly, the current ranges should be similar or the loads so arranged to avoid damaging any of the zeners.

The zener can also be used as a series element, as shown in Fig. 3-12. This circuit is used only where a small voltage drop is required. The shunt circuit of Fig. 3-11 is used for large voltage drops. In the series circuit of Fig. 3-12, the drop across the zener is 5.6 V, lowering the 28-V input to 22.4 V. Note that the entire load current, plus the current through R_S, passes through the series diode. Thus, the total current must be used in calculating the required power-handling ability (wattage rating) of the zener.

Several zener diodes can be connected in series to act as a divider to obtain several regulated voltages, as shown in Fig. 3-13. Using the four diodes shown, it is possible to obtain the following 10 voltages across corresponding terminals: 3.9*DE*, 6.8*CD*, 10.7*CE*, 12*BC*, 18.8*BD*, 22.7*BE*, 27*AB*, 39*AC*, 45.8*AD*, and 49.7*AE*.

It is also possible to arrange zener diodes so as to produce a regulated output at a voltage lower than that of the diode. Such a circuit is shown in Fig. 3-14, where two zeners are used. The output

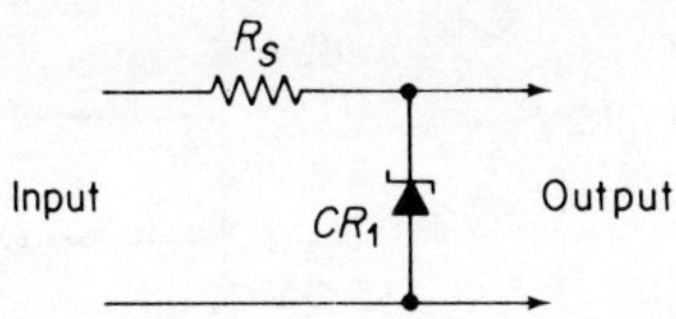

Output voltage = Zener voltage

Zener power dissipation = Zener voltage × Zener current

Minimum input voltage > 1.4 × Zener voltage

Zener voltage ≈ 0.7 × lowest input voltage

Safe Zener power ≈ 3 × load power

$$R_S \text{ (in ohms)} \approx \frac{(\text{max input voltage} - \text{Zener voltage})^2}{\text{safe power dissipation}}$$

Input voltage variation < 30 % of max input voltage

Load current = Zener voltage/load resistance

Load power = Zener voltage × load current

Safe Zener current ≈ 3 × load current

Figure 3-10. Basic zener-diode regulator circuit.

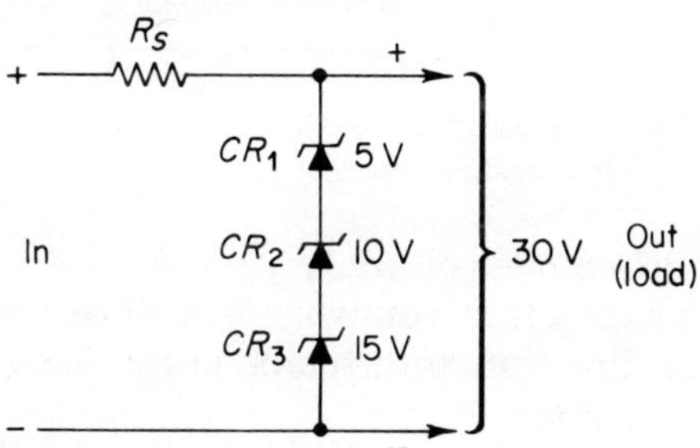

Figure 3-11. Basic zener-diode circuit providing regulation of a voltage higher than individual zener voltages.

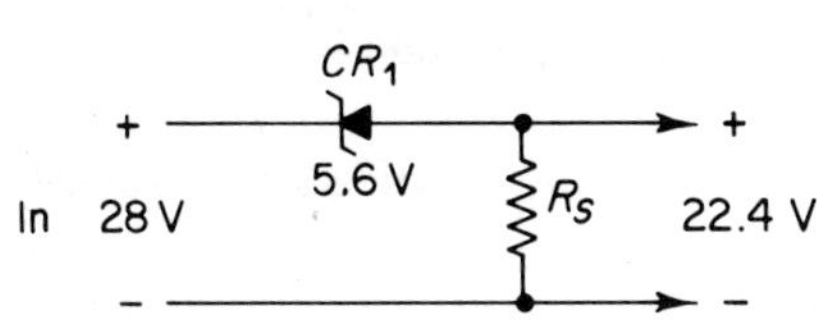

Figure 3-12. Basic zener-diode series circuit to provide regulation where small voltage drops are required.

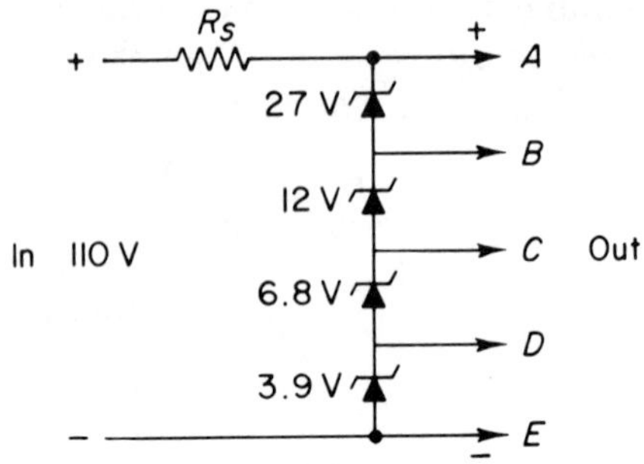

Figure 3-13. Basic zener-diode circuit used as a voltage divider.

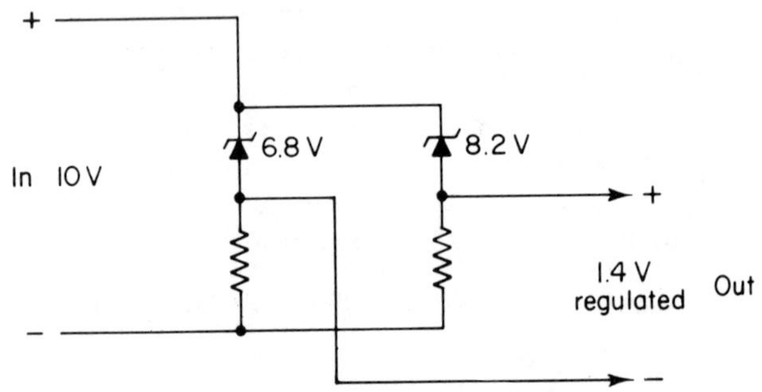

Figure 3-14. Basic zener-diode circuit used as a difference supply for low voltages.

is the regulated difference voltage (8.2 − 6.8 = 1.4). This arrangement provides good temperature compensation since both zeners tend to drift in the same direction with temperature changes, thus maintaining the difference voltage.

Zeners can also be arranged to provide an adjustable regulated voltage as shown in Fig. 3-15. Any combination of diodes can be used to obtain the desired range of output voltage.

3-6-3 Extending Zener Regulation

The voltage control ability of a zener diode can be increased if the diode is used to control the operating point of a transistor or group of transistors. There are two basic types of transistor regulators: shunt and series. The shunt regulator is placed across the power-supply output, while the series regulator is placed in series with the power-supply output.

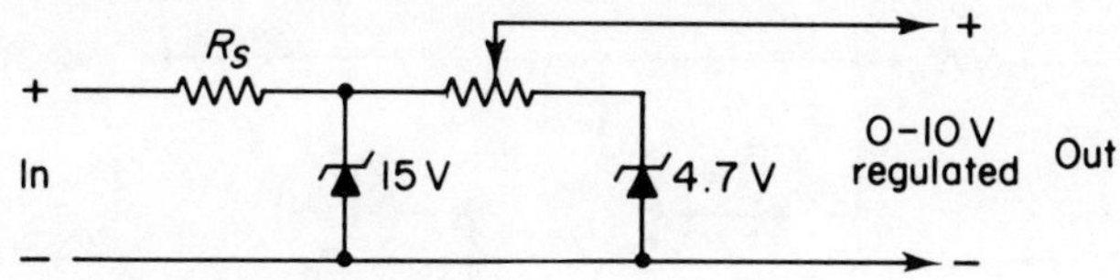

Figure 3-15. Basic zener-diode circuit used to provide an adjustable regulated output.

3-6-4 Shunt Voltage Regulators

The simplest form of shunt voltage regulator using a transistor is shown in Fig. 3-16. Transistor Q_1 appears across the power-supply output as a variable "bleeder" resistor, with current flowing between emitter and collector. Base current flows through zener CR_1. Both of these currents, as well as the load current, flow through the series resistor R_1.

If the power supply load should increase, more current is drawn through R_1 and the power-supply output voltage tends to drop. Under these conditions, less current is drawn through CR_1, the forward bias on Q_1 is lowered, and less emitter-collector current is drawn from the power supply. This causes a lower drop across R_1 and tends to increase the power-supply output voltage, thus offsetting the initial drop in voltage.

When the power supply is subject to large current changes, shunt regulators are often connected in cascade to increase their effectiveness. A typical cascade shunt regulator is shown in Fig. 3-17. Transistors Q_1 and Q_2 are placed across the power-supply output and act as variable resistors. Base current for Q_1 flows through CR_1; base current for Q_2 flows through R_2. The voltage across R_2 is determined by current flowing through R_2. All of these currents, as well as load current, flow through the series resistor R_1.

As power-supply load decreases, less current is drawn through R_1 and the output voltage rises. More current is drawn through CR_1

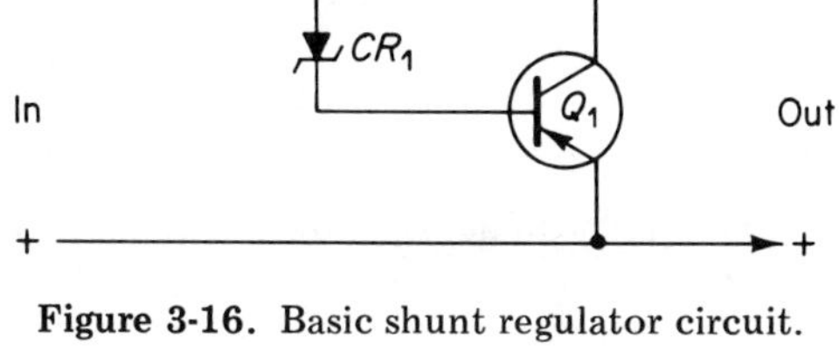

Figure 3-16. Basic shunt regulator circuit.

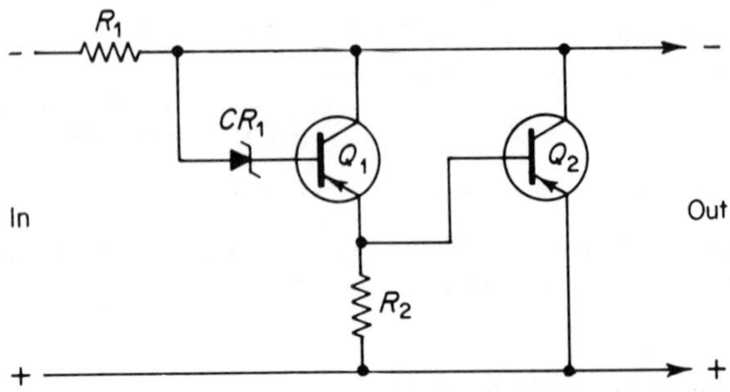

Figure 3-17. Basic cascade shunt regulator circuit.

and through the base-emitter of Q_1. This raises the forward bias of Q_1, causing more emitter-collector current to be drawn from the power supply. Also, more current is passed through emitter resistor R_2, causing the voltage drop across R_2 to increase. The forward bias on Q_2 increases and more Q_2 emitter-collector current is drawn from the power supply. The increased current through Q_1 and Q_2 causes a higher drop across R_1, which tends to decrease the power-supply output voltage and thus offset the initial rise in voltage.

The transistor shunt regulator can be adapted to supply voltage lower or higher than the supply voltage.

Figure 3-18 shows a shunt regulated supply for outputs higher than the zener voltage. Neglecting the effect of R_S or assuming that the source voltage is at the junction of R_S and R_1, the output voltage is determined by the ratio of R_1/R_2, or

$$\text{output voltage} = \frac{R_1 + R_2}{R_2}$$

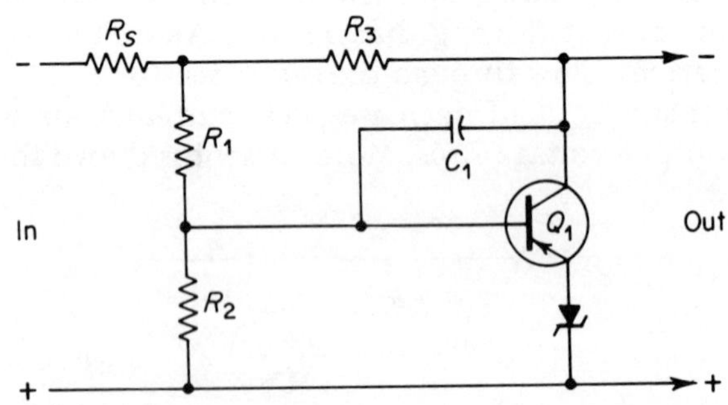

Figure 3-18. Basic shunt regulator for voltages higher than the zener voltage.

For example, if R_1 and R_2 are of the same value, the output voltage is twice that of the zener voltage. Resistor R_3 compensates for variations in the supply to the regulator circuit. A high resistance of R_3 results in overcompensation, while an R_3 with too low a resistance produces undercompensation. The value of R_3 is often found by trial and error using a variable resistance.

Figure 3-19 shows a shunt regulated supply for outputs lower than the zener voltage. Potentiometer R_2 acts as a variable voltage divider and sets the regulated output voltage. The forward bias on Q_1 is determined by the voltage drop across R_1 (or the current through R_1). If the supply voltage tends to increase, more current flows through CR_1 and R_1, causing an increase in forward bias on Q_1. The increase in Q_1 collector-emitter current passing through R_S causes a greater voltage drop across R_S, offsetting the initial rise in supply voltage.

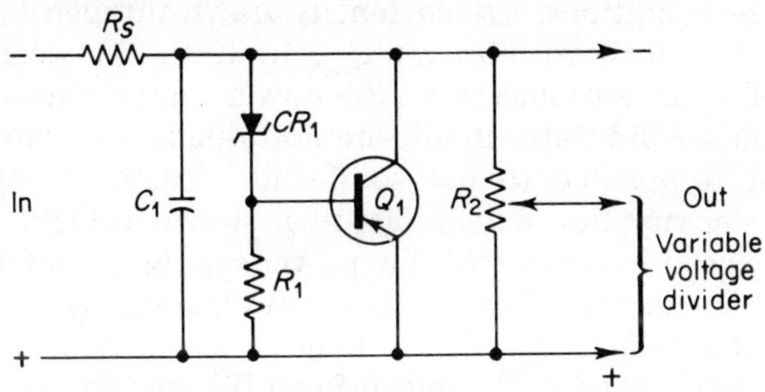

Figure 3-19. Basic shunt regulator for voltages lower than the zener voltage.

3-6-5 Series Voltage Regulators

The series circuit is usually used where higher currents must be regulated. The basic series transistor voltage regulator is shown in Fig. 3-20. Transistor Q_1 is placed in series with the power-supply output, together with series resistor R_1. Transistor Q_1 acts as a variable "series resistor" with current flowing from the collector to emitter. Current also flows through resistor R_2 and zener CR_1 to establish a voltage at the base of Q_1. The Q_1 base voltage is determined by current flow through R_2. The base voltage remains fixed in relation to the positive terminal of the power supply but varies in relation to the negative terminal.

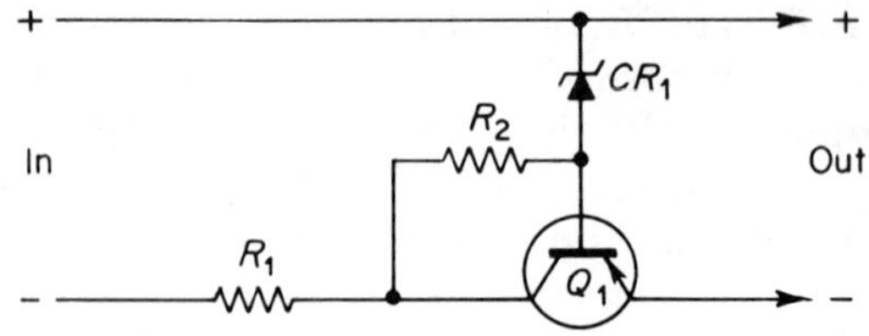

Figure 3-20. Basic series regulator circuit.

When the forward bias of the base-emitter circuit is increased, the emitter-collector "resistance" in series with the power-supply output is lowered. This causes a lower voltage drop across the emitter-collector resistance and raises the power-supply output voltage.

For example, if the power-supply load should increase, more current is drawn through the series resistor as well as the Q_1 emitter-collector resistance and the power-supply output voltage drops. Under these conditions, less current is drawn through CR_1 and R_2, increasing the forward bias on Q_1. In turn, the decrease in Q_1 emitter-collector resistance raises the power-supply output voltage.

Although solid-state circuits are not usually associated with high voltages, it is possible to use solid-state circuits to regulate high-voltage power supplies. Such a circuit is shown in Fig. 3-21. Transistor Q_1 is placed in series with the power-supply output and acts as a variable series resistor. Current also flows through R_1 to establish a voltage at the base of Q_1. This voltage is determined by the current flowing through resistor R_1 and through the emitter-collector circuit of Q_2.

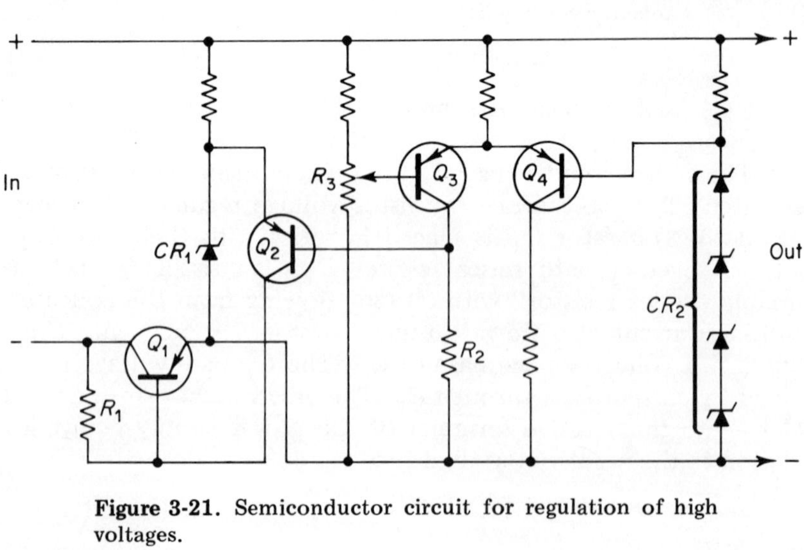

Figure 3-21. Semiconductor circuit for regulation of high voltages.

When the forward bias of Q_1 is decreased, the Q_1 emitter-collector resistance in series with the power supply is raised. This causes a higher voltage drop across the emitter-collector resistance, and lowers the power-supply output voltage. The emitter-collector current of Q_2 is determined by the forward bias voltage obtained across R_2. This voltage is determined by the emitter-collector current of Q_3. The emitter of Q_2 is held at a fixed voltage by the action of zener CR_1.

Transistors Q_3 and Q_4 are used to establish a control voltage for the regulation circuit. The base of Q_4 is held at a fixed voltage by the action of CR_2. The emitter-collector current of Q_4 remains constant and the emitter of Q_4 remains at a fixed voltage. Since the emitter of Q_3 is connected directly to the emitter of Q_4, the emitter of Q_3 also remains at a fixed voltage. However, the voltage at the base of Q_3 varies with the power-supply output voltage.

If the power supply output voltage changes, the forward bias on Q_2 changes, causing a change in the voltage drop across R_2. This changes the forward bias on Q_2, causing a change in emitter-collector current. Since the emitter-collector current of Q_2 passed through R_1, the voltage drop across R_1 changes, making the base of Q_1 more or less negative. This changes the forward bias on Q_1 and changes the emitter-collector resistance in series with the power-supply output. In turn, this causes a higher or lower voltage drop across the emitter-collector resistance and changes the power-supply output voltage to offset the initial rise or fall in voltage.

The power supply output voltage is set by adjustment of potentiometer R_3, which determines the bias voltage on the base of Q_3. In practice, the power supply is adjusted by connecting a voltmeter across the power-supply output terminals (usually with the load connected) and setting R_3 for the desired voltage.

3-6-6 High-Current Regulators

If a very high current must be regulated and all of this current passes through a single series transistor, the transistor might not be able to dissipate all of the heat satisfactorily, even though heat sinks are used. This can result in damage to the transistor. To overcome the problem, several transistors can be connected in parallel with each other. Then the string of transistors is connected in series with the power-supply output, as shown in Fig. 3-22.

Transistors Q_3 through Q_6 are connected in series with the power-supply output, and in parallel with each other, to act as variable series "resistors." Current is divided equally among all four so that each transistor dissipates one-fourth of the total heat.

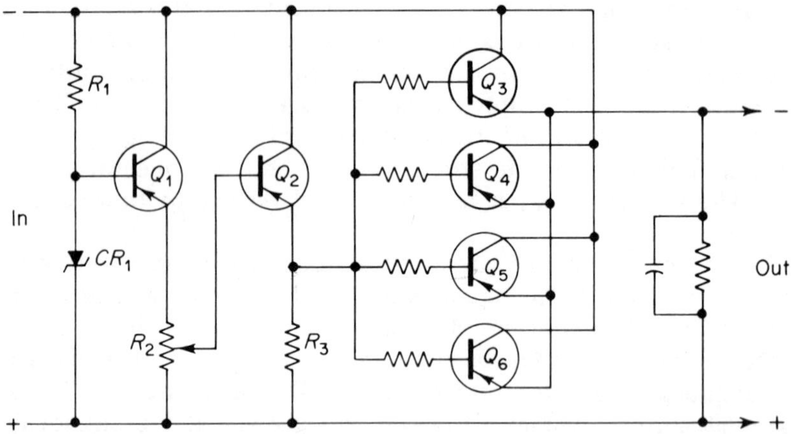

Figure 3-22. Semiconductor circuit for regulation of voltages where high currents are present.

The base of Q_1 is held at a fixed voltage (in relation to the emitter) by the action of zener CR_1. However, the voltage across the emitter-collector circuit of Q_1 varies with the power-supply voltage. If the power-supply voltage changes for any reason, the voltage across R_2 varies and changes the forward bias on Q_2. In turn, this changes the current through R_3. Since the voltage across R_3 determines the forward bias on all four transistors (Q_3 through Q_6), the voltage across R_3 determines the amount of resistance offered by transistors Q_3 through Q_6. A change in the Q_3 through Q_6 voltage drop changes the power-supply output voltage to offset the initial change in voltage. The power-supply output voltage is set by adjustment of R_2, which determines the forward bias on the base of Q_2.

3-6-7 Constant-Current Regulation

Solid-state devices can be used to regulate a power supply so as to produce a *constant current* rather than a constant voltage. The basic circuit is shown in Fig. 3-23. Transistor Q_1 functions as a variable "series resistor" in the power-supply output. Two parallel circuit paths exist. One path is through zener CR_1 in series with bias resistor R_3. The other path is through resistor R_1 and transistor Q_1.

If there is any variation in current at the power-supply output, the current through R_3 varies and results in a change of forward bias on Q_1. In turn, the emitter-collector resistance of Q_1 varies to correct the current flow. The net result is that for every change in R_3 current, there is an equal and opposite change in Q_1 current. Current output of the circuit is set by R_1. The current remains constant

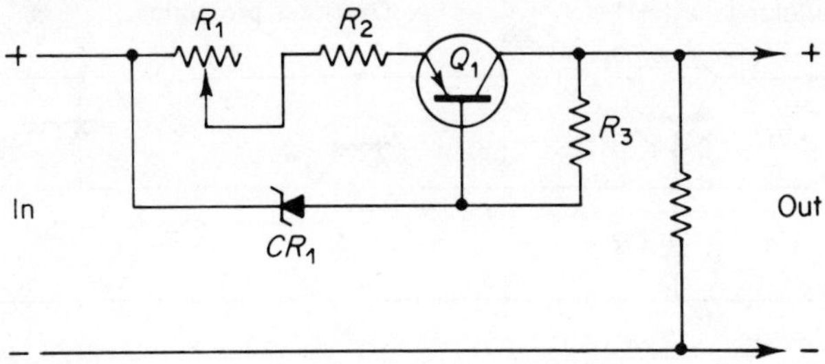

Figure 3-23. Basic semiconductor current regulator circuit.

(within limits) in spite of any load changes. However, the power-supply output voltage varies and the load changes.

3-7 OVERLOAD-PROTECTION CIRCUITS

A series regulator is subject to damage caused by overloads. This can be in the form of a high input voltage or an excessive output load. Either way, the series transistor will burn out. A high input voltage is not likely. Any prolonged period of high input voltage will probably burn out the basic power-supply component first (unless the series regulator design is on the borderline of safety). On the other hand, an excessive output load (low load resistance and high output current) is quite common. (A screwdriver shorted across the regulator output terminals will do the job nicely.) A series regulator can be provided with some form of protection from overloads due to excessive output current.

Figure 3-24 is the working schematic of an overload-protection circuit for series regulators. Operation of the regulator portion of the circuit (Q_1, R_1, CR_1) is the same as described in Sec. 3-6-5. Operation of the overload-protection circuit is controlled by the voltage across R_3. All of the load current passes through R_3 and develops a corresponding voltage. When the load current is below a certain value (at a safe level below the maximum rating of the series regulator), the voltage drop across R_3 is not sufficient to forward bias Q_2. Thus, Q_2 remains nonconducting as long as the load current is at a safe level.

If diode CR_2 is silicon, there will be about a 0.5-V drop from the output terminal to the emitter of Q_2. Similarly, if Q_2 is silicon, another 0.5-V drop is required before Q_2 turns on. Thus, the drop across R_3 must be about 1 V or more (typically 1.1 V) before Q_2 turns on. The value of R_3 is chosen so that the drop is about 1.1 V when the output load current is at the maximum safe level. With Q_2 turned on, part of the current through R_1 is passed through Q_2,

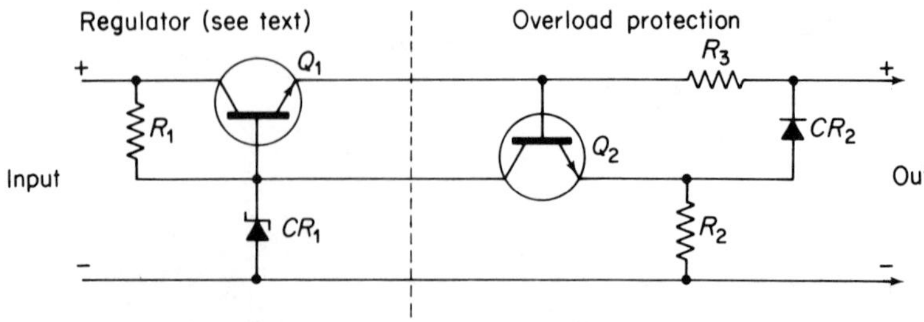

With Q_2 and CR_1 silicon, drop across $R_3 \approx 1.1$ V, at maximum current output
Output voltage ≈ output from regulator − 1 V
Output current and power ≈ same as regulator
$R_2 \approx 10 \times$ load resistance, at maximum current
Load resistance ≈ output voltage / load current
R_3 (in ohms) ≈ 1.1/maximum load current
R_3 (in watts) ≈ (maximum load current)^2 × R_3 resistance

Figure 3-24. Overload protection for series regulator.

thus drawing current away from the base of Q_1. Transistor Q_1 is cut off (or partially cut off), thus preventing (or limiting) current flow to the output load.

When the load is returned to normal, the drop across R_3 is below 1.1 V and transistor Q_2 is turned off. Transistor Q_2 must be capable of withstanding the full output voltage. However, the current (or power) requirements for Q_2 are not high in comparison to those of Q_1. Transistor Q_2 is off and dissipates no power except in case of an overload. When an overload occurs, current through Q_2 is limited by emitter resistor R_2.

3-8 POWER-CONVERSION CIRCUITS

Solid-state devices can be used effectively in power-conversion circuits. Such solid-state circuits have replaced the vibrator-type circuits for converters and inverters. Usually, the term *inverter* refers to a circuit used to change d-c power into a-c power. If the a-c output is further rectified and filtered to provide direct current (say at a higher voltage than the source), the overall circuit is called a *converter.*

The *push-pull switching inverter* is the most widely used type of power-conversion circuit. For inverter applications, the circuit provides a square-wave a-c output. When the inverter is used to provide d-c to d-c conversion, the square-wave voltage is usually applied to a bridge rectifier and filter.

Figure 3-25 is the circuit of a one-transformer converter used to supply power in the range 15 to 55 W. Any voltage imbalance in the

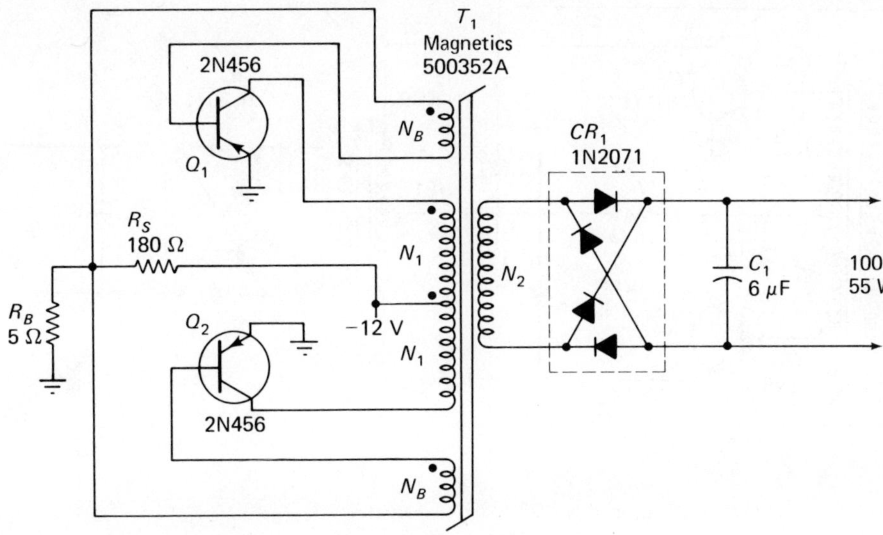

N_1 = 29 turns of No. 17 AWG
N_2 = 275 turns of No. 24 AWG
N_B = 6 turns of No. 24 AWG

Figure 3-25. One-transformer converter with 12 V input and 100 V, 55 W output.

circuit causes one of the transistors (Q_1, for example) to conduct a small amount of current. Regeneration turns Q_2 off and Q_1 is driven into saturation. The collector current of Q_1 increases. As the T_1 core becomes saturated, the Q_1 collector current increases rapidly and is limited only by the resistance in the collector circuit and the transistor characteristics. When the T_1 core is saturated, the induced voltage in the winding is zero. The resulting lack of base drive caused Q_1 to turn off and the collector current drops to zero. When the core approaches negative saturation, Q_2 is turned off and the current drops to zero, turning Q_1 on. A new cycle is initiated each time one of the two transistors reaches saturation.

Resistors R_B and R_S are added to place a bias on the bases of both Q_1 and Q_2. This bias provides a starting current and reduces the effect of base-emitter voltage variations. The frequency of oscillation is determined by the design of transformer T_1. The secondary output voltage is rectified by the bridge-connected diodes CR_1 and filtered by capacitor C_1.

Figure 3-26 is the circuit of a two-transformer converter used to supply power in the range 100 to 250 W. Although two transformers

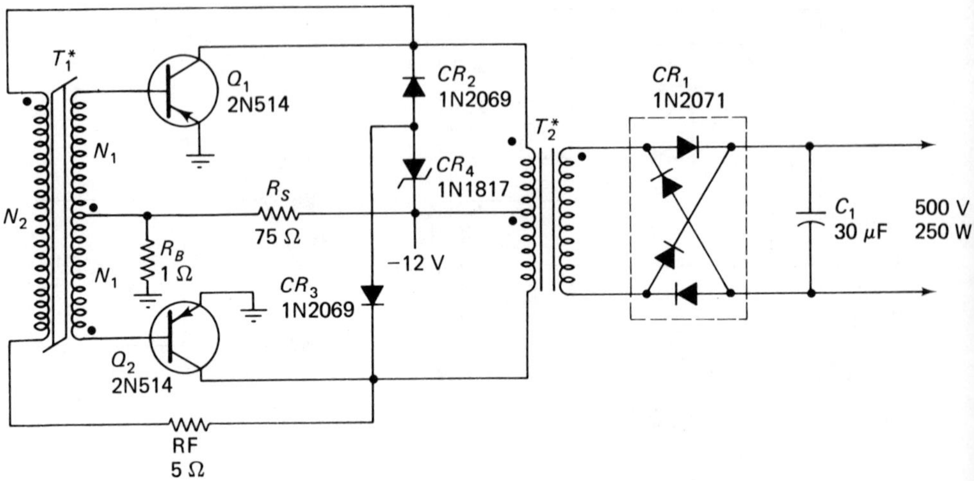

T_1: N_1 = 35 turns of No. 20 AWG
N_2 = 140 turns of No. 24 AWG

T_2: laminations E1–125; primary turns 28, secondary turns 623; turns layers, primary 7/4, secondary 63/10; primary wire size, 2 parallel No. 13 bifilar wound; secondary wire size, No. 24; lamination dimensions per diagram.

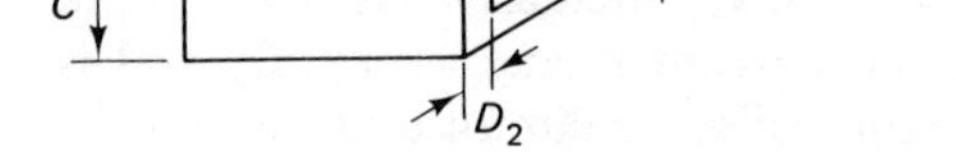

$A = 1\frac{3}{16}$ in.
$B = 3\frac{3}{4}$ in.
$C = 3\frac{1}{8}$ in.
$D_1 = \frac{5}{8}$ in.
$D_2 = \frac{5}{8}$ in.
$D_3 = 1\frac{7}{8}$ in.

Figure 3-26. Two-transformer converter with 12 V input and 500 V, 250 W output.

are used, only transformer T_1 saturates. Thus, the extra current necessary at saturation is small compared to the load current. This allows the use of a small drive transformer T_1 (with a saturable core) to drive a much larger, relatively inexpensive, power transformer T_2 (with a conventional, nonsaturable core) which steps up the output voltage to the required value.

When one of the transistors (Q_1, for example) conducts, the collector swings from the supply voltage to near zero (saturation). The voltage builds up across the primary of T_2 and is applied to the primary of T_1 through feedback resistor R_F. Transistor Q_2 is biased off, while Q_1 is biased into conduction. As soon as the core of T_1

reaches saturation, the increasing current causes an additional voltage drop across R_F, thus aiding the regenerative action (Q_2 off; Q_1 on). Transistor Q_2 continues in this state until reverse saturation of the transformer is reached. The circuit then switches back to the initial state and the cycle is completed.

The collector current of the conducting transistor rises to the value of the load current, plus the magnetizing current of T_2 and the feedback current needed to produce the drive. The magnetizing current of T_2 is never more than a fraction of the rated load current, since T_2 is not allowed to saturate.

Resistors R_B and R_S are added to place a bias on the bases of Q_1 and Q_2. This bias provides a starting current and reduces the effect of base-emitter voltage variations. The frequency of oscillation is determined by the design of transformer T_1 and the value of feedback resistor R_F.

Chapter 4

Oscillator Circuits

This chapter is devoted entirely to oscillator circuits. These circuits include *LC*, crystal-controlled, *RC*, and blocking oscillators, as well as multivibrators.

4-1 *LC* AND CRYSTAL-CONTROLLED OSCILLATORS

LC oscillators are those that use inductances (coils) and capacitors as the frequency-determining components. Typically, the coils and capacitors are connected in series- or parallel-resonant circuits and adjusted to the desired operating frequency. Either the coil or capacitor can be variable. *LC* oscillators are used at higher frequencies (RF). Both the Hartley and Colpitts oscillators are used in present-day solid-state equipment. (The Hartley and Colpitts are classic vacuum-tube oscillator circuits.) However, the Colpitts is generally the most popular.

Transistor *LC* oscillators can be crystal-controlled. That is, a quartz crystal can be used to set the frequency of operation, with an adjustable *LC* circuit used to "trim" the oscillator output to an exact frequency. In addition to the Hartley and Colpitts, there are a number of other crystal oscillator circuits suitable for transistors. These include the Pierce oscillator, harmonic or overtone oscillators, and oscillators that use two transistors to provide the necessary feedback required for oscillation.

4-1-1 Basic Solid-State *LC* Oscillators

Figure 4-1 shows two arrangements of the Hartley oscillator circuit. The circuit of Fig. 4-1a uses a bypassed emitter resistor to provide proper operating conditions. The circuit of Fig. 4-1b uses a

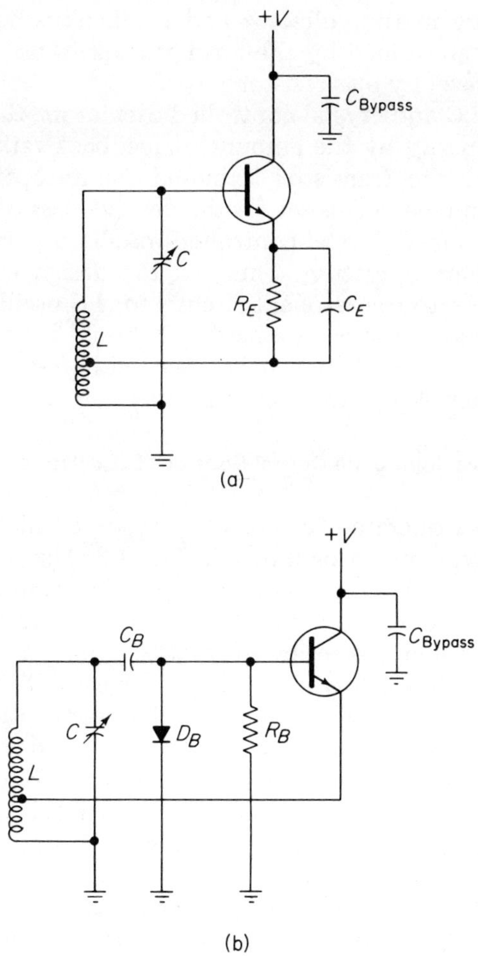

Figure 4-1. Basic solid-state Hartley oscillator circuits.

base resistor and biasing diode. The amount of feedback in either circuit depends upon the position of the coil *L* tap. Output from the Hartley circuits can be obtained through inductive coupling to the coil or through capacitive coupling to the base.

Figure 4-2 shows two forms of the Colpitts oscillator circuit. Colpitts circuits are more commonly used in VHF and UHF equipment than the Hartley circuit because of the mechanical difficulty involved in making the tapped coils required at high frequencies. Feedback is controlled in the Colpitts circuit by the ratio of capacitance C' to C''. Note that the *LC* circuits in the basic configuration of Fig. 4-2 are in the base circuit. Feedback is between emitter and base, through C' and C''. In most practical Colpitts circuits, the *LC* components are in the collector and feedback is between collector and emitter (determined by the ratio of capacitors). The base is held at a fixed d-c level by bias resistors.

In most *LC* and crystal-controlled oscillators, the class of operation is set primarily by the amount of feedback rather than the bias point. That is, the transistor is biased for an optimum operating point, and then feedback is set for the desired class of operation.

LC and most crystal-controlled oscillators require resonant circuits for their operation. Thus, all the design considerations of Chapter 2 apply to the resonant circuits for *LC* oscillators. Similarly, the design considerations of Chapter 2 should be studied when RF oscillators of any kind are to be used with frequency multipliers and/or power amplifiers to form transmitter circuits.

4-1-2 Basic Solid-State Crystal-Controlled Oscillators

Transistors operate efficiently in crystal oscillator circuits such as the Pierce-type oscillator shown in Fig. 4-3. The Pierce oscillator is very popular because of its simplicity and minimum number of components. No *LC* circuits are required for frequency control. Instead, the frequency is set by the crystal.

At frequencies below about 2 MHz, a capacitive voltage divider is sometimes used across the crystal. The connection between the voltage-divider capacitors must be grounded so that the voltage developed across the capacitors is 180° phase-inverted.

It is frequently desirable to operate crystals in communications equipment at their *harmonic*, or *overtone*, frequencies. Figure 4-4 shows two circuits designed for overtone operation. Additional feedback is sometimes obtained for the overtone crystal by means of a capacitive divider, acting as the *LC* circuit bypass. Most third-overtone crystals operate satisfactorily without this additional feedback, but for the fifth and seventh harmonics, the extra feedback is

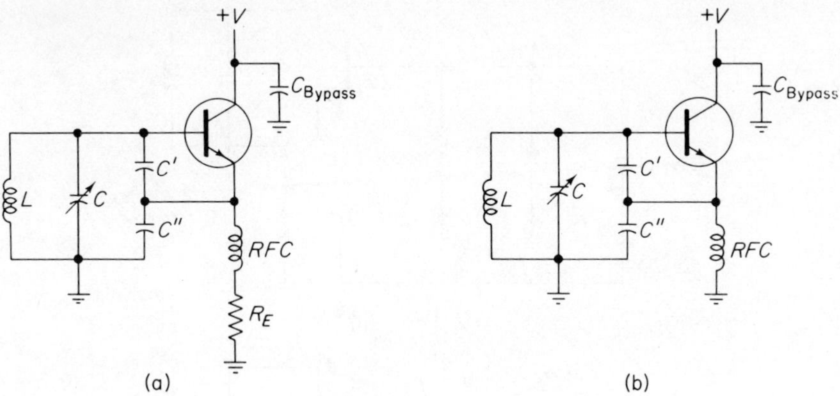

Figure 4-2. Basic solid-state Colpitts oscillator circuits.

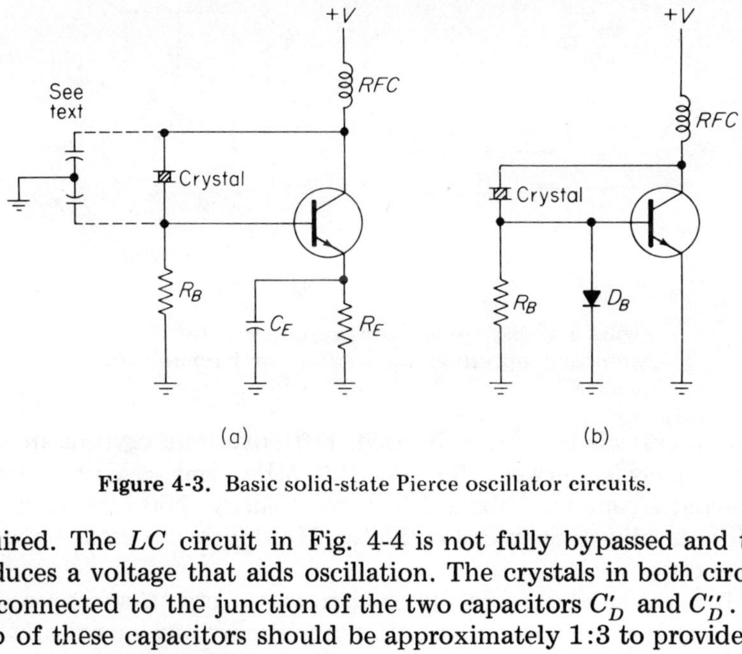

Figure 4-3. Basic solid-state Pierce oscillator circuits.

required. The *LC* circuit in Fig. 4-4 is not fully bypassed and thus produces a voltage that aids oscillation. The crystals in both circuits are connected to the junction of the two capacitors C'_D and C''_D. The ratio of these capacitors should be approximately 1:3 to provide the required amount of additional feedback.

The circuit of Fig. 4-5 operates well at *low frequencies*. The crystal is located in the feedback circuit between the emitters of the two transistors and operates in the series mode. Capacitor C_2 is used for precise adjustment of the oscillator frequency. A reduction in C_2 capacitance increases the frequency slightly.

The practical limit of crystal fundamental resonance is about 25 MHz for transistor crystal oscillators. From 20 to 60 MHz, third-

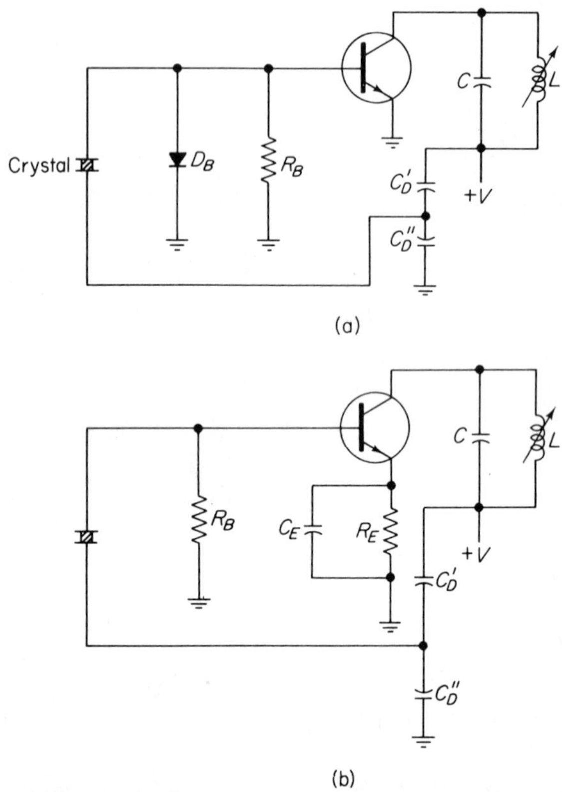

Figure 4-4. Basic solid-state crystal oscillator circuits permitting operation at overtone or harmonic frequencies.

overtone crystals are generally used. Fifth-overtone crystals are used for frequencies between 60 and 120 MHz, and seventh-overtone crystals at frequencies above 120 MHz. Usually, 150 MHz is the top limit for a solid-state crystal oscillator. Multipliers are used above this frequency.

In most solid-state oscillator circuits, the crystal series-resonant frequency is most appropriate. At "overlapping" frequencies, where either a fifth or seventh overtone can be used (such as 115 to 125), the power output is about the same for either crystal. Of course, the seventh overtone may present other problems, such as more critical adjustments and a tendency to oscillate at the fifth overtone.

In general, the oscillator *LC* circuit is tuned to the crystal overtone. A better viewpoint is that the reactive elements of the *LC* circuit and crystal form a coupling network between the transistor output and input, which produces the necessary phase shift. The

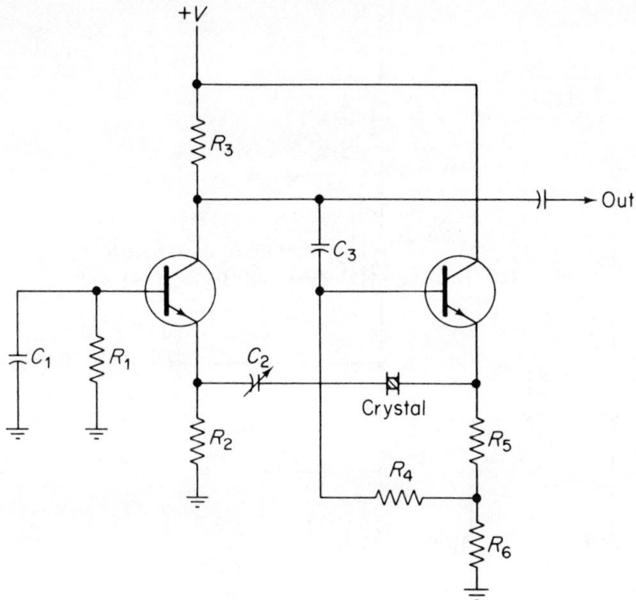

Figure 4-5. Basic solid-state low-frequency crystal oscillator circuit.

function of the crystal is to introduce an additional reactive element capable of causing large phase-shift changes for a very small frequency change. To maintain the 360°-loop phase shift, changes in reactances in the circuit can be compensated for by only tiny frequency shifts, due to the extreme high Q of the crystal, and usually excellent temperature stability. Where extreme stability is required, the crystal is usually housed in a temperature-stable environment (crystal oven).

Crystals reduce oscillator efficiency as a result of losses in the crystal. These losses are represented by their series resistance. Typically, crystal series resistance is on the order of 20 to 100 Ω. Oscillator efficiency may be increased by reducing signal currents in all dissipative elements. Generally, this is done by including RF chokes (RFCs) in the transistor element leads. The values for such chokes are determined by frequency, as is discussed in the following paragraphs.

4-1-3 Practical Solid-State Crystal Oscillators

Figure 4-6 is the working schematic of a crystal-controlled oscillator. This circuit is one of the many variations of the Colpitts oscillator. The output frequency is fixed and controlled by the

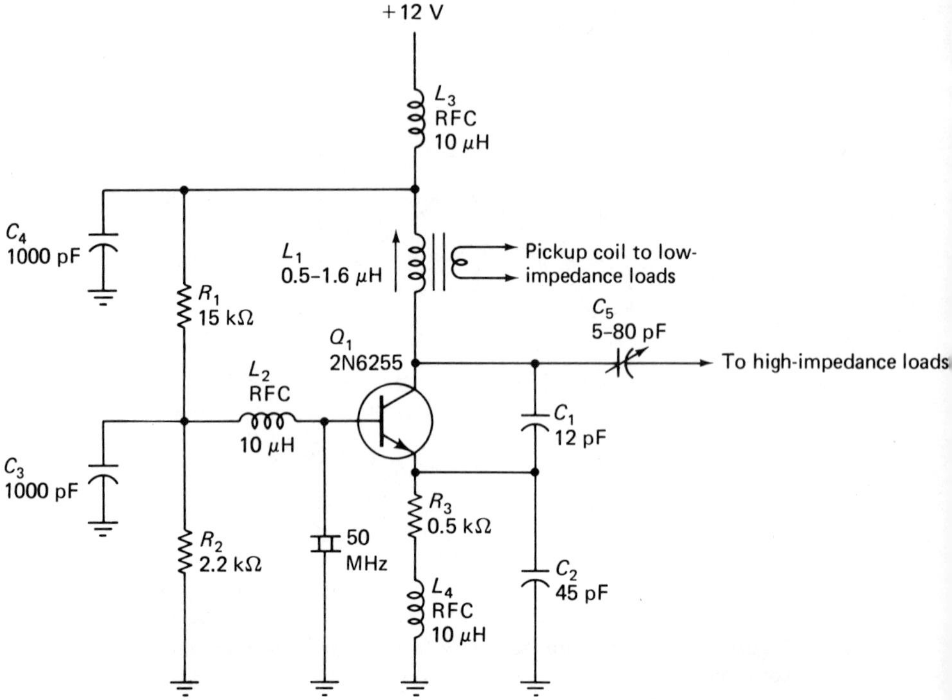

Figure 4-6. 50-MHz crystal-controlled oscillator circuit.

crystal. The circuit can be operated over a narrow range by L_1 (which is slug-tuned). Many factors must be considered in the design of such circuits. For example, the frequency-determining components must be temperature-stable, and mechanical movement of the individual components should not be possible. The following notes describe the most important design and operating characteristics for the oscillator circuit of Fig. 4-6.

Oscillator-frequency stability. Many factors affect oscillator-frequency stability. For example, there is usually some optimum bias value and supply voltage that produces maximum frequency stability over a given range of operating temperature. However, the one factor that can be controlled by the designer is the percentage of feedback. (Note that this percentage refers to feedback versus output voltage.) The lowest practical feedback level is about 10 percent. The best feedback level is usually about 15 percent. Rarely is more than 25 to 30 percent ever required, although some higher-frequency oscillators are operated at 40 percent.

The percentage of feedback is also increased if the tuning circuits are made high-C (large values of capacitance, with corresponding low values of inductance). Such high-C circuits produce sharper resonant tuning. A large value of inductance (with a corresponding lower value of capacitance) introduces more resistance into the resonant circuit, thus lowering the circuit Q to produce broader tuning, as described in Chapter 2.

Oscillator efficiency. For maximum efficiency, the resonant circuit should be at the same frequency as the crystal. If reduced efficiency is acceptable, the resonant circuit can be at a higher frequency (multiple) of the crystal frequency. However, as a guideline, the resonant circuit should not be operated at a frequency higher than the fourth harmonic of the crystal frequency. Figure 4-7 shows the decrease in power output or efficiency of an oscillator operated at harmonic frequencies. Figure 4-7 also shows that efficiency is reduced when overtone crystals are used (instead of fundamental crystals).

Bias circuits. The bias-circuit components (R_1, R_2, R_3) are selected to produce a given current flow under no-signal conditions. The bias circuit is calculated and tested on the basis of normal operating point, even though the circuit will never be at the operating point. A feedback signal is always present and the transistor is always in a state of transition. The collector current should be set at a value to produce the required power output. With the correct bias-feedback relationship, the output power of the oscillator is about 0.3 times the input power. Typically, the voltage drop across L_1 and L_3 is very small, so that the collector voltage equals the supply voltage. Thus, to find a correct value of current for a given power output and

Crystal frequency of operation	RF power output as % of d-c power input			
		Harmonics		
	Fund.	Second	Third	Fourth
Fundamental	27	15	10	5
Third overtone	23	15	10	5
Fifth overtone	22	12	7	3
Seventh overtone	22	12	7	3

Figure 4-7. Typical RF power output of crystal oscillator circuits.

supply voltage, divide the desired output by 0.3 to find the required input power. Then divide the input power by the supply voltage to find the collector current flow.

Feedback signal. The signal output appears at the collector, and with the proper bias-feedback relationship, the output voltage is about 80 percent of the supply voltage. The amount of feedback is set by the ratio of C_1 and C_2. For example, if C_1 and C_2 are of the same value, the feedback signal is one-half of the output signal. If C_2 is made about three times the value of C_1, the feedback signal is about 0.25 of the output signal voltage.

It may be necessary to change the value of C_1 in relation to C_2, in order to get a good bias-feedback relationship. For example, if C_2 is decreased in value, the feedback increases and the oscillator operates nearer the class C region. An increase in C_2, with C_1 fixed, decreases the feedback and makes the oscillator operate as class A. Keep in mind that any change in C_2 (or C_1) also affects frequency. Thus, if the C_1-C_2 values are changed, it will probably be necessary to change the value of L_1.

As a first trial value, the amount of feedback should be equal to, or greater than, cutoff. That is, the feedback voltage should be equal to or greater than the voltage necessary to cut off collector current flow. Under normal conditions, such a level of feedback should be sufficient to overcome the fixed bias (set by R_1 and R_2) and the variable bias set by R_3. As discussed, feedback is generally within the limits of 10 and 40 percent, with the best stability in the range 15 to 25 percent.

Frequency. Frequency of the circuit is determined by the resonant frequency of L_1, C_1, and C_2 and by the crystal frequency. Note that C_1 and C_2 are in series, so that the total capacitance must be found by the conventional series equation

$$C = \frac{C_1 \times C_2}{C_1 + C_2}$$

Also note that the output capacitance of the transistor must be added to the value of C_1. At low frequencies, the output capacitance can be ignored since the value is usually quite low in relation to a typical value for C_1. At higher frequencies, the value of C_1 is lower, so the output capacitance becomes of greater importance.

Capacitor C_1 can be made variable. However, it is generally easier to make L_1 variable, since the tuning range of a crystal-controlled oscillator is quite small. Typically, the value of C_2 is about three times the value of C_1 (or the combined values of C_1 and the transistor output capacitance, where applicable). Thus, the signal voltage (fed back to the emitter terminal) is about 0.25 of the total output signal voltage (or about 0.2 of the supply voltage, when the usual bias-feedback relationship is established).

Resonant circuits. Any number of L and C combinations can be used to produce the desired frequency. That is, the coil can be made very large or very small, with corresponding capacitor values. Often, practical limitations are placed on the resonant circuit (such as available variable inductance values). In the absence of some specific limitations, and as a starting point for resonant circuit values, the capacitance should be 2 pF per meter. For example, if the frequency is 30 MHz, the wavelength is 10 m and the capacitance is 20 pF.

Output circuits. Output to the following stage can be taken from L_1 by means of a pickup coil (for low-impedance loads) or coupling capacitor (for high-impedance loads). Generally, the most convenient output scheme is to use a coupling capacitor (C_5) and make the capacitor variable. This makes it possible to couple the oscillator to a variable load (a load that changes impedance with changes in frequency).

Bypass and coupling capacitors. The values of bypass capacitors C_3 and C_4 should be such that the reactance is 5 ohms or less at the crystal operating frequency. A higher reactance (200 Ω) can be tolerated in some circuits. However, when crystal output is low, the lower reactance is preferred.

The value of C_5 should be approximately equal to the combined parallel output capacitance of the transistor and C_1. Often, this is the midrange value of C_5 (if C_5 is variable).

Radio-frequency chokes. The values of RFCs L_2, L_3, and L_4 should be such that the reactance is between 1000 and 3000 Ω at the operating frequency. The minimum current capacity of the chokes should be greater (by at least 10 percent) than the maximum anticipated direct current. Note that a high reactance is desired at the operating frequency. However, at high frequencies, this can result in very large chokes that produce a large voltage drop (or are too large physically).

4-1-4 Practical Variable-Frequency Oscillators

Figure 4-8 is the working schematic of a variable-frequency oscillator. This circuit is also one of the many variations of the Colpitts oscillator. The circuit is chosen here for maximum stability at frequencies up to about 0.5 MHz. Oscillation is sustained by feedback to the emitter from the junction of C_1 and C_2 (as is the case for the crystal oscillator).

All the design considerations for the variable-frequency oscillator are the same as for the crystal-controlled oscillator (Sec. 4-1-3) with the following exceptions.

Generally, C_1 is made variable to tune across a given frequency range. However, L_1 can be made variable if required.

The values of coupling and bypass capacitors C_3, C_4, and C_5 (if used) should be such that the reactance is 200 Ω at the *lowest operating frequency* (when variable capacitor C_1 is at full value). Note that capacitor C_4 and the output capacitance of the transistor may add to the C_1 capacitance. This tends to lower the resonant frequency of the L_1, C_1, C_2 circuit slightly from the calculated value. However, since C_1 is variable, there should be no problem in tuning to a desired frequency.

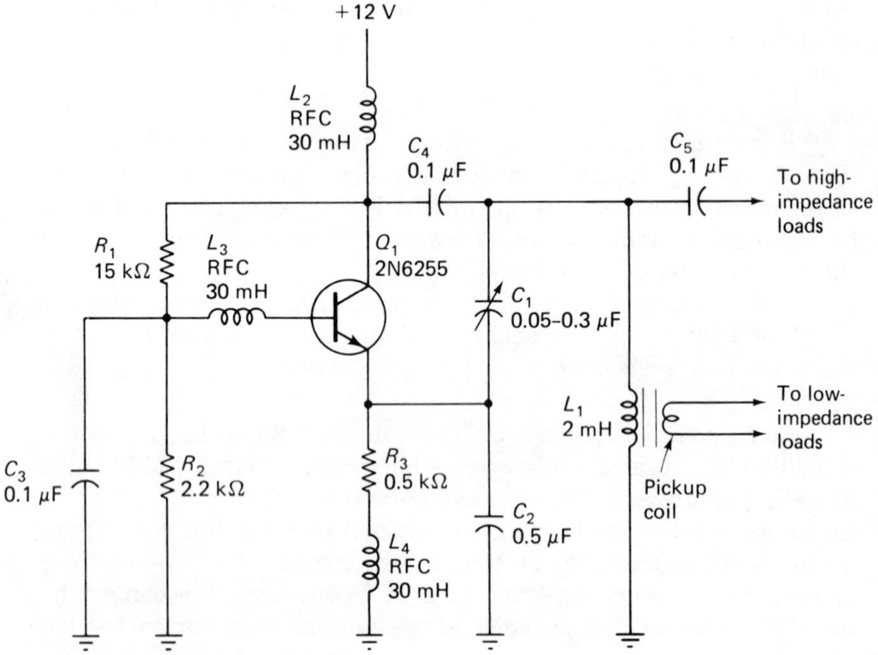

Figure 4-8. 10-60 kHz variable-frequency oscillator circuit.

4-2 *RC* OSCILLATOR CIRCUITS

Figure 4-9 is the working schematic of an *RC* (resistance-capacitance) oscillator. Such oscillators are used at audio frequencies instead of the *LC* (inductance-capacitance) oscillators described in Sec. 4-1. *RC* oscillators avoid the use of inductances, which are not practical in the audio-frequency range. *RC* oscillators are usually operated class A, thus producing good waveforms. The feedback principle is used in *RC* oscillators. The collector signal is fed back through an *RC* network to the base. The time constants of the *RC* network determine the oscillator output frequency. In the circuit of Fig. 4-9, each of the three identical *RC* networks shift the phase by about 60°, resulting in a total phase shift of 180°.

4-2-1 Design Considerations

The following specific points should be considered when designing the circuit of Fig. 4-9.

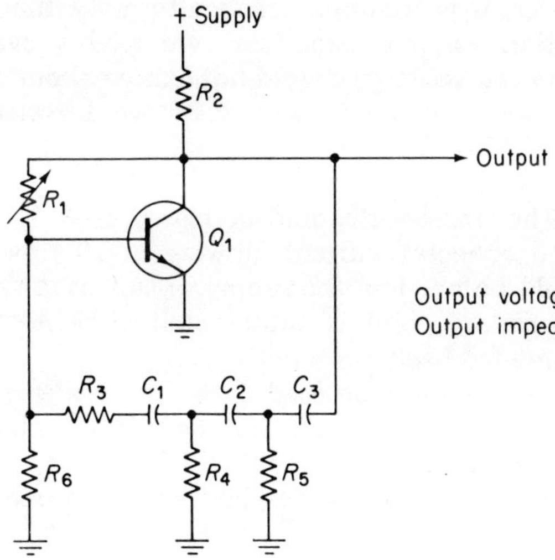

Output voltage ≈ 0.3 × supply
Output impedance ≈ R_2

Frequency $\approx \frac{1}{18\,RC}$, where $R = R_3 = R_4 = R_5$ and $C_1 = C_2 = C_3$

$C_1 = C_2 = C_3 = \frac{1}{18\,R \times \text{frequency}}$
R_1 = 500 kΩ variable
R_2 and R_6 = see text
$R_3 = R_4 = R_5$ = 10 kΩ maximum

Figure 4-9. RC oscillator.

Power output. RC oscillators do not have the efficiency of LC oscillators, since RC oscillators are operated class A. Also, there is considerable power loss in the RC network. Typically, RC oscillators are never more than 30 percent efficient. Generally, power output is not a major design consideration in RC oscillators. Instead, voltage output is the prime factor. Typically, the voltage output is about 30 percent of the supply voltage. Thus, the transistor collector voltage limits (or the available supply) set the maximum output voltage.

Output frequency. The oscillator frequency is determined by the RC time constant. To simplify design, the same values are used in all three RC networks. Thus, the output frequency is about one-third of the typical time-constant frequency relationships of $1/(6.28RC)$. Thus, the output frequency is about $1/(18RC)$. A more exact frequency calculation cannot be made in practical design, since the transistor capacitance and resistance values are added to the RC network. However, the $1/(18RC)$ relationship is satisfactory for trial values.

If a variable output frequency is required, either C or R can be made variable. However, it is common practice to make the C variable, since three-section variable capacitors are readily available. Typically, the RC network resistors should not exceed about 10 kΩ. With this value, the lowest audio frequency (above 1 Hz) can be obtained with a C of less than 6 μF.

Bias network. The transistor should be biased class A. That is, there should be some collector current flowing at all times. The collector voltage should be one-half the supply voltage at the operating point. A true operating point is never obtained since there is always some signal being fed back.

An emitter resistor need not be used since there is always some reverse bias supplied by the feedback signal (one-half of each cycle). The base is set at the correct voltage by R_1. Usually, R_1 is quite large (typically 0.25 to 0.5 MΩ), since base current is on the order of 0.1 mA or less.

Transistor selection. The transistor must be capable of oscillating at the desired frequency. This presents no particular problem since most RC oscillators are used at frequencies below 100 kHz (generally below 20 kHz). However, the transistor must have a gain of about 60 at the operating frequency to overcome the power loss introduced by the RC networks. If transistor gain is high, resistor R_6 is added.

There is a design trade-off relationship between transistor gain and the collector resistor (R_2) value. If the value of R_2 is equal to the value of RC network resistors R_3, R_4, and R_5, a transistor gain of about 60 is required. If the value of R_2 is increased to about twice the value of R_3, R_4, and R_5, the required gain can be reduced to about 45. However, an increase in the value of R_2 lowers the operating-point collector voltage, thus lowering the available output voltage swing.

4-3 TWIN-T *RC* OSCILLATOR CIRCUITS

Figure 4-10 is the working schematic of a twin-T RC (resistance-capacitance) oscillator. Such circuits are used at audio frequencies where a highly stable output at a fixed frequency is desired.

4-3-1 Design Considerations

The following specific points should be considered when designing the circuit of Fig. 4-10.

Power output. Typically, twin-T RC oscillators are never more than about 30 percent efficient. Similarly, the voltage output is about 30 percent of the supply voltage. Thus, the transistor collector voltage limits (or the available supply) set the maximum output voltage.

Frequency. The oscillator frequency is determined by the RC time constants of the twin-T network. This network is essentially a filter that has a sharp null, or balance, at the resonant frequency, $F = 1/(6.28RC)$. By decreasing the value of shunt resistor R_3 slightly from the balance point, the output of the twin-T network is a small, in-phase signal that is rapidly changing in phase at the balance frequency. When the network is adjusted off-balance (the normal oscillating condition) by R_3, the approximate frequency is found by $F = 1/(5RC)$.

Bias network. Both transistors should be forward-biased so that they operate class A. This will ensure a good waveform. The base bias for Q_1 (an emitter follower with no voltage gain) is supplied through R_4, network resistors R_1 and R_2, and collector resistor R_5. As a rule, R_4 should equal the total series resistance of R_1 and R_2. The collector of Q_1 is connected directly to the supply, while the Q_1 emitter is returned through the base resistance of Q_2. Base resistor R_6 should be approximately 10 times the value of R_1. Transistor Q_2 should be biased similar to class A. However, considerable voltage gain is required to overcome the network loss. Therefore, R_5 (collector) should be approximately 60 times the value of R_7 (emitter) for maximum voltage gain. For a realistic R_5 value, the value of R_7 should be less than 100 Ω, typically about 50 Ω.

Any number of *RC* combinations can be used to produce the desired operating frequency. However, since R_1 and R_2 form part of the bias network, design of the *RC* network should start with these resistors. As first trial values, R_1 and R_2 should be 1000 times the value of R_7.

Transistor selection. The transistors must be capable of oscillating at the desired frequency. This presents no particular problem, since most *RC* oscillators are used at frequencies below 100 kHz (generally below 20 kHz). Both Q_1 and Q_2 can be the same type, if convenient. However, Q_2 must have a gain of about 100 at the operating frequency to overcome the power loss.

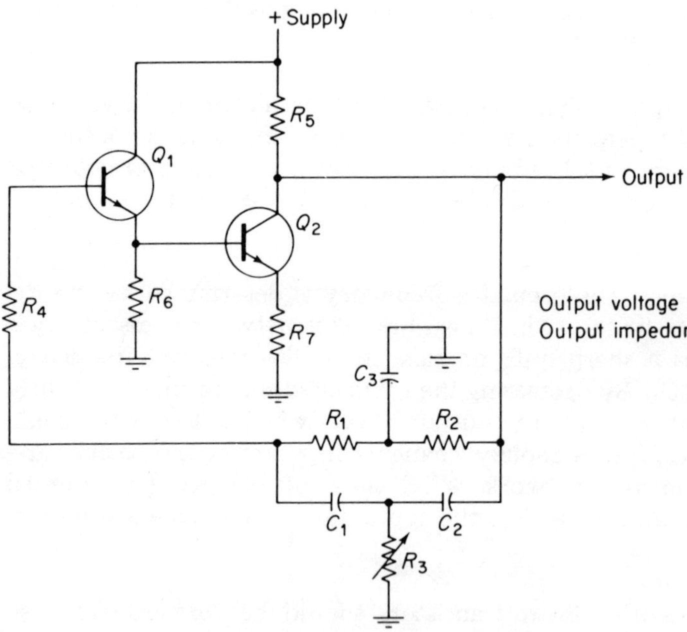

Output voltage ≈ 0.2 × supply
Output impedance ≈ R_5

Frequency $\approx \frac{1}{5\,RC}$, where $R = R_1 = R_2$, $R_3 = 0.5R_1$ and $C = C_1 = C_2$, $C_3 = 2C$

$R_4 \approx R_1 + R_2$ $\quad R_5 \approx 60\,R_7$

$R_6 \approx 10\,R_7$ $\quad R_7 \approx 50-100\,\Omega$

$R_1 = R_2 = 1000\,R_7$ $\quad C_1 = C_2 = \frac{1}{5\,R_1 \times \text{frequency}}$

$R_3 = 0.5\,R_1$ $\quad C_3 = 2C_1$

Figure 4-10. Twin-T RC oscillator.

4-4 BLOCKING OSCILLATOR CIRCUITS

Figure 4-11 is the working schematic of a blocking oscillator. Note that two versions of the circuit are shown, one with a tapped transformer. The blocking oscillator is one of the simplest solid-state oscillators and operates on the relaxation principle. The transistor is initially forward-biased to conduct through the transformer primary winding. This causes a signal to be fed back to the base through C_1. The base is driven very hard in the forward direction, so that C_1 is rapidly charged through the forward-biased emitter-base junction. The output pulse at the T_1 secondary is generated by the rapid turn-on of the collector current, and the steepness of the pulse wavefront is limited only by the leakage inductance of T_1. The top of the pulse is flattened by collector-current saturation. When saturation is reached, the feedback signal drops and is no longer able to forward bias the transistor. Capacitor C_1, in discharging, reverse-biases Q_1, and discharges slowly through R_1. The cycle is restarted by forward bias. The time constant R_1C_1 determines the off-time between pulses and thus determines the oscillator frequency.

The output of a blocking oscillator is similar to that of Fig. 4-12. Blocking oscillators are not suited for radio frequencies, or an application where a sine wave (or anything approaching a sine wave)

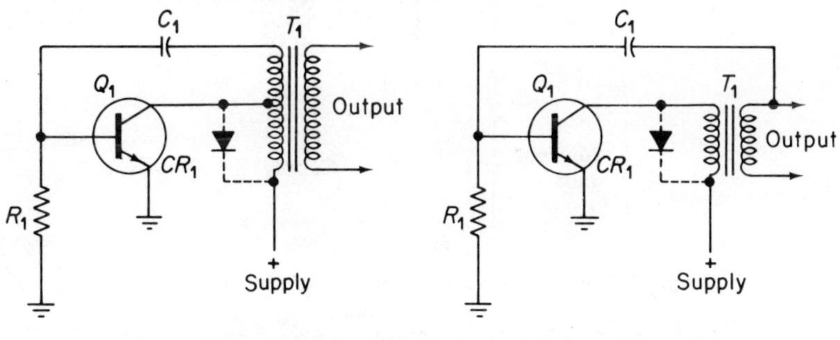

Output voltage (peak pulse) ≈ supply, where $T_1 = 1{:}1$

$T_1 = 5{:}1$ typical

Frequency $\approx \dfrac{1}{R_1 C_1}$

Maximum pulse power $\approx \dfrac{\text{output voltage}^2}{\text{load impedance}}$

Average power dissipated ≈ maximum pulse power × duty cycle

$C_1 \approx 0.1 - 10\ \mu F$

$R_1 \approx 0.5\ M\Omega$ or less

Figure 4-11. Free-running blocking oscillator.

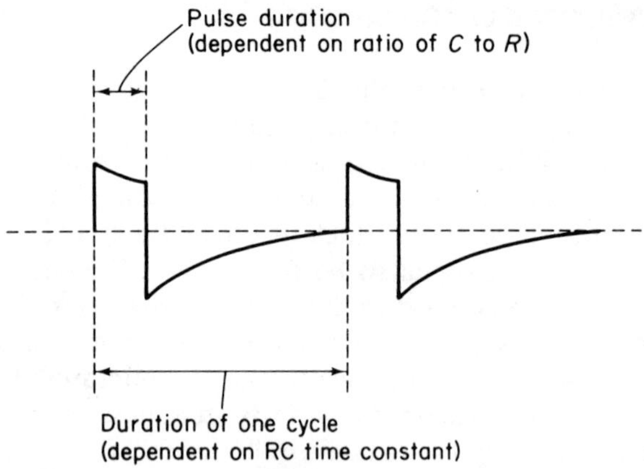

$$\text{Duty cycle} = \frac{\text{pulse duration}}{\text{duration of one cycle}}$$

Figure 4-12. Typical blocking oscillator output waveform.

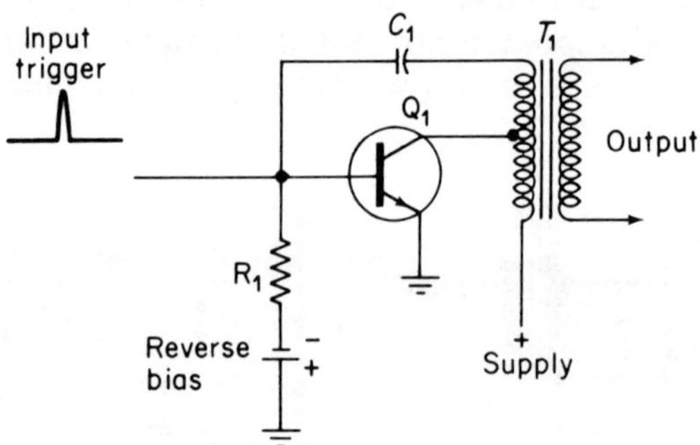

Output voltage (peak pulse) $\approx$ supply, where $T_1 = 1{:}1$

$$\text{Frequency} \approx \frac{1}{R_1\, C_1} \qquad \begin{matrix}\text{Maximum}\\ \text{pulse}\\ \text{power}\end{matrix} \approx \frac{\text{output voltage}^2}{\text{load impedance}}$$

$C_1 \approx 0.1\text{–}10\ \mu\text{F}$

$R_1 \approx 0.5\ \text{M}\Omega$ or less $\approx$ input impedance of oscillator

$$\begin{matrix}\text{Average}\\ \text{power}\\ \text{dissipated}\end{matrix} \approx \begin{matrix}\text{maximum}\\ \text{pulse}\\ \text{power}\end{matrix} \times \text{duty cycle}$$

Reverse bias $\approx$ peak trigger input $\times$ 0.5

$$C_1 \approx \frac{1}{\text{freq.} \times R}$$

Figure 4-13. Triggered or driven blocking oscillator.

is needed. However, blocking oscillators are excellent sources of the steep wavefront pulses required in switching applications. A pulse rise time of 0.1 μs is not uncommon for blocking oscillator outputs.

Blocking oscillators can be used as driven oscillators where a single output pulse is produced in response to a trigger pulse. This requires that the transistor emitter-base be reverse-biased as shown in Fig. 4-13. With this arrangement, the transistor is initially cut off so that a trigger signal moves the operating point into the active region just long enough to start a pulse cycle.

One of the major advantages of a blocking oscillator is that very little current is drawn between pulses. Therefore, high pulse currents can be drawn from the output for short durations, without exceeding the power capability of the transistor.

4-4-1 Design Considerations

The following specific points should be considered when designing the circuits of Figs. 4-11 and 4-13.

Voltage output. If the primary and secondary windings of T_1 have the same number of turns (a 1:1 turns ratio), the output voltage (pulse peak) is approximately equal to the supply voltage. Generally, the secondary winding of T_1 has fewer turns than the primary to provide a voltage stepdown. When the untapped transformer version of the circuit (Fig. 4-13) is used (where C_1 is connected to the secondary), the full output voltage is applied to the base during the brief pulse peak. In some cases, this can result in damage to the transistor. As a guideline, a 5:1 turns ratio is used, so that the output voltage is about one-fifth of the supply voltage.

Transformer selection. Any transformer can be used with a blocking oscillator, provided that the primary can withstand the voltage and current, and the secondary is at the desired output impedance. There is a momentary voltage surge across the primary winding equal to approximately twice the supply voltage. Often, special-purpose transformers are designed for use with blocking oscillators. Such transformers are supplied with design data for the blocking oscillator circuit. This information should be followed when available.

Transistor selection. The transistor must be capable of oscillating at the desired frequency and must be capable of withstanding the full supply voltage continuously. The momentary voltage buildup across the transformer primary is also applied to the transistor. Thus, the collector may be at about twice the supply voltage during the pulse peak. This condition can be corrected by the addition of diode CR_1 across the primary, as shown in dashed form on Fig. 4-11. Diode CR_1 must be capable of withstanding twice the supply voltage without breakdown. Since any diode has some leakage, as well as some forward voltage drop, the addition of CR_1 to the circuit can result in a drop of the output voltage.

Current through the transistor, and the resultant power dissipation, is difficult to calculate in a blocking oscillator. As a rough approximation, divide the square of the anticipated output voltage (across the transformer secondary) by the anticipated load impedance to find the maximum pulse power. The average power dissipated by the transistor then depends on the duration of the pulse in relation to the spacing between pulses.

Bias requirements. The free-running version of the blocking oscillator (Fig. 4-11) is initially forward-biased through R_1. The amount of bias is not critical. Once the circuit begins to oscillate, the emitter-base junction is driven into full forward bias and full reverse bias by the charge and discharge of C_1. The driven version of the blocking oscillator (Fig. 4-13) requires a fixed reverse bias on the emitter-base junction. This bias must be less than the available trigger source. As a first trial, make the reverse bias voltage one-half of the trigger voltage.

Operating frequency. The operating frequency is determined by the time constant of R_1C_1 and is approximately equal to the reciprocal of the time constant. The exact frequency is difficult to calculate, since both the transistor and transformer characteristics can affect the charge and discharge functions. Also, the supply voltage can have some effect on frequency. However, blocking oscillators are fairly stable with respect to power-supply variations.

Various combinations of R and C can be used to produce a given time constant (and thus a given frequency). However, the following rules are generally applied. The value of C_1 is between 0.1 and 10 μF, with a value of less than 0.5 MΩ, for a free-running blocking oscillator in the audio-frequency range. When a larger value of C is used (with an R of corresponding low value) to produce a given RC time constant, the pulse duration is longer in relation to the complete cycle. That is, the on-time (or duty cycle) is longer. This increases the average power dissipation, as well as the average power output.

4-5 MULTIVIBRATOR CIRCUITS

Figure 4-14 is the working schematic of a multivibrator (sometimes known as a *square-wave oscillator*). The circuit is of the *high-current* type. That is, the emitter resistors are about half the collector resistor values, resulting in very high switching currents through the transistors. This requires transistors with a higher current capability (and higher power dissipation) but produces a circuit with high frequency stability. That is, the circuit will maintain frequency to about 1 part in 10^4, in spite of large changes in supply voltage.

The same circuit can also be used as a *low-current* type, where the collector resistors are about 10 times the emitter-resistance value. This requires a lower current (and power dissipation) capability for the transistors, but results in frequency changes with variations of the supply voltage. This is not always an undesirable condition. For

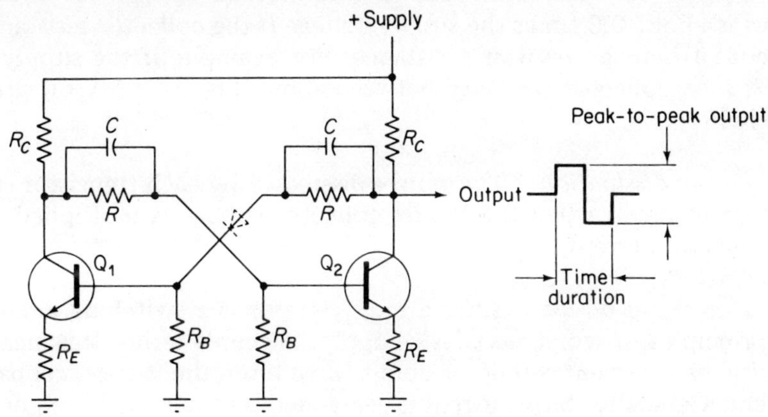

Output voltage (peak-to-peak) $\approx 0.6 \times$ supply $\approx$ operating point collector voltage

Q_1 and Q_2 power dissipated $\approx$ collector voltage $\times$ collector current

$R \approx 10 \times R_C$ Frequency $\approx \dfrac{1}{RC}$ Time duration $\approx RC$ time constant

Lowest collector voltage $\approx 0.3 \times$ supply

$$\text{Maximum collector current} \approx \frac{\text{power dissipation limit of } Q_1 \text{ and } Q_2}{\text{lowest collector voltage}}$$

$$R_C \approx \frac{\text{supply} - \text{lowest collector voltage}}{\text{maximum collector current}}$$

$R_E \approx 0.5 \times R_C$ $C \approx \dfrac{1}{\text{freq.} \times R}$

R_B = see text

Figure 4-14. High-current multivibrator.

example, in telemetry circuits, a low-current type multivibrator is used as a voltage-to-frequency converter to transmit voltage information. However, for stability, high-current type is far superior.

In either type, the output is a symmetrical square wave. Both the positive and negative portions of the cycle are of the same duration and amplitude. The output can be taken from either half of the circuit. Similarly, either half of the circuit can be triggered, if desired. The circuit can be either free-running, where the frequency is determined by the *RC* time constant, or driven by a trigger source.

4-5-1 Design Considerations

The following specific points should be considered when designing the circuit of Fig. 4-14.

Output voltage. The peak-to-peak output voltage (or square wave) is about 0.6 times the supply voltage if the collector resistance is about twice the emitter resistance. For example, if the supply is 10 V, each collector will vary between about 3.5 and 9.5 V on alternate half-cycles.

Power dissipation. The power dissipated by each transistor can be approximated when the lowest collector voltage is multiplied by the collector current.

Worst-case design. Since a multivibrator is a switching circuit, the principles of worst-case design apply. In simple terms, this means that the base current should be about three times the theoretical base current. Generally, base current is calculated on the basis of required collector current divided by gain. In switching circuits, use three times the calculated base current.

When the worst-case design principle is applied, the transistor is always overdriven. While this is undesirable in most circuits, it is required for switching circuits. Under normal operating conditions, a switching circuit is driven into full saturation, or full cutoff, in the shortest possible time.

Switching time. The total switching time of each transistor must be far less than the duration of one output cycle. Total switching time of a transistor is considered as the rise time, fall time, delay time, and storage time all added together. As a rule, switching time must be less than 0.1 of the pulse duration (for a complete cycle). For example, if a transistor has a total switching time of 1 μs, the pulse duration should be at least 10 μs. Thus, the maximum multivibrator operating frequency would be 100 kHz under these conditions. Switching-time information is found on the data sheets for switching transistors (or transistors that could be used in switching applications).

Symmetrical output. If both halves of the circuit are symmetrical, the output will be symmetrical. In some free-running circuits, a slight unbalance is introduced to ensure starting. A multivibrator is inherently stable and may not start in the free-running circuit. An alternative method for starting a free-running multivibrator is to connect a diode in series with either of the feedback resistances, as shown in dashed form on Fig. 4-14. The circuit is then asymmetrical until it starts and reaches full operation.

Operating frequency. The operating frequency is determined by the time constant of the feedback RC and is approximately equal to the reciprocal of the time constant. The exact frequency is difficult to calculate, since the transistor characteristics can affect the charge and discharge function. Various combinations of R and C can be used to produce a given time constant (and thus a given frequency). However, the value of R should be approximately equal to 10 times the value of the collector resistor. Capacitors of corresponding values can then be selected to produce the desired frequency.

Bias relationships. The bias circuit can be calculated, and tested, on the basis of normal operating point, even though the circuit is never at the operating point. A feedback signal is always present, and the transistors are always in a state of transition between full saturation and full cutoff. As guidelines, the collector should be 0.6 of the supply voltage, with the emitter less than 0.2 of supply voltage. The base should be about 0.5 V positive (for silicon NPN transistors) or negative (for silicon PNP transistors) in relation to the emitter.

4-5-2 Schmitt Trigger Circuits

The Schmit trigger is another bistable circuit similar to the multivibrator. However, the Schmitt trigger is not a true oscillator since it returns to the steady state in the absence of a trigger signal. A typical Schmitt trigger circuit is shown in Fig. 4-15. This circuit provides an output only in the presence of an input of sufficient amplitude.

With no trigger input, capacitor C_2 charges to the value of the Q_2 collector. Transistor Q_2 is forward-biased by the positive voltage at the junction of R_2 and R_3. Transistor Q_1 is reverse-biased by the

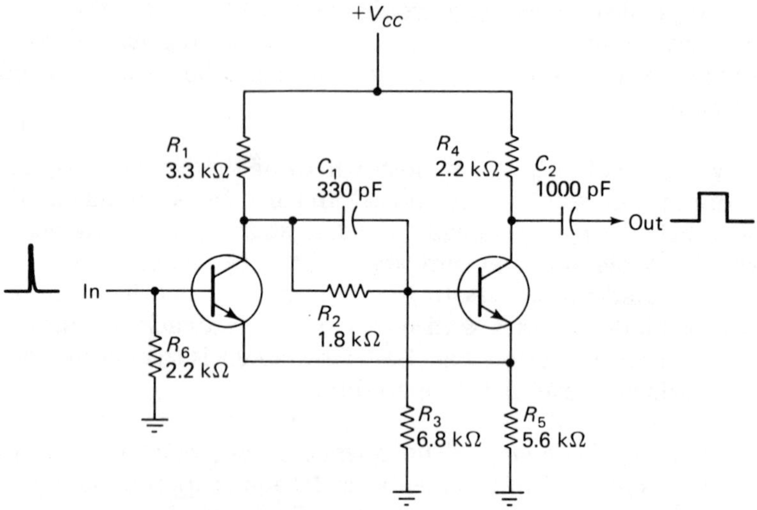

Figure 4-15. Schmitt trigger circuit.

voltage developed across R_5 (which makes the NPN emitter more positive than the base).

With a trigger of sufficient amplitude applied, the reverse bias on Q_1 is overcome, and Q_1 conducts. This drops the voltage on the collector of Q_1 and the base of Q_2, removing the forward bias from Q_2. Transistor Q_2 stops conducting, and an output pulse is produced at the Q_2 collector.

The Schmitt trigger is used where it is necessary to reshape pulses. Typically, the Schmitt trigger is used to convert inputs with any waveform (pulse spikes, a-c waveforms, etc.) into rectangular or square output pulses. The amplitude and duration of the output pulses are determined by the values of the Schmitt trigger rather than the input waveforms. Thus, the Schmitt trigger is sometimes known as a *squaring* circuit.

Chapter 5

Filter, Attenuator, and Waveforming Circuits

This chapter is devoted entirely to filter, attenuator, and waveforming or wave modification circuits. These circuits include *RC* filters, *LC* filters, active filters, attenuators and pads, clippers, slicers, limiters, clamps, and d-c restorers.

5-1 FILTER BASICS

The primary purpose of a filter is to discriminate against the passage of certain groups of frequencies while simultaneously passing other groups or portions of the frequency spectrum. Although filter circuits range from the very simple to the very complex, there are only two basic types of filters, *active* and *passive*. The active filters described here are essentially a solid-state frequency-selective amplifier stage. The passive filters are either *RC* (resistor-capacitor), used primarily for low- or audio-frequency applications, or *LC* (coil-capacitor), for use at higher frequencies. Filters can also be classified by the frequencies that they are designed to pass or reject. The four most common filter classifications are as follows:

A *low-pass filter* is one that passes all frequencies below a selected value and attenuates higher frequencies. The low-pass filter is also known as a *high-cut filter*.

A *high-pass filter* is one that passes all frequencies above a selected value and attenuates lower frequencies. The high-pass filter is also known as a *low-cut filter*.

A *band-elimination filter*, also known as a band-stop, band-rejection, or band-suppression filter, is one that suppresses a selected band of frequencies while passing all lower and higher frequencies.

A *bandpass filter* is one that passes a selected band of frequencies.

5-2 *RC*-FILTER CIRCUITS

The use of resistance-capacitance (*RC*) filters is the simplest way to filter audio-frequency signals. Low frequencies are removed by a series capacitor and shunt resistor (Fig. 5-1), and high frequencies are filtered by interchanging the resistor and capacitor (Fig. 5-2). The filters in Figs. 5-1 and 5-2 are essentially *RC* circuits with the following three factors: resistance, reactance, and impedance. The amount of attenuation produced by *RC* filters is dependent on the ratio of resistance or reactance to impedance.

In the high-pass (low-frequency attenuation) circuit of Fig. 5-1, input voltage is applied across both the resistor and capacitor. Output voltage appears across the resistance. As frequency decreases, reactance increases, as does the impedance. Since the input voltage remains constant, the current through the circuit decreases ($I = E/Z$) with an increase in impedance. Thus, the current through the resistance decreases and the output voltage drops. In a high-pass *RC* circuit, the attenuation is calculated by

$$\text{output voltage} = \text{input voltage} \times \frac{R}{Z}$$

Since the value of R is fixed, for practical purposes, and the value of Z varies inversely with frequency, the voltage output varies directly with frequency (output voltage drops with decreases in frequency).

In the low-pass (high-frequency attenuation) circuit of Fig. 5-2, the input voltage is applied across both the resistor and the capacitor. Output voltage is taken across the capacitor. As frequency increases, the reactance decreases, as does the impedance. However, since the impedance is composed of both reactance and fixed resistance

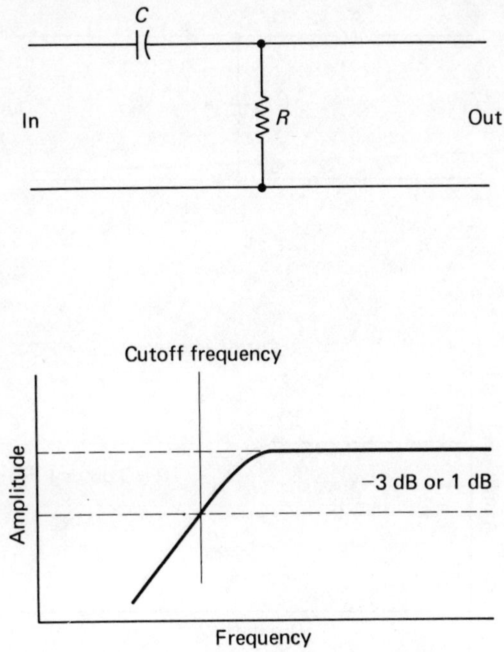

3-dB cutoff frequency (hertz) $\approx 1/(6.28RC)$
3-dB R (ohms) $\approx 1/(6.28FC)$
3-dB C (farads) $\approx 1/(6.28FR)$
1-dB cutoff frequency (hertz) $\approx 1/(3.2RC)$
1-dB R (ohms) $\approx 1/(3.2FC)$
1-dB C (farads) $\approx 1/(3.2FR)$

Figure 5-1. Basic high-pass (low-frequency attenuation or cut) RC filter circuit.

the impedance does not decrease as much as the reactance. Thus, the decrease in reactance (in relation to impedance) causes the voltage output to decrease. In a low-pass *RC*-filter circuit, the attenuation is calculated by

$$\text{output voltage} = \text{input voltage} \times \frac{X_C}{Z}$$

Since the reactance varies at a greater rate than the impedance, the voltage output varies directly with frequency (voltage drops with increases in frequency).

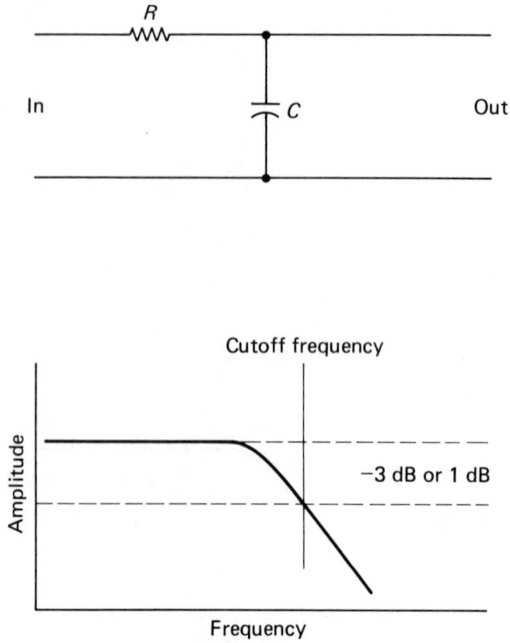

3-dB cutoff frequency (hertz) ≈ 1/(6.28*RC*)
3-dB *R* (ohms) ≈ 1/(6.28*FC*)
3-dB *C* (farads) ≈ 1/(6.28*FR*)
1-dB cutoff frequency (hertz) ≈ 1/(3.2*RC*)
1-dB *R* (ohms) ≈ 1/(3.2*FC*)
1-dB *C* (farads) ≈ 1/(3.2*FR*)

Figure 5-2. Basic low-pass (high frequency attenuation or cut) RC filter circuit.

5-2-1 Design Considerations for *RC* Filters

Filter attenuation is rated in terms of decibel drop at a given frequency. Generally, *RC* filters are designed to produce an approximate 3-dB drop (to 0.707 of input) at a selected cutoff frequency. However, some *RC* filters are designed to produce a 1-dB drop at the cutoff frequency. The equations for both the 1-dB and 3-dB drops are given in Figs. 5-1 and 5-2.

In any *RC* filter, either the capacitor or the resistor value can be assumed to find the other value for a given cutoff frequency. In practical design, the resistance value is usually assumed because the resistor is chosen to meet the other circuit requirements. For example, the resistance in a high-pass filter may also form the circuit input or output impedance.

5-2-2 Bandpass *RC* Filters

Both the high-pass (low-cut) and low-pass (high-cut) filter circuits can be combined to provide a bandpass *RC* filter, as shown in Fig. 5-3. As a guideline, the high-frequency limit (F_H) must be at

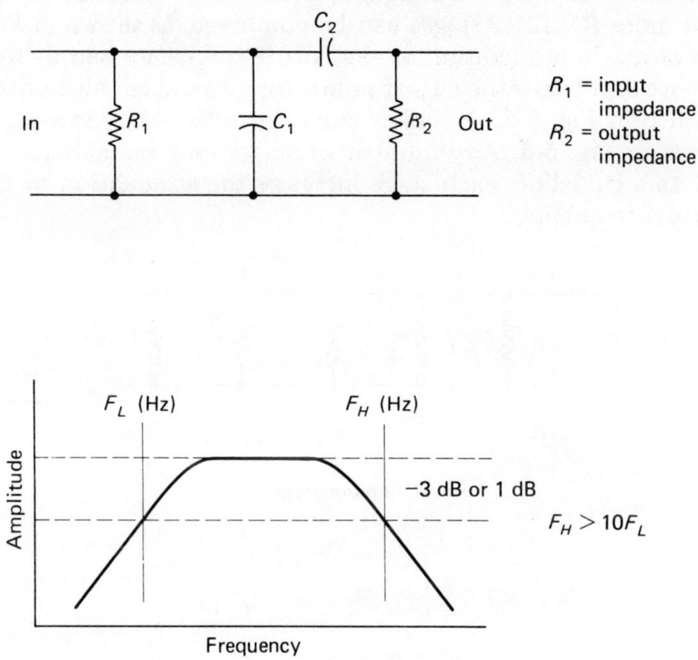

$$F_L \approx \frac{1}{6.28C_2(R_1 + R_2)} \quad \text{(for 3 dB)}$$

$$F_H \approx \frac{R_1 + R_2}{6.28C_1 R_1 R_2} \quad \text{(for 3 dB)}$$

$$C_1 \text{ (farads)} \approx \frac{R_1 + R_2}{6.28F_H R_1 R_2} \quad \text{(for 3 dB)}$$

$$C_2 \text{ (farads)} \approx \frac{1}{6.28F_L(R_1 + R_2)} \quad \text{(for 3 dB)}$$

For 1 dB, use 3.2 instead of 6.28

Figure 5-3. Basic band-pass RC filter.

least 10 times the low-frequency limit (F_L) for such a bandpass circuit to be effective. If the high- and low-frequency limits are not greater than 10:1, there is considerable interaction between the combined circuits. Even with a 10:1 ratio, some interaction may occur. Thus, the equations of Fig. 5-3 are approximate.

5-2-3 Multistage *RC* Filters

A single *RC* filter provides a gradual transition from the passband to the cutoff region. If a rapid transition is necessary for design, two or more *RC* filter stages can be combined, as shown in Fig. 5-4. The increase in attenuation at the cutoff frequency and at frequencies above and below the cutoff point, for a two-stage high-pass filter, are shown in Fig. 5-5. A similar curve for a two-stage low-pass filter is shown in Fig. 5-6. Any number of stages may be added to an *RC* filter. As a guideline, each stage increases the attenuation by 6 dB at the cutoff frequency.

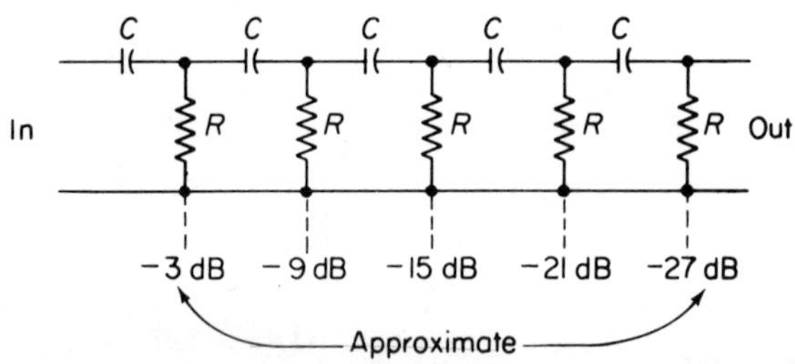

$$\text{Cutoff frequency} \approx \frac{1}{6.28RC}$$

$$R\,(\text{in ohms}) \approx \frac{1}{6.28FC}$$

$$C\,(\text{in farads}) \approx \frac{1}{6.28FR}$$

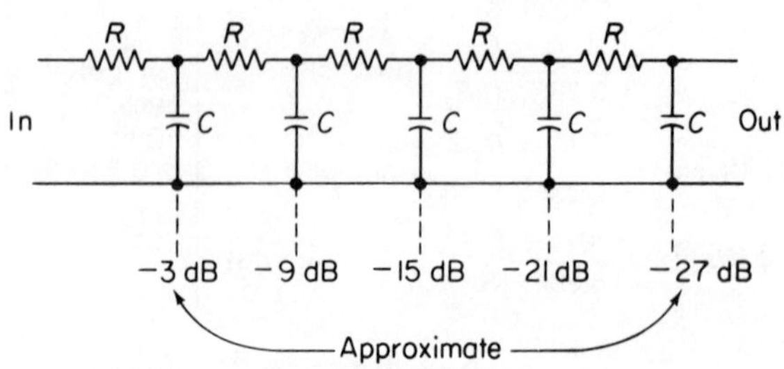

Figure 5-4. Basic multistage RC filters.

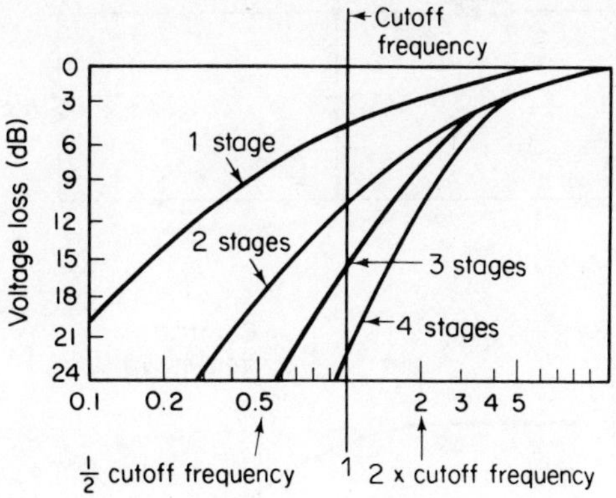

Figure 5-5. Voltage loss curve for high-pass RC filter stages.

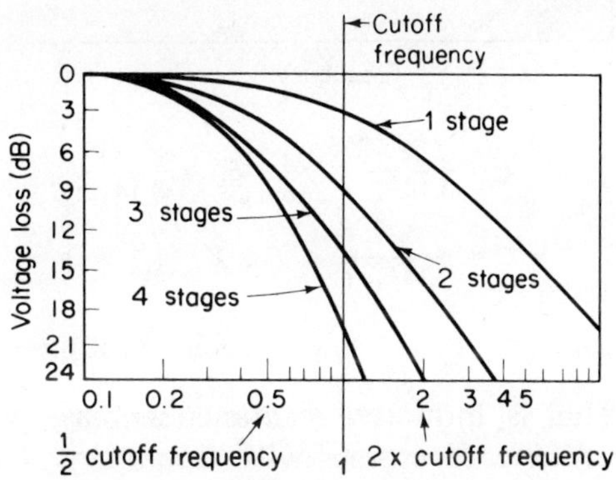

Figure 5-6. Voltage loss curves for low-pass RC filter stages.

5-3 *LC*-FILTER CIRCUITS

The most precise filters are composed of inductors (coils) and capacitors. An *LC* filter can also include resistors. The input and output connections of most *LC* filters are terminated in source and load resistances or by impedances equal in value to the impedance of the filter.

5-3-1 Low-Pass *LC* Filter

Figure 5-7 shows a typical low-pass *LC* filter in its simplest form, the *basic L configuration*. All *LC* filters take advantage of the fact that capacitors and coils operate inversely in the presence of alter-

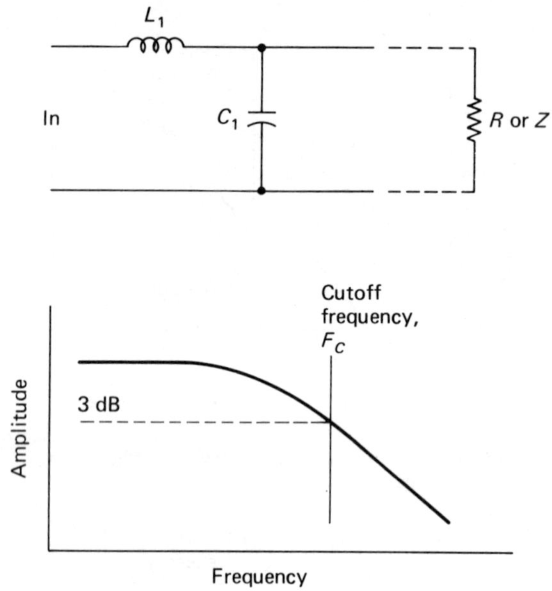

$$L \approx \frac{R}{3.14 F_C} \qquad F_C \approx \frac{1}{3.14\sqrt{LC}}$$

$$C \approx \frac{1}{3.14 F_C R} \qquad R \text{ or } Z \approx \sqrt{\frac{L}{C}}$$

Figure 5-7. Basic L low-pass constant-K filter.

nating current. That is, inductive reactance increases with frequency, and capacitive reactance decreases with frequency. Thus, in the low-pass *LC* filter, the parallel unit (a capacitor across the line) has a decreasing reactance as frequency is increased. This acts to bypass high frequencies but has an increasing reactance to low frequencies. The series unit (a coil in the line) has an increasing reactance to high frequencies but passes the low frequencies.

In most *LC* filters, the product of the impedances presented by the capacitance and inductance remains constant with changes in frequency (because of the inverse change in reactance). For example, if capacitive reactance goes down for a given increase in frequency, the inductive reactance goes up by a corresponding amount; hence the name *constant-K filter*.

One of the problems with the basic L-type filter is that it does not provide sharp frequency change. To increase frequency sharpness, another coil can be added to the basic L configuration, as shown in Fig. 5-8. Such a filter is known as the T type because of its appearance. In a T-filter, the value of capacitor *C* does not change from that of a basic L configuration, and the basic equations are the same. The

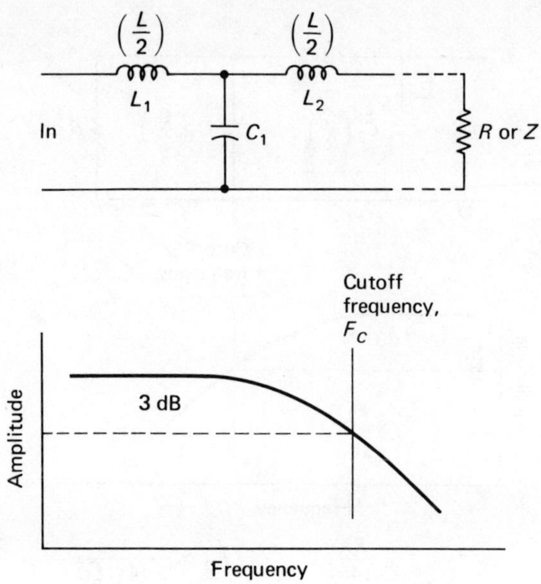

$$\text{Total } L \approx \frac{R}{3.14F_C} \qquad F_C \approx \frac{1}{3.14\sqrt{LC}}$$

$$C_1 \approx \frac{1}{3.14F_C R} \qquad R \text{ or } Z \approx \sqrt{\frac{L}{C}}$$

Figure 5-8. T-type low-pass constant-K filter.

total inductance L_1 and L_2 must be equivalent to that of the single coil in a basic L configuration. Usually, the desired total inductance is divided evenly between the two coils, so that each coil in a T-type low-pass filter has one-half the total desired inductance.

Frequency sharpness can also be increased by adding another capacitor, as shown in Fig. 5-9. Such a filter is known as the *pi type.* In a pi-type *LC* filter, the value of *L* does not change, but the total capacitance of C_1 and C_2 must be equivalent to that of the single capacitor in a basic L configuration. Usually, the desired total capacitance is divided evenly between the two capacitors, so that each capacitor in a pi-type low-pass filter has one-half the total desired capacitance.

5-3-2 High-Pass *LC* Filter

Figure 5-10 shows a typical high-pass filter in its simplest form, the basic L configuration. High-pass filters also take advantage of the fact that the capacitors and coils operate inversely in the presence of alternating current. Thus, in a high-pass *LC* filter, the series unit (a

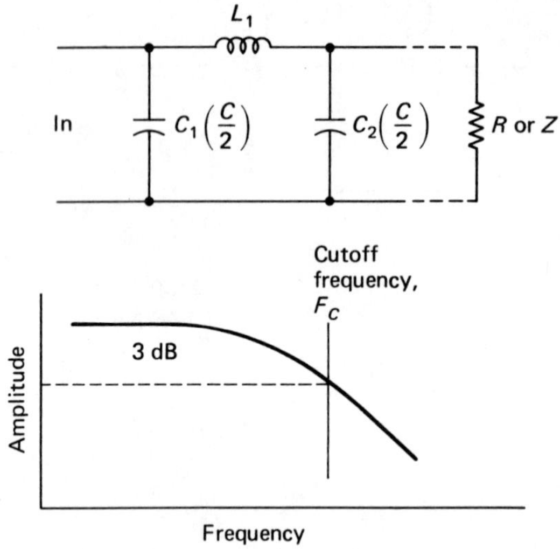

$$L \approx \frac{R}{3.14F_C} \qquad F_C \approx \frac{1}{3.14\sqrt{LC}}$$

$$\text{Total } C \approx \frac{1}{3.14F_C R} \qquad R \text{ or } Z \approx \sqrt{\frac{L}{C}}$$

Figure 5-9. Pi-type low-pass constant-K filter.

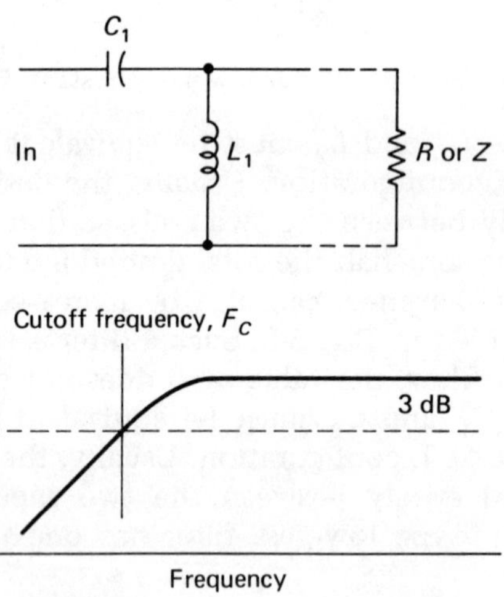

$$L \approx \frac{R}{12.6F_C} \qquad F_C \approx \frac{1}{12.6\sqrt{LC}}$$

$$C_1 \approx \frac{1}{12.6F_C R} \qquad R \text{ or } Z \approx \sqrt{\frac{L}{C}}$$

Figure 5-10. Basic L high-pass constant-K filter.

capacitor in the line) has a decreasing reactance as the frequency is increased. The series unit passes high frequencies but offers an increasing reactance to low frequencies. The parallel unit (a coil across the line) bypasses low frequencies but has an increasing reactance to high frequencies. Most high-pass *LC* filters are of the constant-*K* type.

To increase frequency sharpness, another capacitor can be added to the basic L configuration, as shown in Fig. 5-11. Such a filter is known as the T-type. In the T-filter, the value of coil *L* does not change from that of the basic L configuration, and the basic equations are the same. The total capacitance of C_1 and C_2 must be equivalent to that of the single capacitor in the basic L configuration. Usually, the desired total capacitance is divided evenly between the two capacitors, so that each capacitor in the T-type high-pass *LC* filter has twice the total desired capacitance.

Frequency sharpness can also be increased by adding another coil, as shown in Fig. 5-12. Such a filter is known as the pi type. In a pi-type *LC* filter, the value of *C* does not change, but the total inductance of L_1 and L_2 must be equivalent to that of the single coil in the basic L configuration. Usually, the desired total inductance is divided evenly between the two coils, so that each coil in the pi-type high-pass *LC* filter has twice the total desired inductance.

5-3-3 Band-Elimination *LC* Filter

The band-elimination filter takes advantage of the different impedance characteristics of series- and parallel-resonant *LC* circuits. A parallel *LC* circuit has maximum impedance at the resonant frequency. A series *LC* circuit has minimum impedance at the resonant frequency. These two *LC* circuits are combined in the band-elimination filter circuit of Fig. 5-13. The series arm has minimum impedance at the center frequency of the desired band. The impedance increases on either side of resonance. This acts to bypass the center frequency. The parallel arm has maximum impedance at the center frequency, with impedance decreasing on either side of resonance. This prevents passage of the center frequency as well as a band of frequencies on either side.

5-3-4 Bandpass *LC* Filter

The bandpass filter takes advantage of the differing impedance characteristics of series- and parallel-resonant circuits. A parallel *LC* circuit has maximum impedance at the resonant frequency. A series *LC* circuit has minimum impedance at the resonant frequency. These

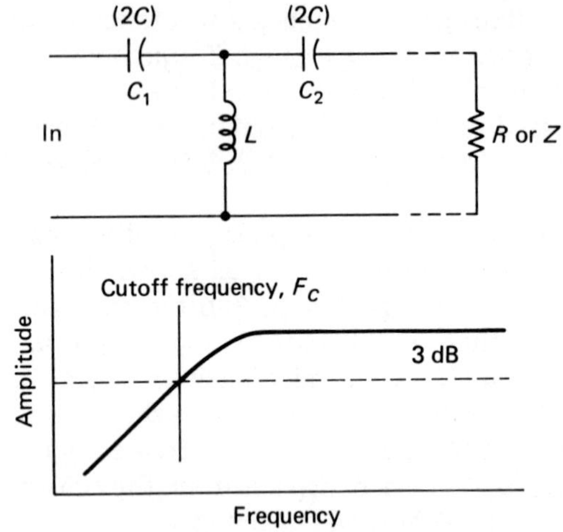

$$L \approx \frac{R}{12.6F_C} \qquad F_C \approx \frac{1}{12.6\sqrt{LC}}$$

$$C \approx \frac{1}{12.6F_C R} \qquad R \text{ or } Z \approx \sqrt{\frac{L}{C}}$$

Figure 5-11. T-type high-pass constant-K filter.

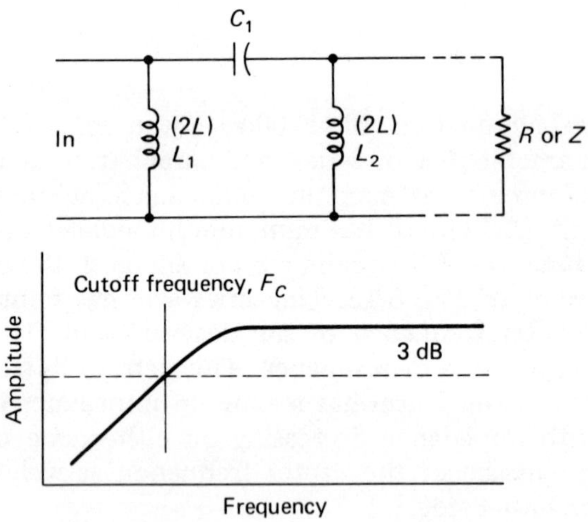

$$L \approx \frac{R}{12.6F_C} \qquad F_C \approx \frac{1}{12.6\sqrt{LC}}$$

$$C \approx \frac{1}{12.6F_C R} \qquad R \text{ or } Z \approx \sqrt{\frac{L}{C}}$$

Figure 5-12. Pi-type high-pass constant-K filter.

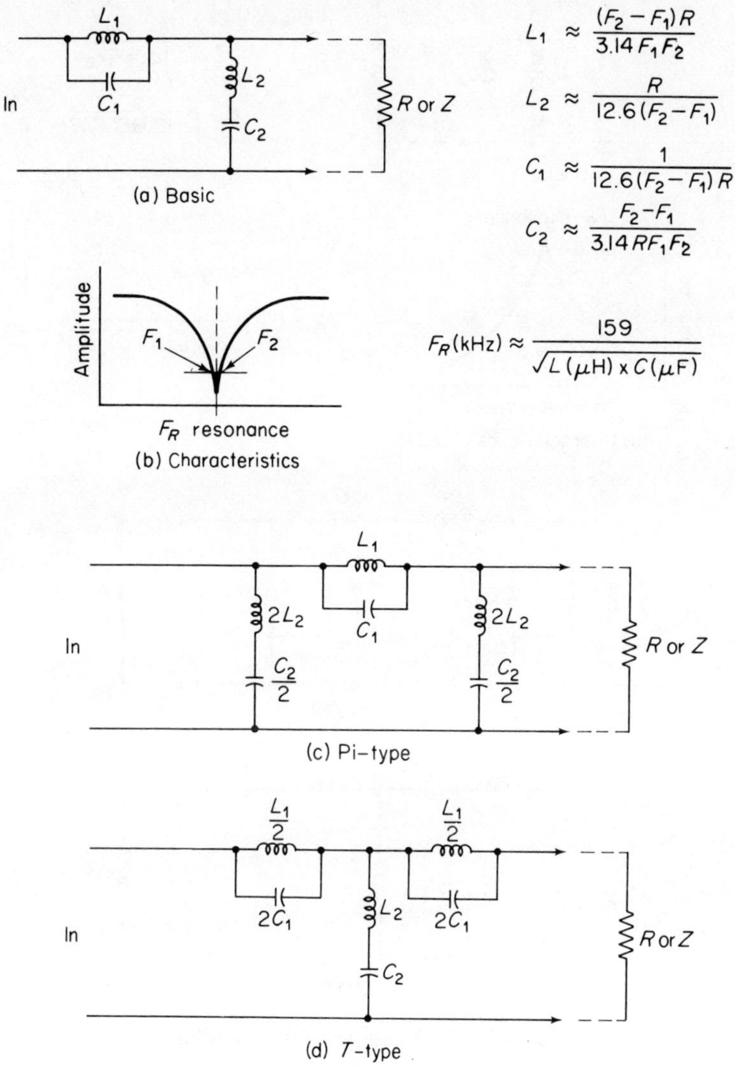

Figure 5-13. Band-elimination filter circuits.

two LC circuits are combined in the bandpass filter of Fig. 5-14. The series arm has minimum impedance at the center frequency of the desired band. The impedance increases on either side of resonance. This acts to pass the center frequency and to reject frequencies above and below the center frequency. The parallel arm has maximum impedance at the center frequency, and the impedance decreases on either side of resonance. This acts to bypass frequencies above and below the center. Therefore, both the series and parallel arms provide

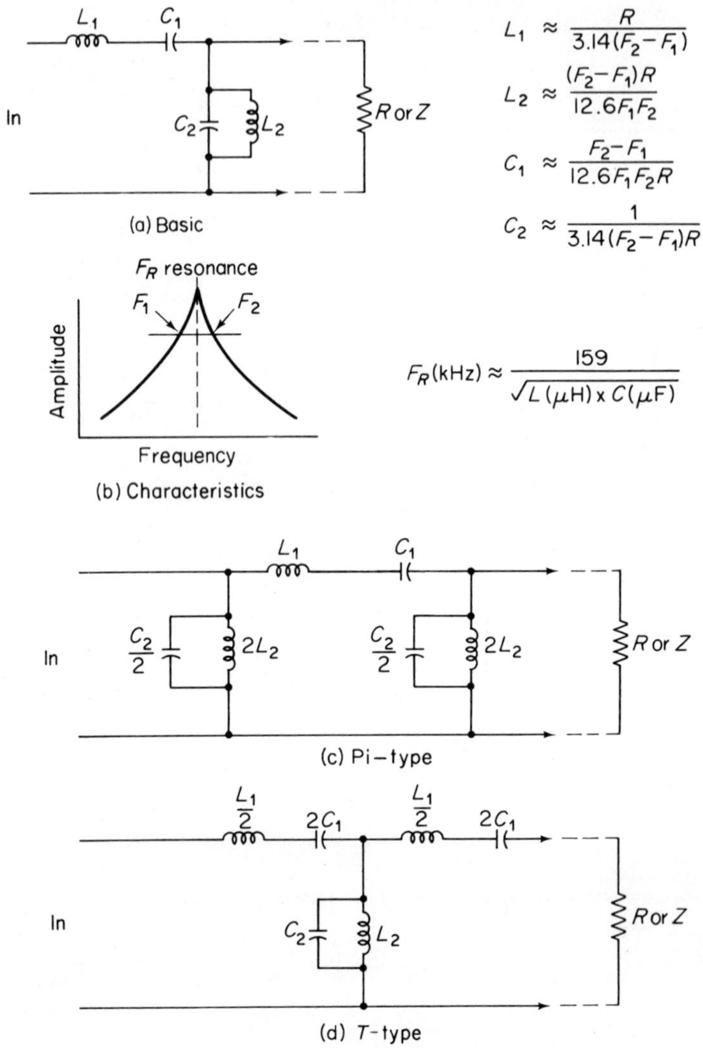

Figure 5-14. Band-pass filter circuits.

passage of the center frequency as well as a band of frequencies on either side of center.

5-4 ACTIVE-FILTER CIRCUITS

In addition to passive filters (either *LC* or *RC*), it is possible to use amplifiers to form active filters. There are two prime advantages in using these active filters. First, it is possible to obtain the equivalent of an inductive reactance without actually using a heavy and bulky inductance usually required for a typical *LC* audio filter. Second, the use of an active filter eliminates the signal loss normally associated with passive filters (either *RC* or *LC*).

5-4-1 Active Low-Pass (High-Cut) Filters

Figure 5-15 shows the basic circuit of an active low-pass filter together with the corresponding characteristic curves for several sets of component values. Note that these values are approximate and will usually require trimming to achieve an exact curve. The typical voltage gain is slightly less than 1 (unity) for transistors with a minimum beta of 20.

The amount of gain and the shape of the curves are set by the amount of feedback in relation to signal (which, in turn, is set by component values). Note that the feedback is positive and thus adds

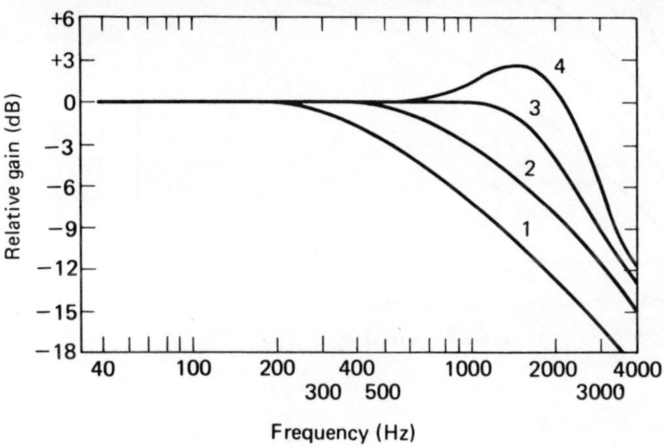

Curves

	C_F	C_1
1	0.003	0.015
2	0.007	0.010
3	0.015	0.005
4	0.05	0.002

(C_F, C_1 in μF)

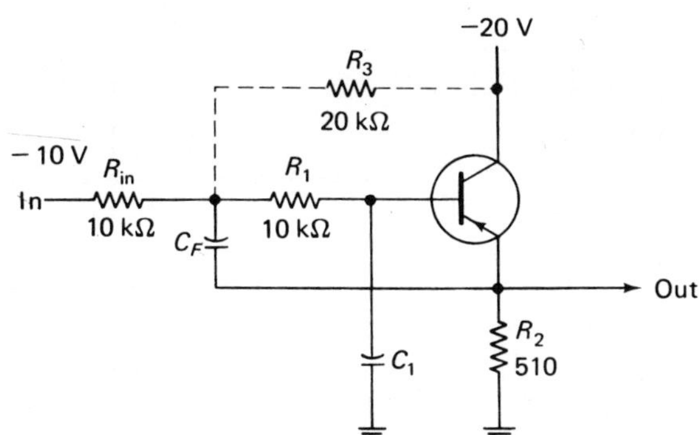

Figure 5-15. Active low-pass (high-cut) filter and corresponding response curves.

to the signal. However, the feedback amplitude (across the entire frequency range) is just below the point necessary for oscillation. The circuit of Fig. 5-15 is an emitter follower, which typically has no voltage gain.

The circuit of Fig. 5-15 requires a bias of approximately −10 V (one-half the −20-V supply) at the input. This can be obtained from a previous stage. If no such stage exists, the bias can be obtained by the addition of the 20-kΩ resistor (shown in phantom as R_3) and by changing the value of R_{IN} to 20 kΩ. Such an arrangement introduces a loss of about 6 dB. Thus, it is better to operate the circuit in Fig. 5-15 by direct coupling from the output of a previous stage.

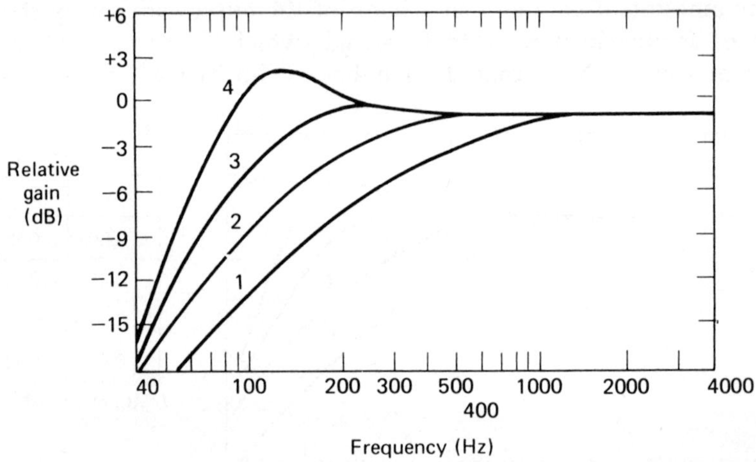

Curves

	C_1	C_2	R_F
1	0.3	0.1	15 kΩ
2	0.3	0.3	5100
3	0.5	0.5	1500
4	1.0	1.0	510

(C_1, C_2 in μF)

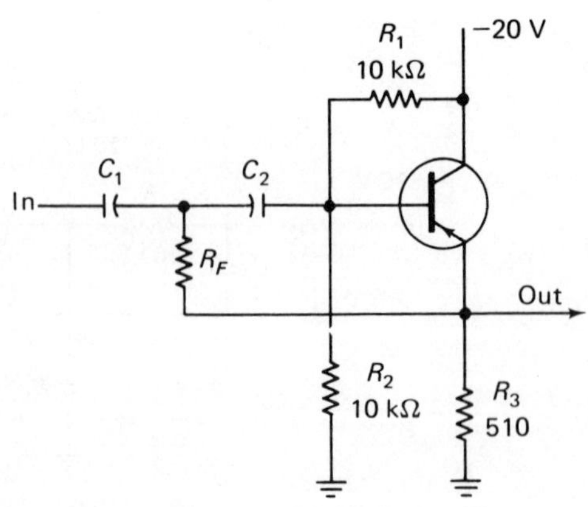

Figure 5-16. Active high-pass (low-cut) filter and corresponding response curves.

5-4-2 Active High-Pass (Low-Cut) Filters

Figure 5-16 shows the basic circuit of an active high-pass filter, together with characteristic curves. The circuit of Fig. 5-16 is the inverse of the Fig. 5-15 circuit. That is, the Fig. 5-16 circuit uses capacitors in series with the base, with feedback obtained through R_F rather than C_F. The gain and shape of the curves are set by the amount of feedback (determined by circuit values).

5-4-3 Active Bandpass Filters

The circuits in Figs. 5-15 and 5-16 can be cascaded to provide a bandpass filter. Any of the curves can be used. However, curve 3 is

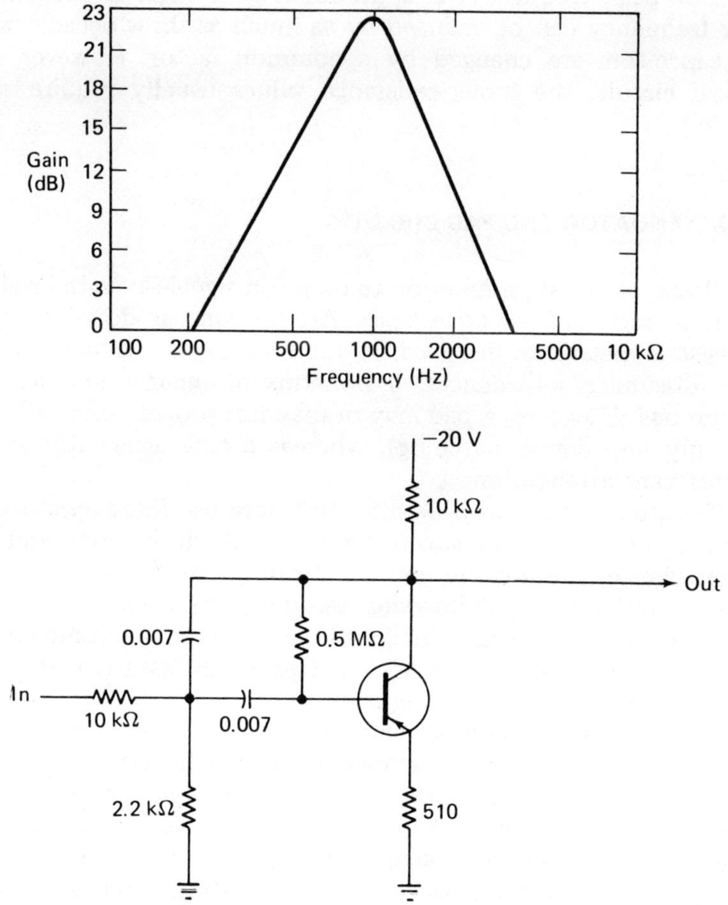

Figure 5-17. Active peaking filter and corresponding response curve.

the most satisfactory because it has the sharpest break at cutoff. Curves 1 and 2 have considerable slope with no sharp break, and curve 4 produces some peaking at the breakpoints. If the circuits are cascaded, the low-pass filter (Fig. 5-15) should follow the high-pass filter (Fig. 5-16). This provides the necessary bias at the input of the low-pass filter (−10 V from the emitter of the high-pass filter).

5-4-4 Active Peaking Filters

The circuits in Figs. 5-15 and 5-16 can be cascaded to provide a peaking filter with the proper selection of components. However, a single-stage tuned amplifier produces the same results. Such a circuit is shown in Fig. 5-17. As shown by the characteristic curve, the center, or peak frequency, is approximately 1 kHz. If desired, the center frequency can be changed by as much as three decades when both capacitors are changed by a common factor. However, in a practical circuit, the input resistance values usually require some trimming.

5-5 ATTENUATOR AND PAD CIRCUITS

Like filters, attenuators function to diminish signal-strength level but not on a frequency-selective basis. Attenuators, as described here, are resistive networks designed to diminish signal strength and to match dissimilar impedances. The terms *attenuator* and *pad* are synonymous. However, a pad may or may not provide some attenuation (only impedance matching), whereas a true attenuator always provides some attenuation.

The attenuators and pads discussed here use fixed resistors and may be of two types: the *asymmetrical,* in which the input and output impedances are not the same, and the *symmetrical*, which have identical input and output impedances. In general, asymmetrical pads are used for impedance matching, but may also provide some attenuation. The symmetrical pads are used primarily as attenuators between two devices with equal impedances.

Many of the attenuators and pads described here are available commercially, in variable-resistance form. Such variable pads are made up of variable resistances (usually potentiometers) ganged together on a common shaft. All of the resistances vary, in proportion, as the shaft is turned. This keeps the input/output impedance relationships contant even though the amount of attenuation is varied.

Although there are any number of attenuator, or pad, circuits, they are generally versions of a few basic types. These basic types include the L, U, T, O, H, and pi types. In general, the L- and U-pads are asymmetrical (for impedance matching), and the T, O, H, and pi types are symmetrical (for attenuation).

5-5-1 Asymmetrical L- and O-pads

The basic L-type pad shown in Fig. 5-18 is an asymmetrical pad used to match impedance of an input source to that of an output source. This provides a smooth transition between devices of differ-

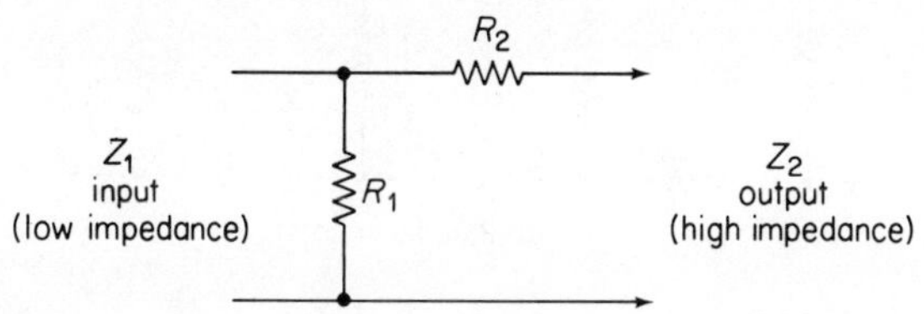

$$R \approx \frac{Z_1}{\sqrt{1 - \frac{Z_1}{Z_2}}} \qquad R_2 \approx Z_2\sqrt{1 - \frac{Z_1}{Z_2}}$$

Figure 5-18. Basic L-type pad and calculations.

ent characteristic impedances, with each device encountering a total impedance equal to its own characteristic impedance. Although a certain amount of signal attenuation is usually required to make this transition, the pad in Fig. 5-18 introduces a minimum of loss while performing the impedance-matching function. If Z_1 (input) has a lower impedance with respect to Z_2 (output), the relationships and equations are as shown in Fig. 5-18. In practice, the standard resistor values that are closest to the calculated values are used for R_1 and R_2.

A balanced arrangement of the L-type pad is shown in Fig. 5-19. Here, each series resistor has one-half the value of R_2, shown originally in Fig. 5-18. When the pads are placed in cascade, as in Fig. 5-19b, the network is sometimes referred to as a *ladder pad*. The basic pad of Fig. 5-19 is often known as a U-pad.

5-5-2 Symmetrical T- and H-pads (Attenuators)

The T-pad, shown in Fig. 5-20, is a symmetrical pad where the impedance of the input device matches that of the output device. The only function of the T-pad is attenuation of the signal. Since impedance matching is not involved, the values of all R_1 resistors are not identical, and the values of R_1 and R_2 are chosen to provide

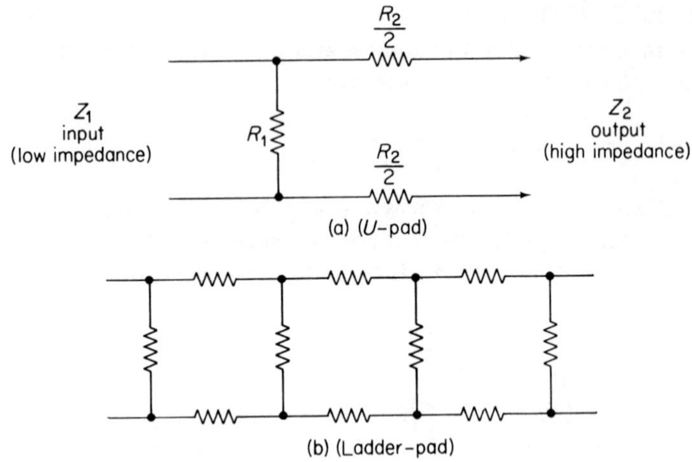

$$R_1 \approx \frac{Z_1}{\sqrt{1 - \frac{Z_1}{Z_2}}} \qquad R_2 \approx Z_2 \sqrt{1 - \frac{Z_1}{Z_2}}$$

Figure 5-19. Balanced minimum-loss U-pad and ladder-pad.

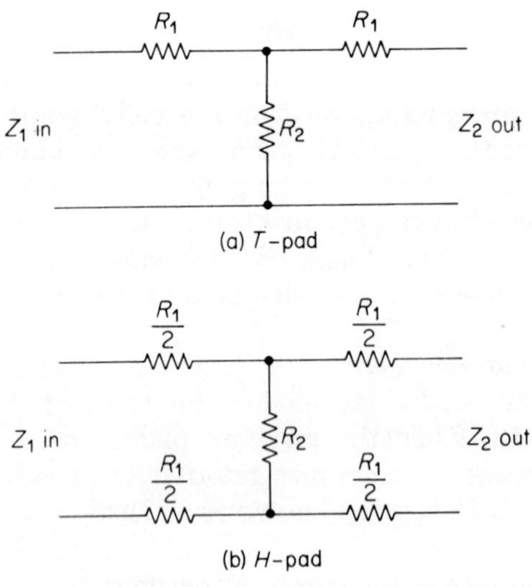

$$R_1 \approx Z\,\frac{(V - 1)}{(V + 1)} \qquad R_2 \approx Z\left[\frac{2V}{(V + 1)(V - 1)}\right]$$

V = input voltage divided by output voltage

Figure 5-20. T-pad and H-pad circuits for attenuation.

the degree of attenuation desired. The T-pad in Fig. 5-20a is an unbalanced type. The balanced version (sometimes known as the H-pad) is shown in Fig. 5-20b. In the balanced version, the R_1 values are cut in half.

The values of R_1 and R_2 are related to the ratio of signal-voltage or signal-current attenuation that is desired. It is usually simpler to use a voltage ratio. For example, if a signal voltage having an amplitude of 100 V is to be attenuated to produce a 10-V output signal, the voltage ratio is 10.

5-5-3 Symmetrical Pi- and O-pads (Attenuators)

The pi-type pad shown in Fig. 5-21 is a symmetrical unbalance type. The balanced version (sometimes known as an O-pad) is shown in Fig. 5-21b. Since the input impedance is the same as the output impedance, no impedance matching is involved. The resistor values are chosen to provide the degree of attenuation desired. As with the T- and H-pads, the equations for solving the resistor values must consider the voltage attenuation ratio.

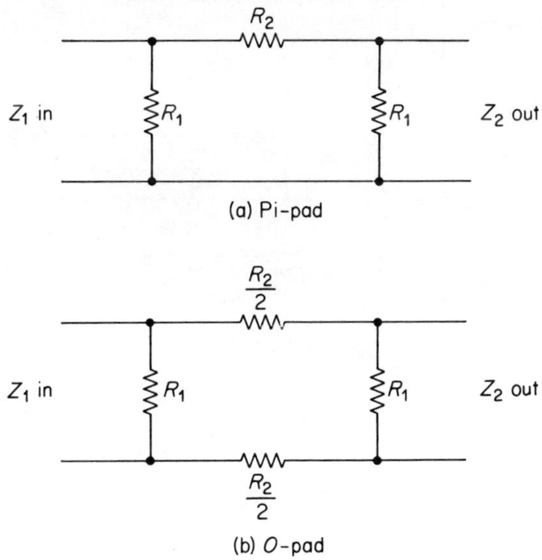

$$R_1 \approx Z\frac{(V+1)}{(V-1)} \qquad R_2 \approx Z\left(\frac{V^2-1}{2V}\right)$$

V = input voltage divided by output voltage

Figure 5-21. Pi-pad and O-pad circuits for attenuation.

5-5-4 Symmetrical Bridged T- and H-pads

An addition resistor can be shunted across the series resistors of the T- and H-pads to form what is known as a *bridged pad*. The bridged T- and H-pads are shown in Fig. 5-22a and b, respectively. The ohmic values of R_1 and R_2 are chosen so that each has a resistance equal to the ohmic value of the impedance; therefore, only the resistance of R_3 and R_4 need be calculated.

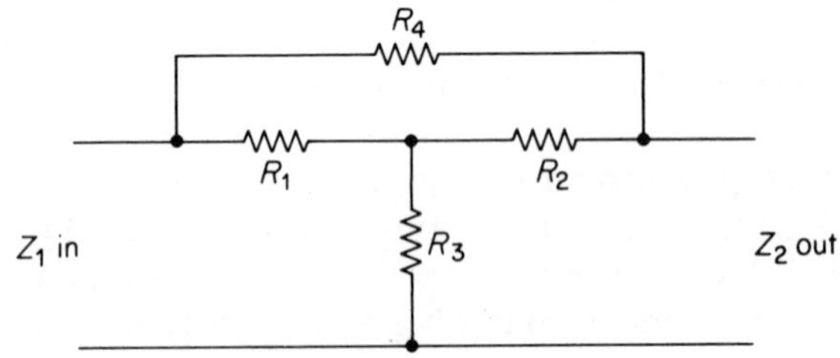

(a) Bridged *T*-pad

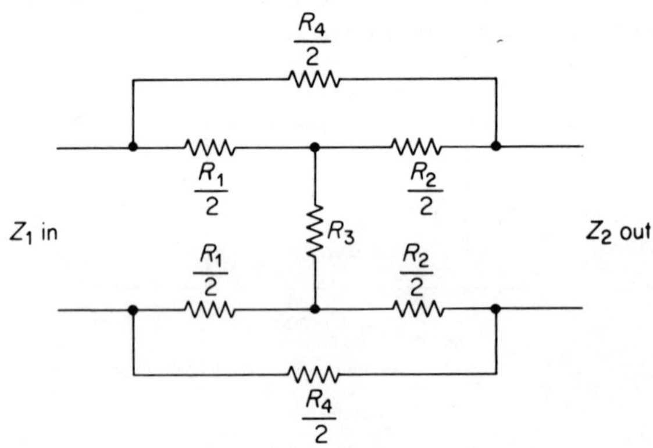

(b) Bridged *H*-pad

$$R_1 = R_2 = Z \qquad R_3 \approx \frac{Z}{V-1} \qquad R_4 \approx Z(V-1)$$

V = input voltage divided by output voltage

Figure 5-22. Bridged T-pad and bridged H-pad circuits.

5-6 CLIPPER CIRCUITS

A clipper circuit clips a particular waveform so that the amplitude of the output signal is proportional to those values of input waveform exceeding a certain critical value.

5-6-1 Basic Diode Clippers

A basic diode clipper circuit is shown in Fig. 5-23. The diode is normally reverse-biased by the bias voltage (battery), and no current flows in the absence of an input signal. If a positive input is applied, current does not flow until the input positive amplitude rises above the bias voltage (4.5 V). The output amplitude is then proportional to the amplitude of the input that exceeds the 4.5-V bias. No current flows for a negative input signal. Therefore, the circuit clips those portions of the signal that are below the positive 4.5-V bias value.

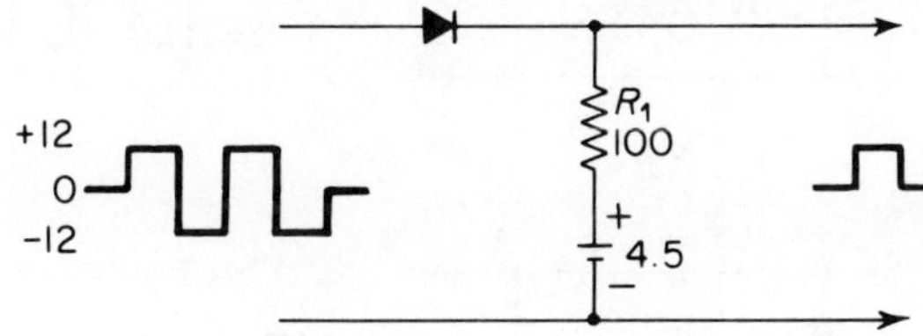

Figure 5-23. Series diode clipper.

5-6-2 Shunt Diode Clippers

Figure 5-24 shows the four basic types of shunt clippers. In all four circuits, the diode is forward-biased during a portion of the input waveform and conducts. During conduction, the low forward resistance of the diode shunts the load resistance R_1, and no output is produced. Thus, that portion of the input waveform is clipped. The diode is reverse-biased during the remainder of the input waveform, so that the remainder of the waveform is reproduced across the output load resistor R_1.

In Fig. 5-24b-d, the diode is reverse-biased by a fixed voltage (represented by a battery). When the input signal exceeds this bias, the diode conducts, and no output voltage is developed across R_1. Thus, that portion of the input signal above the bias voltage is clipped. With negative input signals, the diode does not conduct and the output is proportional to the input.

5-6-3 Parallel Clippers (Slicers)

Two shunt clippers can be combined to produce a parallel clipper (sometimes known as a slicer), as shown in Fig. 5-25a. The same effect can be produced by an overdriven, common-emitter, class A transistor amplifier as shown in Fig. 5-25b. In either circuit, the purpose is to clip the peaks of an a-c input signal so that the out-

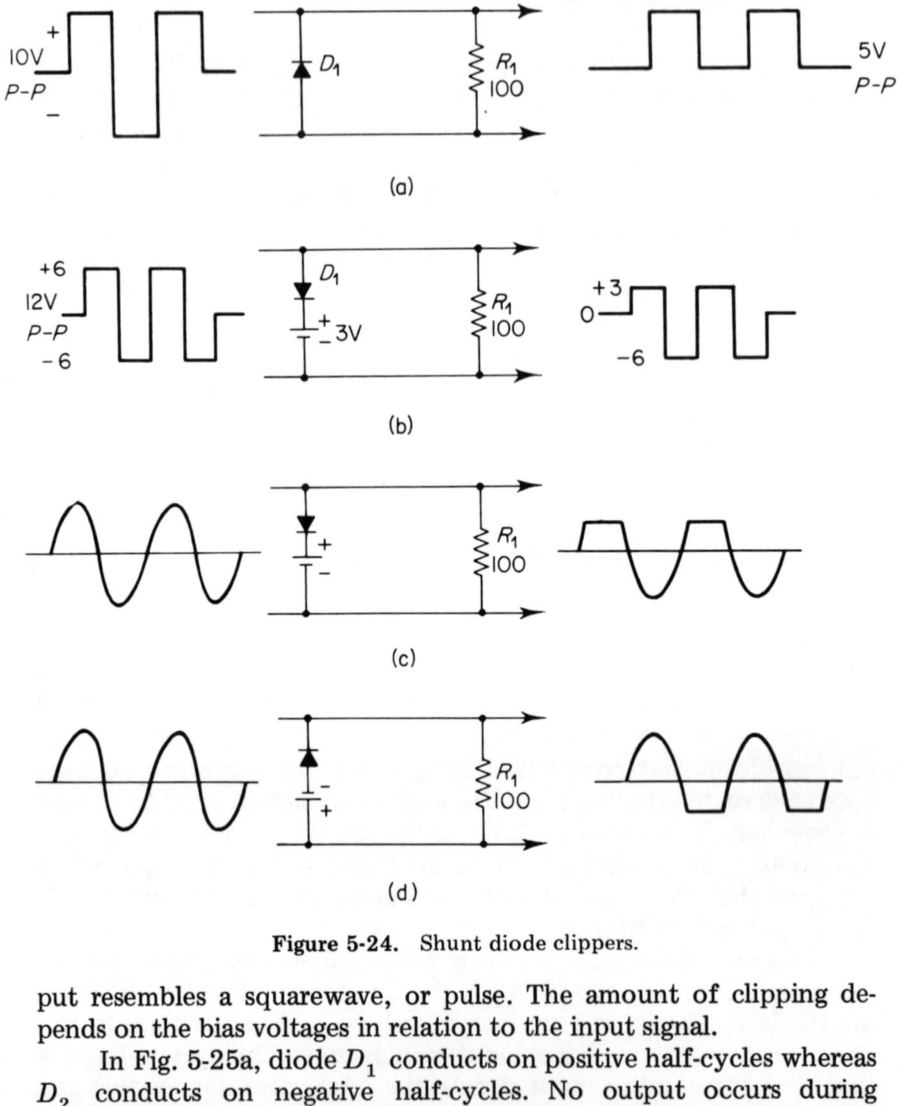

Figure 5-24. Shunt diode clippers.

put resembles a squarewave, or pulse. The amount of clipping depends on the bias voltages in relation to the input signal.

In Fig. 5-25a, diode D_1 conducts on positive half-cycles whereas D_2 conducts on negative half-cycles. No output occurs during conduction of either diode, so the peaks are clipped. The greater the bias voltage, the lower the output amplitude. In Fig. 5-25b, base-emitter bias must be of a value that permits the input signal to drive the transistor into saturation, and into cutoff, on positive and negative peaks. When the transistor is driven into saturation, a further increase in input signal causes no further increase in output. When in cutoff, a further input increase produces no further change in output. Therefore, both the positive and negative peaks of the input waveform are clipped (or sliced).

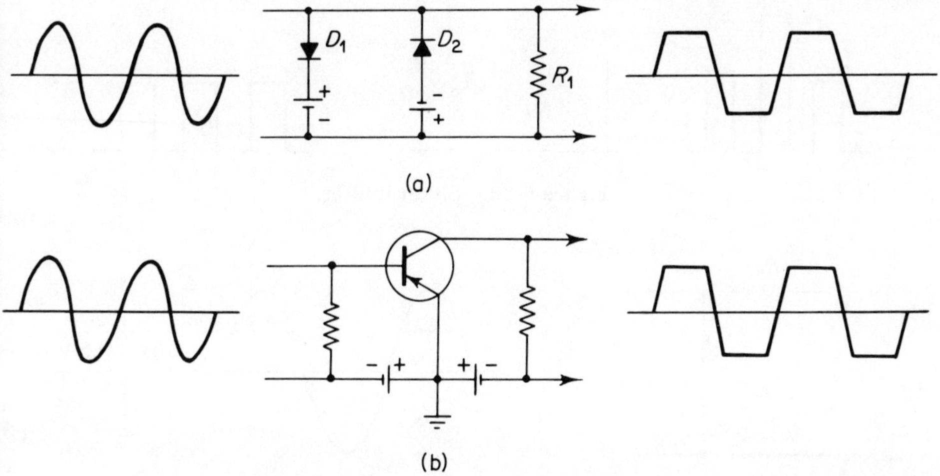

Figure 5-25. Parallel clippers (slicers).

5-6-4 Shunt Limiters

The biased shunt clipper can be used as a *limiter circuit* as shown in Fig. 5-26. Here, all input signals that exceed the bias voltage are limited to the value of the bias voltage. The diode is reverse-biased by the fixed voltage (battery) and does not conduct on any input pulses equal to, or below, the bias value. Any input pulses that exceed the bias value cause the diode to conduct and remove the output. Such circuits are often used to remove transients such as overshoots. The circuit has no effect on pulses of amplitudes below that of the bias voltage.

5-6-5 Design Considerations for Clippers

Figure 5-27 is the working schematic of a basic diode clipper (sometimes known as a peak limiter). As shown, the output waveform is clipped (held at a peak limit) to a value determined by the bias voltage and the forward-voltage drop of the diode. For example, if the bias is 3 V and the forward-voltage drop is 0.5 V, the output is clipped to 3.5 V. Either the positive or negative peaks can be clipped, depending on the polarity of the bias and diode. Both peaks can be clipped if two diodes and bias sources, with opposite polarities, are used.

The four basic diode characteristics must be considered in clipper circuits. However, maximum reverse voltage (peak inverse) is the prime consideration. The maximum reverse voltage rating of the

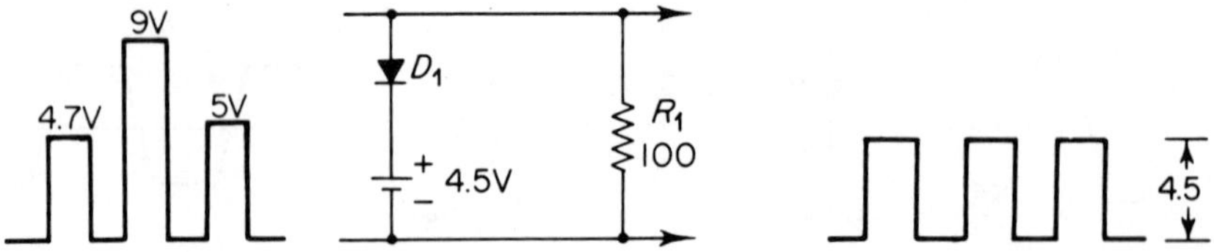

Figure 5-26. Shunt limiter.

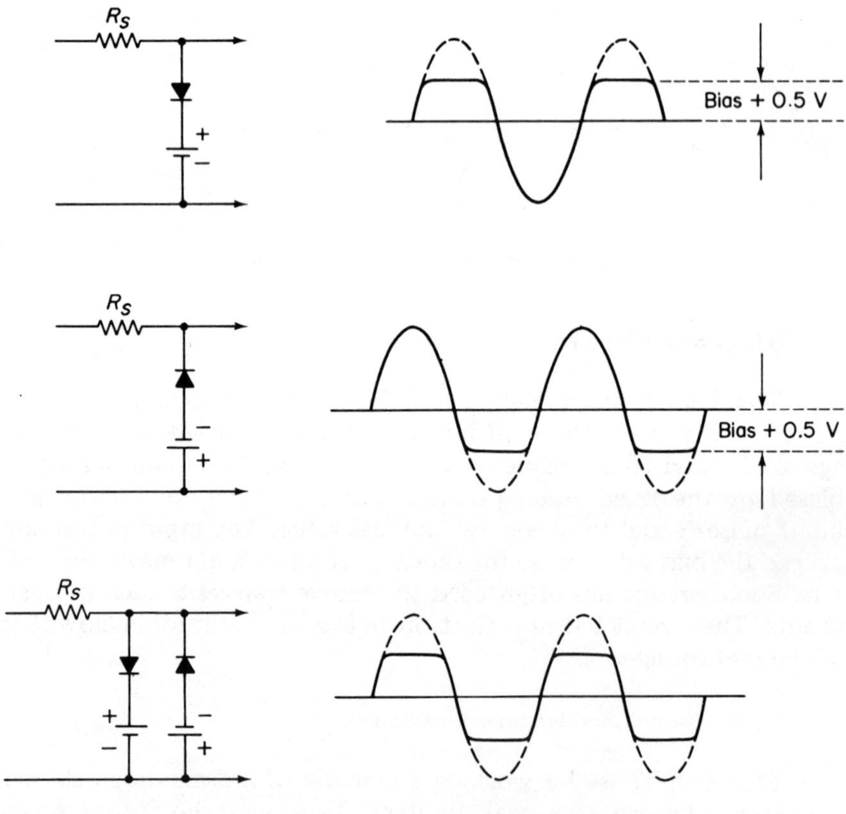

$$R_S \text{ (minimum)} \approx \frac{\text{input voltage}}{\text{max forward current of diode}}$$

Figure 5-27. Biased diode clipper.

diode must exceed the input voltage. The forward-voltage drop should always be minimum and is typically about 0.5 V for silicon diodes and 0.2 V for germanium diodes. Reverse current (leakage) should also be minimum. Any excessive leakage in the diode clips or distorts the waveform. Maximum forward current of the diode is not critical but does set the value of series resistance R_S.

The minimum value of R_S is set by the value of input voltage and maximum forward current rating of the diode. For example, if the input is 10 V and the diode is rated at 0.1 A, the minimum value of R_S is 10/0.1, or 100 (ohms). As a safety factor, the value of R_S should be at least twice the minimum value. This may result in a trade-off. The output current is passed through R_S, resulting in some voltage drop across R_S. This means that the unclipped peak is attenuated. A low value of R_S produces a minimum of attenuation, assuming a given output current. However, the value of R_S must never be less than that required to produce a safe current level in the diode.

Figure 5-28 is the working schematic of an unbiased clipper using a zener diode. The output waveform is clipped to a value determined by the zener voltage and the forward voltage drop of the diode. The advantage of this circuit is that no bias voltage is required. The disadvantage is that zener diodes not not usually have a sharp breakdown below about 4 V. Thus, the circuit should be used at voltages greater than 4 V. The value of R_S is calculated in the same way as for the basic diode clipper except that the maximum current rating of both the zener and clipper diodes must be considered. However, the current rating of the zener is usually greater than that of the clipper, so only the diode current rating need be considered.

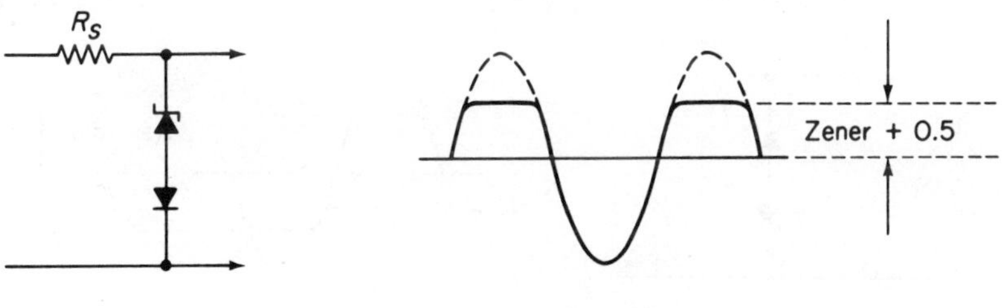

$$R_S \text{ (minimum)} \approx \frac{\text{input voltage}}{\text{max current of both diodes}}$$

Figure 5-28. Zener-diode clipper (unbiased).

Figure 5-29 is the working schematic of a clipper circuit using only a zener diode. As shown, the output waveform is clipped in both polarities. One polarity is clipped at the zener voltage, with the other polarity clipped at about 0.5 V (which is the typical drop across a zener). The value of R_S is calculated in the same way as the other clipper circuits. However, the required zener characteristics are somewhat different than those of the basic diode clipper. The zener voltage rating should be about 0.6 times the input voltage. If a lower zener voltage is used, there may be too much current flowing through the zener or the value of R_S will be excessive. Zener diodes are also rated as to power dissipation. This power is approximately

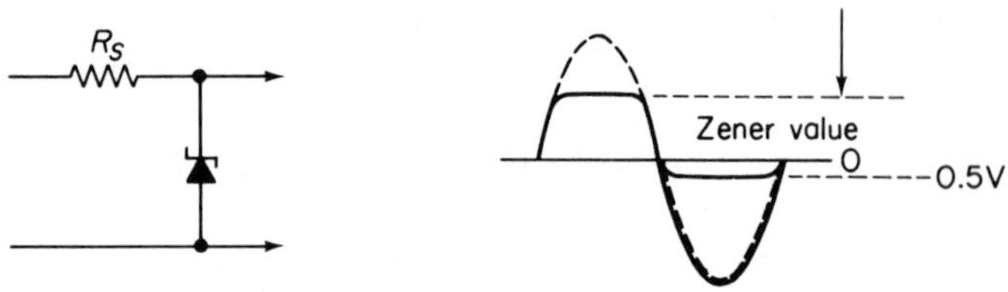

Zener voltage ≈ 0.6 × input voltage

$$R_S \text{ (minimum)} \approx \frac{\text{input voltage}}{\text{max current of zener}}$$

Figure 5-29. Zener clipper (without clipping diode).

equal to the product of the zener voltage and the maximum current (as set by R_S).

Figure 5-30 is the working schematic of a double-peak clipper using two zener diodes. This circuit is similar to that of Fig. 5-29 except that the two zener diodes are connected back to back. As shown, the output waveform is clipped by an equal amount in both polarities. The peaks are clipped to the zener voltage, plus about 0.5 V. The value of R_S is calculated in the same way as the previous clipper circuits.

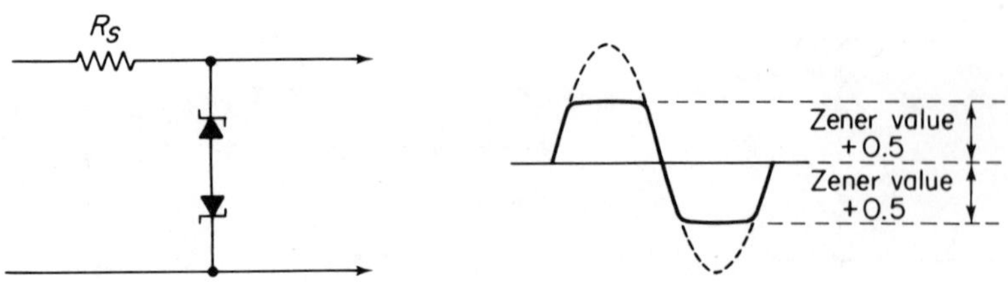

Zener voltage ≈ 0.6 × input voltage

$$R_S \text{ (minimum)} \approx \frac{\text{input voltage}}{\text{max current of Zener}}$$

Figure 5-30. Double-peak clipper (with two zener diodes).

Figure 5-31 is the working schematic of a low-level noise suppression circuit using two diodes. When the input voltage is below the usual diode voltage drop, there is considerable attenuation. If the input voltage exceeds the diode voltage drop, there is little attenuation. As a rule of thumb, if silicon diodes with a 0.5-V drop are used, any input signals below about 0.1 V are attenuated at least 30 dB, and any input signals above about 1 V are attenuated less than 3 dB. As in other clipper circuits, the diodes must be capable of withstand-

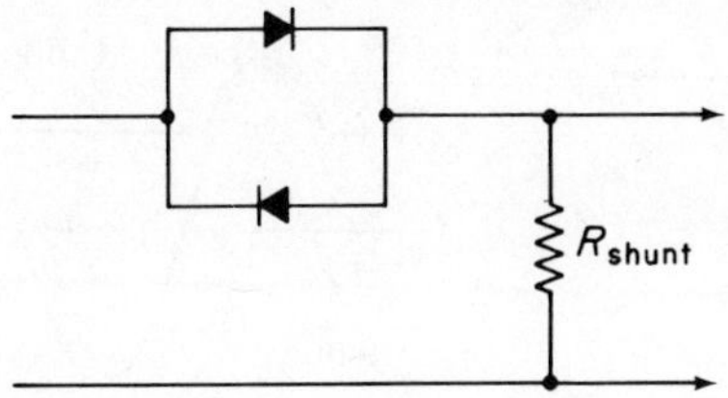

$R_{shunt} \approx 10\ k\Omega$ when minimum input is 0.1 V

Figure 5-31. Low-level noise suppressor (series-diode type).

ing the maximum input voltage and passing the maximum circuit current. The value of shunt resistor R depends upon the minimum input voltage level. As a guideline, the value of R should be about 10 kΩ when the minimum input signals are about 0.1 V.

Figure 5-32 is the working schematic of a low-level clipper using two diodes. The output voltage is limited to the usual diode voltage drop (about 0.5 V for silicon diodes). Any input signals with a peak-to-peak amplitude of less than 0.5 V should pass without distortion or attenuation. The diodes must be capable of withstanding the maximum input voltage. The value of R_S is set by the value of the input voltage and the maximum forward current rating of the diode.

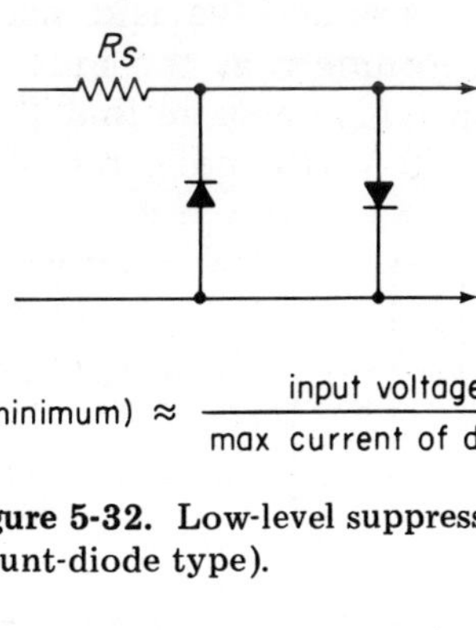

$$R_S\,(\text{minimum}) \approx \frac{\text{input voltage}}{\text{max current of diodes}}$$

Figure 5-32. Low-level suppressor (shunt-diode type).

Figure 5-33 is the working schematic of a high-level noise suppressor, often known as a *base clipper*. The circuit is similar to that of the basic clipper (Fig. 5-27), but with the diode polarity reversed. The output waveform follows peak voltages in excess of the bias voltage. Whenever the peak voltage drops below the bias level or swings in the opposite polarity, the diode conducts and the output voltage returns to the bias level. Thus, any signals below the bias level

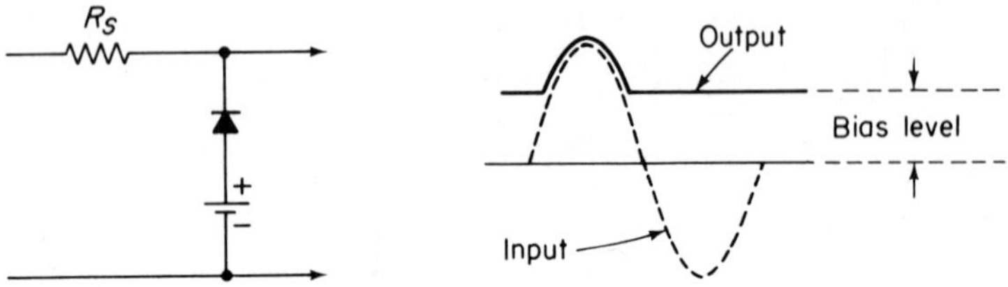

$$R_S\ (\text{minimum}) \approx \frac{\text{input voltage}}{\text{max current of diode}}$$

Figure 5-33. Base clipper.

are suppressed. As in the case of other clipper circuits, the diode must be capable of withstanding the maximum input voltage. The minimum value of R_S is set by the maximum current rating of the diode.

5-7 CLAMP (D-C RESTORER) CIRCUITS

When pulses are passed through capacitor coupling between stages, the d-c component of the pulses is lost. In the circuit of Fig. 5-34, if the clamping diode is not present, the pulses on the output side of capacitor C_1 are of the same amplitude as the input pulses, except that the output pulses are positive and negative with respect to ground. For example, assume that the input pulses (Q_1 collector) have an 18-V peak value. Also assume that the steady-state value on the input side of C_1 is 10 V (the collector of Q_1 is 10 V). Between

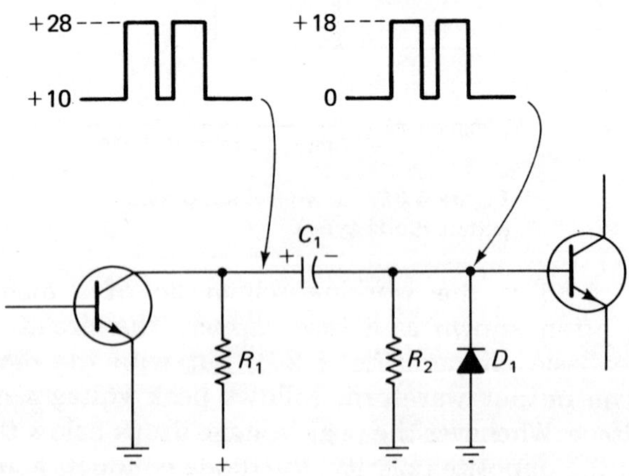

Figure 5-34. Clamp or d-c restorer circuit.

pulses C_1 charges to the steady-state value through R_2. That is, the input side of C_1 is positive with respect to ground, and the output side is negative with respect to ground. Without the diode, when pulses are present, C_1 charges up to the 18-V peak value. On the input side C_1 goes from +10 V up to +28 V. On the output side, C_1 goes from −10 V up to +8 V. Thus, the output pulses are positive and negative with respect to ground, even though the input pulses are always positive with respect to ground.

Diode D_1 prevents this condition by discharging C_1 between pulses. This makes the output side of C_1 zero with respect to ground until a pulse arrives. Then the output side rises to the peak value (in this case, 18 V).

5-7-1 Design Considerations for Clamps

Figure 5-35 is the working schematic for a clamp. As shown, the circuit is essentially a high-pass *RC* filter with a diode across the shunt resistance. The time constant must be long to prevent distortion. As a guideline, the time constant should be on the order of 100 ms for any frequency from about 10 Hz to 1 MHz. Often, it is necessary to select *RC* values on the basis of external circuit considerations. For example, the resistance may have to be a certain value to match impedance. In that case, the capacitance should be of a corresponding value to produce the 100-ms time constant. Preferably,

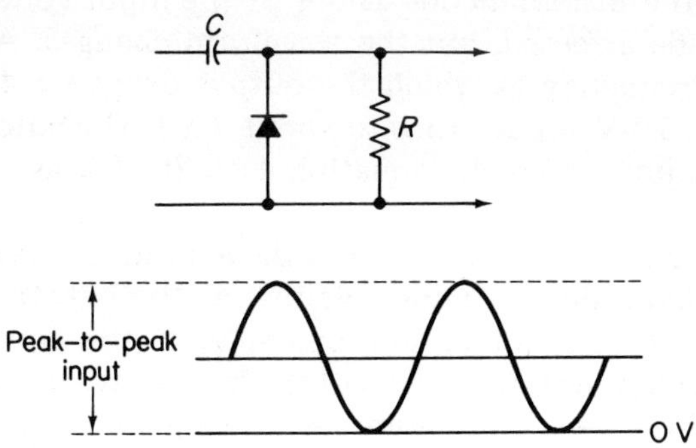

$$RC \approx 100 \text{ msecs}$$

$$C \approx \frac{0.1}{R}$$

$$R \approx \frac{0.1}{C}$$

Figure 5-35. Basic clamp.

the resistance should be as large as possible, thus eliminating the need for frequent recharging.

With the circuit as shown, the negative peaks of an input signal are clamped to ground (at 0 V). If the diode polarity is reversed, the positive peaks are at zero and the negative peaks are below zero. If a fixed bias source is connected in series with the diode, the negative peaks are clamped to the bias value. For example, if the bias is 1 V and the input is 3 V peak to peak, the output negative peak is at 1 V with the positive peak at 4 V.

5-8 MODIFICATION OF SQUARE WAVES

In various pulse circuits, it is often necessary to modify the square-wave output of the circuit (such as the output of a multivibrator) to get a certain waveform. The most common waveforming circuits used with square waves are the high-pass and low-pass *RC* filters described in Sec. 5-2.

5-8-1 Design Considerations

When a square wave is applied to a low-pass *RC* filter, the output is an integrated waveform, as shown in Fig. 5-36. In this case, the *RC* network produces an output voltage proportional to the algebraic sum of the instantaneous values of the input voltage. Not only is the amplitude affected, but the waveform changes. Any *RC* filter has a cutoff frequency at which the output drops by 3 dB from the input value (a 10-V input drops to about 7 V). The cutoff frequency can be approximated by the equation $1/(6.28RC)$, as shown in Fig. 5-36.

The amount of attenuation and the amount of waveform modification are determined by the relation of the cutoff frequency to the square-wave frequency. For example, if the square-wave frequency is much lower than the cutoff frequency, both attenuation and modification are at minimum. As the square-wave frequency increases, the output (an integral of a square wave) is a series of increasing and decreasing ramps. The ramps are linear, but only for a time that is shorter than about half the *RC* time constant. For very accurate integration of the square wave, the integration time should be about 0.1 of the *RC* time constant.

The relationship between frequency and waveform is shown in Fig. 5-37. In Fig. 5-37a, the frequency of the square waves is about one-third of the cutoff frequency. Note that there is little attenuation but slight modification of the waveform. When the square-wave

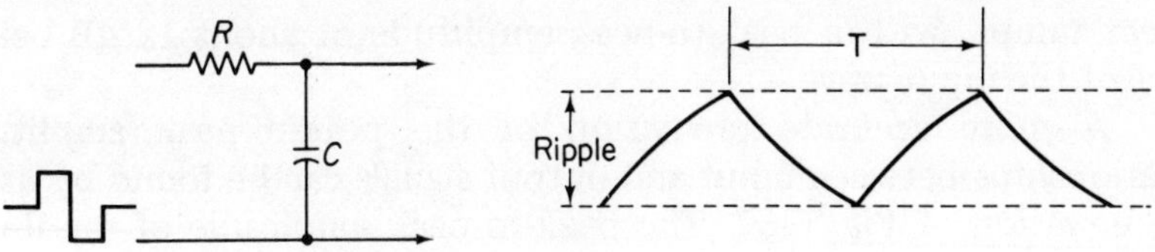

$$\text{Cutoff frequency} \approx \frac{1}{6.28\,RC}$$

$$T \approx RC \qquad \text{Frequency} \approx \frac{1}{T}$$

$$RC \text{ time constant} \approx \frac{100}{4 \times \text{squarewave freq.} \times \text{desired percentage of ripple}}$$

$$C \approx \frac{\text{time constant}}{R}$$

Figure 5-36. Low-pass RC filter with square-wave input and integrated ramp output.

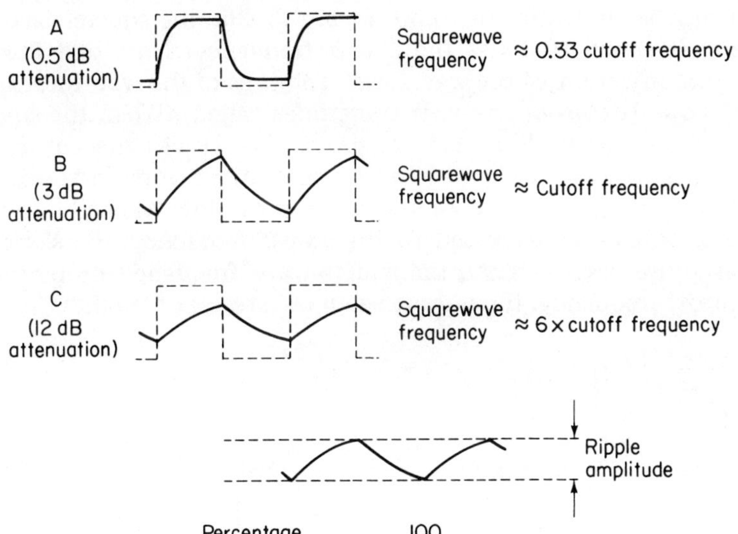

$$\text{Percentage of ripple} \approx \frac{100}{4 \times \text{frequency} \times RC}$$

Figure 5-37. Relationship of frequency and waveform in output of low-pass RC filter (with square waves at the input).

frequency is increased to the cutoff frequency, as shown in Fig. 5-37b, the peak-to-peak waveform is attenuated by 3 dB, and the waveform has the shape of an almost complete exponential rise or fall. When a square-wave frequency is increased to about six times the cutoff frequency (Fig. 5-37c), the output is a series of (almost) linear ramps, with a peak-to-peak amplitude of about 12 dB below that of the input wave.

A more accurate prediction of the peak-to-peak amplitude relationships between input and output signals can be found by using the equation of Fig. 5-37. The peak-to-peak amplitude of the linear ramps, often known as the *ripple*, or *ripple amplitude*, is always less than 100 percent of the square-wave input amplitude. However, the equation applies only when ripple amplitudes are less than 50 percent of the input amplitude.

When a square wave is applied to a high-pass *RC* filter, the output is a differentiated waveform, as shown in Fig. 5-38. In this case, the *RC* network produces an output voltage proportional to the time rate of change of the input voltage.

As in the case of the low-pass (integrator) circuit, the amount and nature of the waveform modification by a high-pass (differentiator) circuit is determined by the relationship of cutoff frequency to square-wave frequency. The relationship between frequency and waveform is shown in Fig. 5-39. In Fig. 5-39a, the square-wave frequency is about 10 times the cutoff frequency. Note that there is slight modification of the waveform. The tops of the waveform show some slope (or droop, as it is sometimes called). When the square-wave frequency is decreased to about three times the cutoff frequency, as shown in Fig. 5-39b, the amount of slope increases and there is some attenuation of the average output. When the square-wave frequency is decreased to the cutoff frequency, the slope becomes quite severe. When the square-wave frequency drops below the cutoff frequency, the output becomes a series of "spikes."

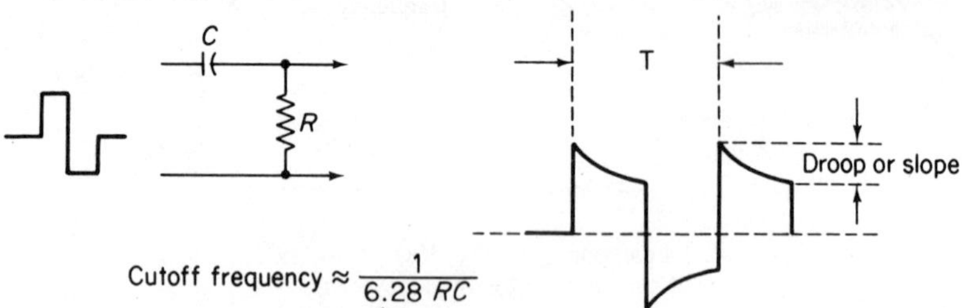

$$\text{Cutoff frequency} \approx \frac{1}{6.28\,RC}$$

$$T \approx RC \quad \text{frequency} \approx \frac{1}{T}$$

$$RC \text{ time constant} \approx \frac{100}{4 \times \text{squarewave frequency} \times \text{desired percentage of slope}}$$

$$C \approx \frac{\text{time constant}}{R}$$

Figure 5-38. High-pass RC filter with squarewave input and differentiated output.

When the percentage of slope is below about 50 percent, the equation of Fig. 5-39 can be used to predict the slope. Note that the equation for slope in a differentiated wave is the same as for ripple in an integrated wave.

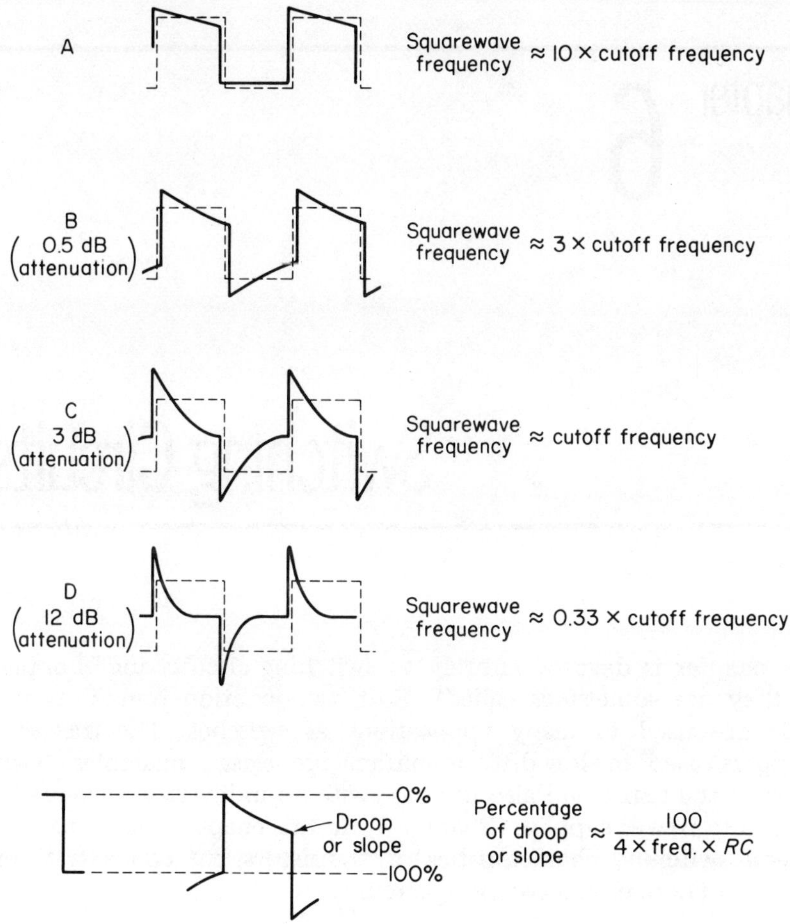

Figure 5-39. Relationship of frequency and waveform in output of high-pass RC filter (with square waves at the input).

Chapter 6

Switching Circuits

This chapter is devoted entirely to switching circuits and choppers (as they are sometimes called). Both two-junction transistors and FETs are used in many applications as switches. The transistor choppers used in low-drift amplifiers are classic examples. Two-junction transistors are also used as switches in inverter or converter solid-state power supplies (Chapter 3). In this chapter, we discuss the general switching characteristics of transistors but concentrate on transistor chopper and switching circuits.

6-1 BASIC CHOPPER CIRCUITS

Chopper circuits are basically of either the *shunt* type or the *series* type, as shown in Fig. 6-1, or a combination of the two. The shunt-type chopper circuit, shown in Fig. 6-1a, operates as follows. When the switch S is opened, a voltage that is directly proportional to the input signal appears across the load. When the switch is closed, all the input signal is shorted to ground. If the switch is opened and closed periodically, the voltage across the load appears as a square

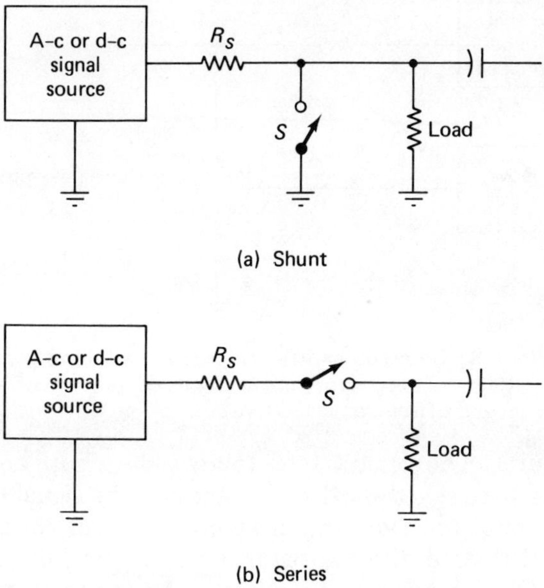

Figure 6-1. Basic chopper circuits.

wave that has an amplitude directly proportional to the input signal. The series chopper circuit, shown in Fig. 6-1b, can also be used to chop signals. The series chopper is particularly useful in telemetry or other systems where a signal source such as a transducer is to be connected periodically to a load.

Early choppers were of the mechanical type and often used relays. Mechanical choppers have generally been replaced by solid-state choppers, but not without introducing certain disadvantages. For example, a transistor cannot match the nearly ideal switching characteristics of a relay (infinite open resistance, and low closed resistance). A two-junction transistor has the additional disadvantages of *offset voltage* and *offset current*. In the case of FETs, particularly MOSFETs, there is a tendency for the chopper drive signal to feed through onto the signal being chopped.

6-2 TWO-JUNCTION TRANSISTOR CHOPPER CIRCUITS

If a common-emitter transistor switch or chopper, such as shown in Fig. 6-2, is in the ON state but the collector current is near zero, the collector-emitter voltage is not zero. It is this *offset voltage* that most seriously limits the performance of two-junction transistors in

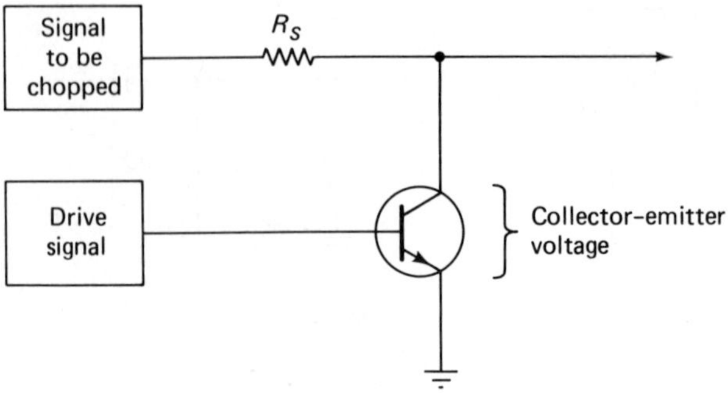

Figure 6-2. Common-emitter transistor switching showing effect of collector-emitter voltage (as an offset voltage and offset current source).

chopper circuits. The circuit that follows the chopper cannot tell the difference between the offset voltage and the signal (particularly if signal is of about the same amplitude as the offset voltage).

In the OFF state, the transistor is not an ideal open circuit and appears as a current source. This *offset current* can induce a voltage in the load, again masking signal levels of comparable level.

The most popular technique to avoid these problems is to use a common-collector (or inverted) circuit rather than a common-emitter circuit. (By inverted circuits it is meant that the control signal or drive signal is applied between the base and the collector instead of between the base and the emitter.) A common-collector version of the basic series chopper circuit is shown in Fig. 6-3. A more sophisticated two-junction transistor chopper circuit is shown in Fig. 6-4a, and an improved version is shown in Fig. 6-4b. In the ON state of

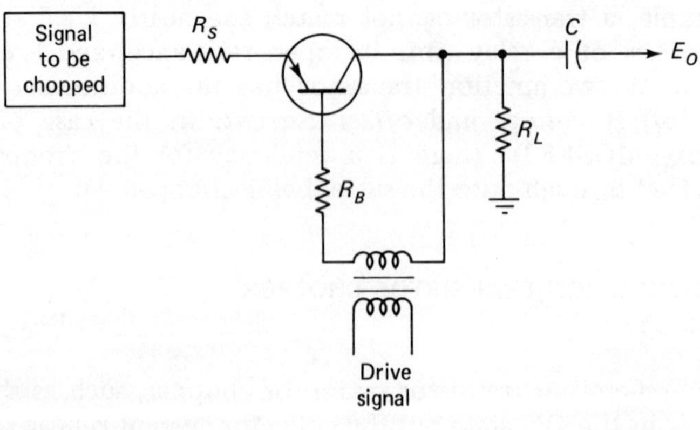

Figure 6-3. Basic two-junction transistor series chopper circuit.

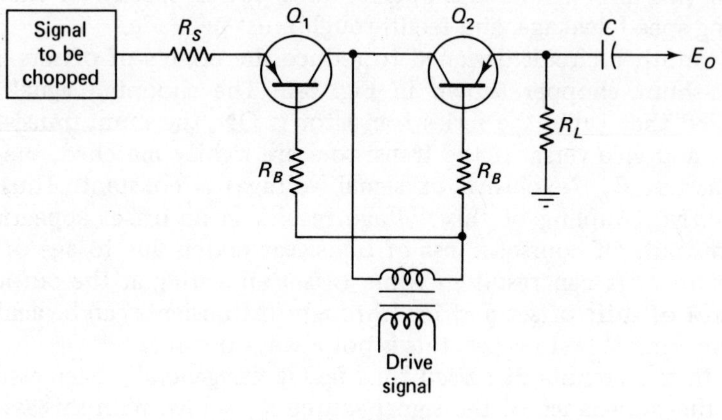

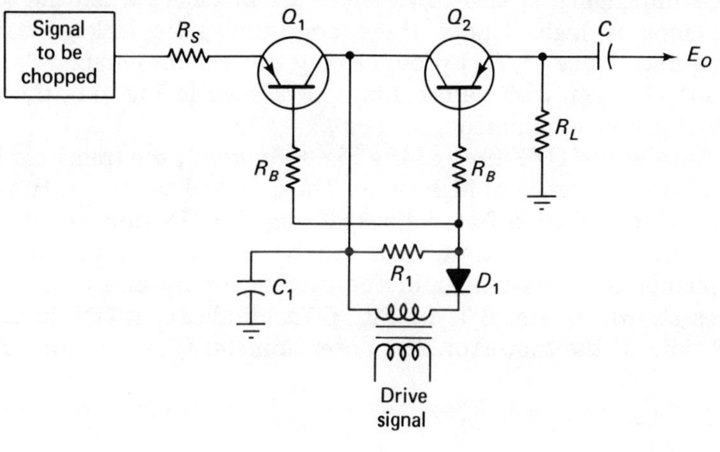

Figure 6-4. Two-junction series-pair chopper circuits.

either Fig. 6-4 circuit, each transistor produces an offset voltage, but the voltages are of opposite polarity and tend to cancel. In the OFF states, the offset currents are of opposite polarity.

In the improved circuit of Fig. 6-4b, the diode D_1 prevents the collector-base junctions from becoming reverse-biased. This eliminates the OFF leakage currents. The diode itself has some leakage, but this is passed by R_1 rather than by the transistors. The small capacitor C_1 forms a capacitive divider with the transistor junction capacitance, to help suppress feedthrough transients. The improvements of Fig. 6-4b are not without disadvantages. The zero off-bias mode and capacitive divider circuits both tend to increase switching times (making the circuit operate at a lower speed). A trade-off among speed, leakage, and feedthrough must be made.

Another circuit designed to reduce the effects of offsets is the series-shunt chopper shown in Fig. 6-5. The chopping signal is so coupled that while the series transistor is ON, the shunt transistor is OFF, and vice versa. If the transistors are rightly matched, the voltage across R_L (exclusive of signal voltage) is constant. Thus, the capacitive coupling of this voltage results in no offset appearing at the output. Of course, a loss of transistor match due to age or temperature drift can result in some offset appearing at the output. A control of drift offset and feedthrough of transients can be achieved by lowering R_L. However, this is not always possible.

In the circuits discussed thus far, it has generally been assumed that the impedance of the signal source R_S is low. With this assumption, any offset voltage due to leakage current flowing through the source impedance is also low. There are instances when the source impedance is high. Under these conditions, the leakage current-source impedance ($I_E R_S$) product may become intolerable. By using a shunt chopper with zero off-bias, as shown in Fig. 6-6, the $I_E R_S$ product can be eliminated.

During the OFF time of the Fig. 6-6 circuit, the transistor looks like a pure resistance of high value. Thus, no leakage flows. However, the presence of an offset voltage during the ON time results in an offset error in the output. This can be removed by introducing a complementary offset later in the system, or by one of two techniques shown in Fig. 6-7. In Fig. 6-7a, diode D_2 is ON during the OFF time of the transistor. The potentiometer is used to introduce a

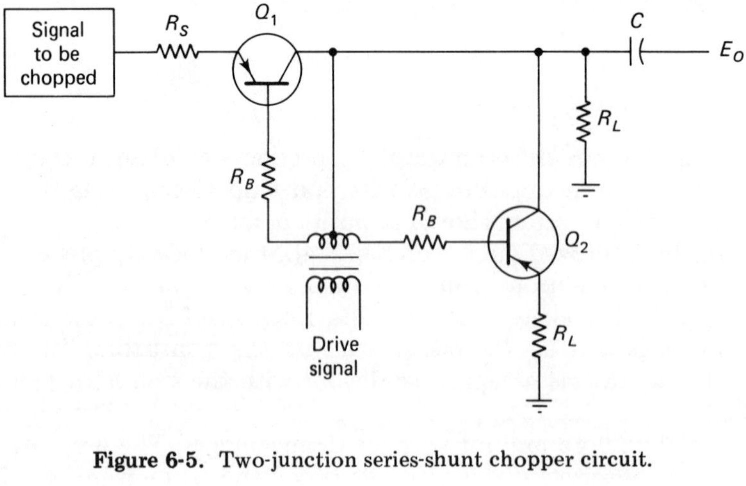

Figure 6-5. Two-junction series-shunt chopper circuit.

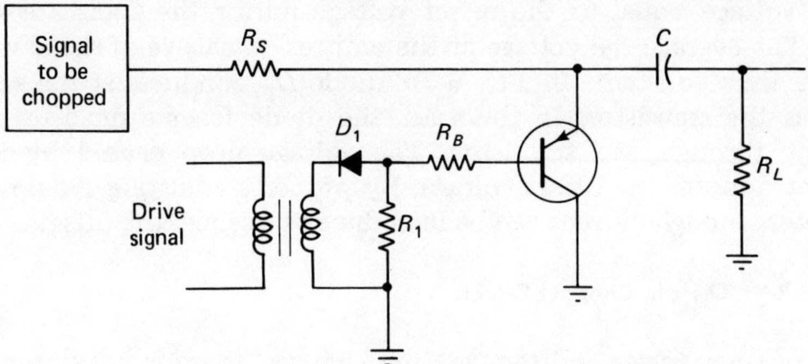

Figure 6-6. Two-junction shunt chopper circuit.

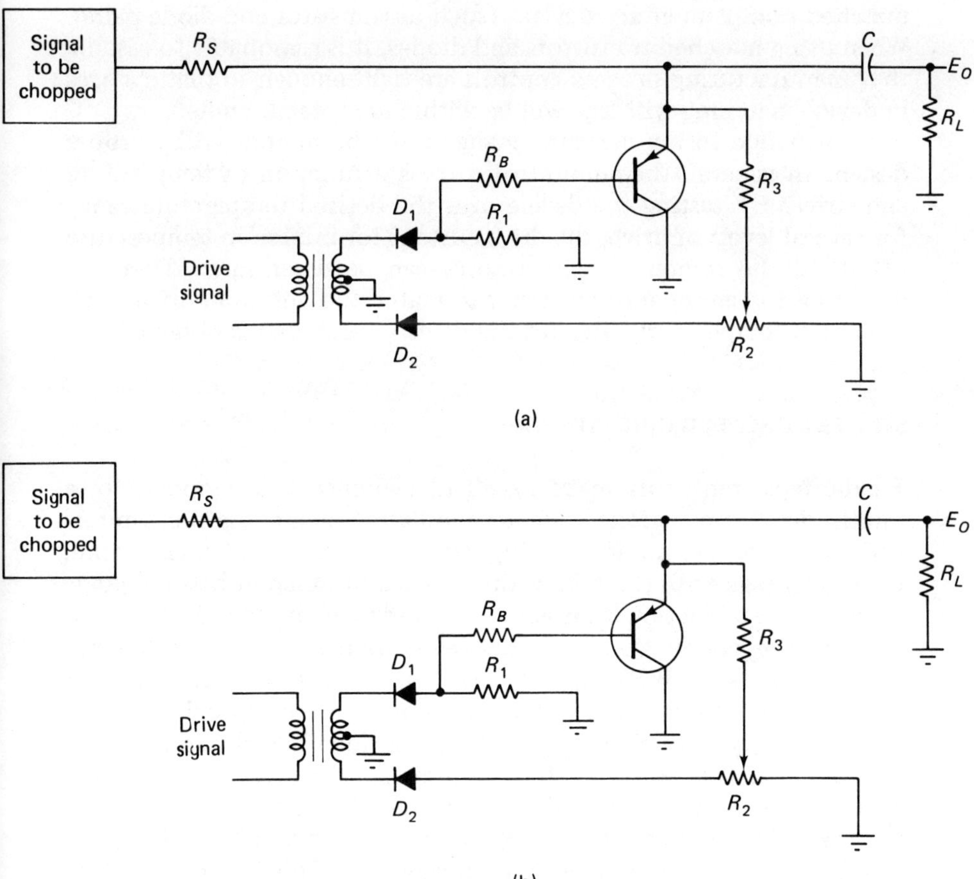

Figure 6-7. Two-junction shunt chopper circuits with offset voltage compensation.

small voltage equal to the offset voltage during the transistor ON time. The average d-c voltage at the emitter (exclusive of signal voltage) is thus constant. In Fig. 6-7b, diode D_2 conducts at the same time as the transistor. In this case, the diode feeds a small reverse current through the transistor. The voltage drop caused by this current opposes the offset voltage. By properly adjusting the potentiometer, enough current can be introduced to cancel the offset.

6-2-1 Drift in Chopper Circuits

As time passes and temperature changes, there is no guarantee that chopper circuits will remain compensated. This results in drift. However, drift can be minimized by proper design and by using well-matched complementary devices (such as transistor and diode pairs). When using matched transistors and diodes, it is reasonable to assume that manufacturing process controls are tight enough so that changes in device matching with age will be within acceptable limits.

Drift due to temperature change must be minimized by proper design. In general, the amount of drift is a function of temperature and drive. By testing the device over the desired temperature range for several levels of drive, the drive current for minimum temperature offset (at the reference temperature) can be determined. Then, by designing the chopper drive circuit to match the optimum drive level, minimum temperature drive for the device being used is obtained.

6-3 FET CHOPPER CIRCUITS

Field-effect transistors make excellent switches and choppers for a multitude of applications, such as modulators, demodulators, sample-and-hold systems, mixing, multiplexing or gating, and many more. One advantage with the FET is the absence of inherent offset voltage found in two-junction transistors (as discussed in Sec. 6-2). This is because the conduction path between drain and source is mostly resistive. In both JFETs and MOSFETs, the conduction channel is either depleted or enhanced by controlling an induced field.

In any type of FET, there is an exceptionally high ratio of OFF resistance to ON resistance for the drain-source channel. The resistance can be as low as several ohms in the ON condition. This results in a very low voltage drop (very low offset voltage). The resistance can be higher than thousands of megohms in the OFF condition, resulting in very little offset current flow.

The extremely high d-c gate input impedance of a FET is also an advantage since little control power is required. In JFETs, the control signal looks into a reverse-biased diode. For MOSFETs, the gate insulation is a high-resistance oxide layer, the properties of which set the impedance. The main limitation of FETs used as switches and choppers is the capacitance between the gate and the drain and the gate and the source. This capacitance feeds through part of the gate control voltage to the signal path. These capacitances are detrimental to high-frequency signal insulation and impose a limitation on response times.

6-3-1 Basic FET Chopper Circuits

Both JFETs and MOSFETs can be used in the three classic chopper configurations: series, shunt, and series-shunt.

The *series chopper* (Fig. 6-8) is the most commonly used. In order to operate in the ohmic region when the FET is ON, the drain current must be limited to a low value. Generally, a large value of load resistance R_L is used to limit the current. In the case of a series chopper, a high value of load resistance also minimizes the ON-voltage error. The leakage of FETs (particularly MOSFETs) is low, and the resultant OFF-voltage error is small. Typically, for a MOSFET chopper with R_L at 100 kΩ, the OFF error is less than 1 μV at room temperature.

The *shunt chopper* (Fig. 6-9) performs the chopping function by periodically shorting the input to ground. This circuit is advantageous where a large source resistance R_S is present.

The *series-shunt chopper* (Fig. 6-10) operates as follows. When Q_1 is ON, Q_2 is OFF, and vice versa. When Q_1 is ON and Q_2 is OFF, the output is similar to that of the series chopper except for the small error introduced by the leakage of Q_2. When Q_1 is OFF and Q_2 is ON, the OFF voltage error due to leakage of Q_1 is reduced since the drain-source resistance is in parallel with R_L.

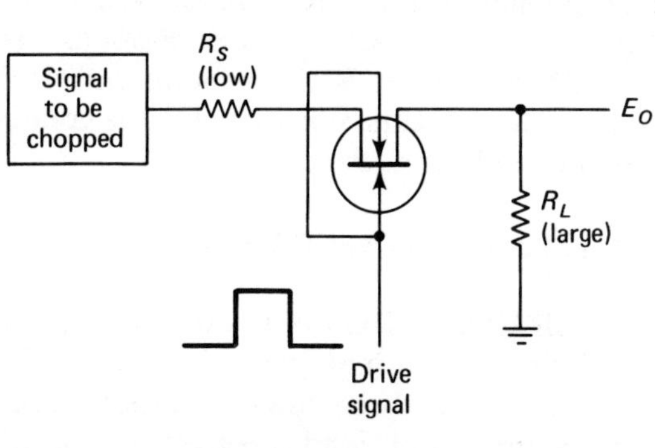

Figure 6-8. Basic FET series chopper circuit.

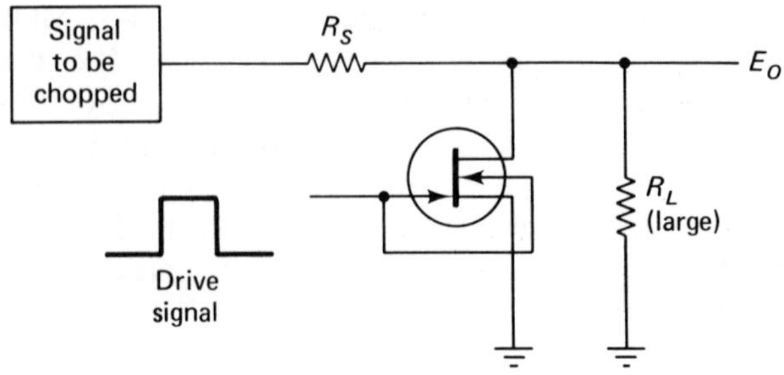

Figure 6-9. Basic FET shunt chopper circuit.

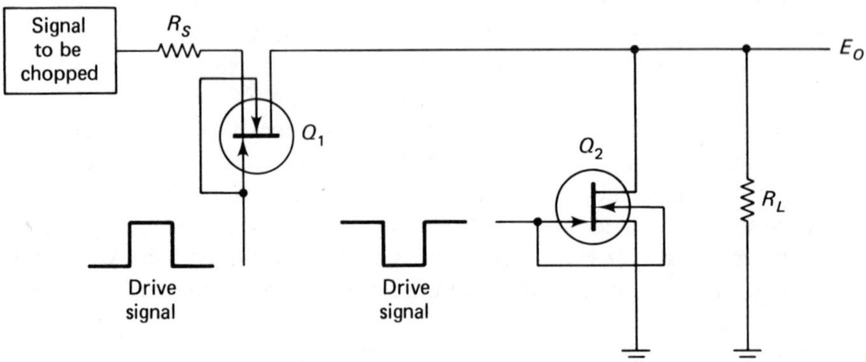

Figure 6-10. Basic FET series-shunt chopper circuit.

Since the leakage current of a MOSFET is quite small, the series-shunt circuit cannot be justified for the sake of minimizing the error due to leakage. However, the series-shunt circuit does have a definite advantage in the area of high-frequency chopping. In the simple series chopper, when the FET is OFF, the FET capacitance must be discharged through the load resistor R_L. The relatively long RC time constant of FET capacitance and load resistance limits the chopping frequency. However, in the series-shunt chopper, every time the series FET turns off, the shunt FET turns on and the low resistance of Q_2 parallels R_L. The RC time constant is thus greatly reduced and the chopping frequency is increased.

6-3-2 Effect of Loads on MOSFET Choppers

Operation of all MOSFET chopper circuits is greatly affected by the magnitude of the source and load resistances. Figure 6-11 lists the output voltages of the three basic chopper circuits for various combinations of source and load resistances. In Fig. 6-11, it is

Source resistance, R_S	Load resistance, R_L	Approximate output voltage, E_O (μV)					
		Shunt chopper		Series chopper		Series–shunt chopper	
		ON	OFF	ON	OFF	ON	OFF
1 MΩ	1 MΩ	0.1	500	500	1	500	0.0001
100 kΩ	1 MΩ	1	900	900	1	900	0.0001
100 Ω	1 MΩ	500	1000	1000	1	1000	0.0001
0 Ω	1 MΩ	1000	1000	1000	1	1000	0.0001
1 MΩ	100 kΩ	0.1	90	90	0.1	90	0.0001
1 MΩ	100 Ω	0.05	0.1	0.1	0.00001	0.1	0.00005
100 kΩ	100 kΩ	1	500	500	0.1	500	0.0001
100 Ω	100 Ω	333	500	333	0.00001	333	0.00005

Input voltage 1 mV; drain-to-source resistance 100 Ω in ON condition, 1000 MΩ in OFF condition.

Figure 6-11. Steady-state chopper output voltage for various source and load resistances.

assumed that the input voltage is 1 mV, and the drain-to-source resistance is 100 Ω in the ON condition, and 1000 mΩ in the OFF condition. The gate leakage resistance of the MOSFET is ignored. The following conclusions can be drawn from the data shown in Fig. 6-11.

1. Only the series or series-shunt circuit should be used when R_S is less than the drain-to-source resistance of the MOSFET.
2. In general, R_L should be high. In any event, R_L should be much greater than the drain-to-source resistance.
3. R_L should always be greater than R_S.
4. Performance of the series-shunt circuit is equal to or better than that of either the series or shunt chopper alone, for any combination of R_S and R_L.

6-3-3 Practical JFET Choppers

Figure 6-12 shows a simple but complete JFET series chopping circuit. The maximum chopping frequency is about 200 kHz. The limitation is primarily due to the long RC time constant for discharging FET capacitances through the 10-kΩ resistors. The time constant can be shortened, and the frequency increased, if the 10-kΩ resistor is reduced in value. However, this results in a greater voltage drop across the divider, consisting of the drain-source resistance and the load resistor.

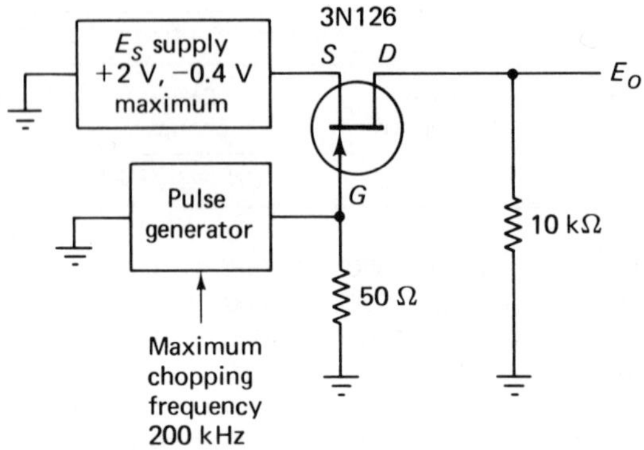

Figure 6-12. Practical series chopper using N-channel JFET.

The maximum allowable input voltages are +2 V and −0.4 V. The reason for these voltage limitations is the ON-OFF requirement of the JFET. For the particular N-channel JFET shown (3N126), the ON condition is 0 V and the OFF condition is a gate-source voltage (V_{GS}) greater than 4 V. The ON condition is obtained by grounding the gate. When the source (E_S supply) starts to go positive, there is a negative potential from gate to source. The negative gate-source voltage causes the drain-source resistance to increase, and the JFET starts to turn off. When the input (source) goes more negative than −0.4 V, the diode from gate to source becomes forward-biased, and the JFET starts to turn ON (thus, the negative limitation of −0.4 V).

In addition to these positive and negative input voltage limitations, there is a minimum input voltage limitation due to feedthrough spikes in the output channel. This feedthrough is caused by FET capacitances, as previously discussed. For inputs less than about 10 mV, the feedthrough spikes become an appreciable part of the output waveform (particularly at high frequencies). There are several circuit techniques useful for minimizing these spikes. First, the control signal (pulse generator) at the gate can be a sloppy square wave, or a sine wave can be used. Next, a capacitor can be connected across the output to filter the spikes. Finally, if a fixed amplitude output is acceptable, a clipper circuit can be connected across the output.

Modified series choppers for large input voltages. When a FET is OFF, large negative input voltages may tend to turn the FET ON again. The circuit of Fig. 6-12 can be modified to get around this input voltage limitation. For example, to overcome the limitation of maximum input voltage when the FET is turned ON, a circuit similar to that of Fig. 6-13 can be used. In this circuit, the FET is ON, the

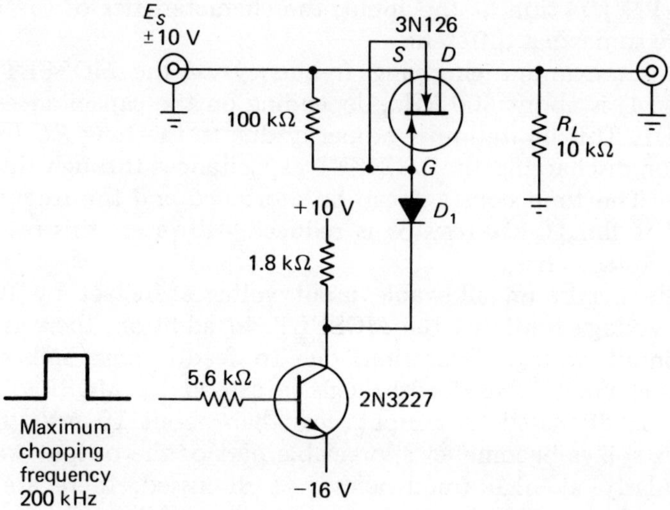

Figure 6-13. JFET chopper circuit for large input voltages.

driver transistor is OFF and a voltage of +10 V appears at diode D_1. This reverse-biases D_1. When the input (source) goes positive, the 100-kΩ resistor from source to gate makes the gate follow in potential as long as the diode does not become forward-biased. If the diode D_1 starts to conduct, a negative voltage appears from gate to source, and the FET starts to turn OFF. For negative input voltages, there is no problem as before, since the gate follows the source until D_1 avalanches.

The circuit of Fig. 6-13 also improves the performance when the FET is OFF. Under these conditions, the driver transistor is turned ON, and when the gate of the FET is at −15 V, the FET remains OFF for inputs up to −10 V. The maximum input voltage of ±10 V is a function of the bias voltage. The real limitation for the input voltage to this circuit is the source-to-gate breakdown voltage. Typical breakdown voltage is 50 V for a JFET (and about 30 V for a MOSFET). With different values of bias, the input can be increased to about +22 V (when both positive and negative inputs are applied) or to about +44 V (when only positive inputs are used). Approximately −6 V is required to keep the 3N126 OFF.

6-3-4 Practical MOSFET Choppers

Figure 6-14 shows a practical MOSFET chopper. Note that this circuit is similar to that of the JFET circuit in Fig. 6-12. Operation of the two circuits is similar. However, since a MOSFET does not have a PN junction at the input, the characteristics of the two circuits are somewhat different.

The maximum chopping frequency of the MOSFET circuit (Fig. 6-14) is about 300 kHz, depending on the capacitances of the MOSFET. This limitation is primarily due to the long *RC* time constant for discharging the MOSFET capacitances through the 10-kΩ resistor. The time constant can be shortened and the frequency increased if the 10-kΩ resistor is reduced. However, this results in a greater voltage error.

The maximum allowable input voltages are set by the gate-source voltage limits of the MOSFET. In addition, there is a minimum input voltage limitation, due to feedthrough spikes in the output channel. This feedthrough is caused by MOSFET capacitances, as discussed. For inputs less than about 10 mV, the feedthrough spikes become an appreciable part of the output waveform (particularly at high frequencies). As discussed, there are several circuit techniques useful for minimizing these spikes. The techniques include sloppy input or control signals, capacitors across the output, and (if a fixed amplitude output is acceptable) a clipper at the output.

Practical MOSFET series-shunt choppers for high-frequency use. Figure 6-15 is a series-shunt high-frequency chopper using complementary enhancement-mode MOSFETs. Using the components shown, the circuit operates satisfactorily at frequencies up to about 5 MHz. An N-channel and a P-channel MOSFET are used as the series and shunting devices, respectively. This allows one drive circuit for both devices.

When the series MOSFET is OFF, the drain-source resistance of the shunting MOSFET is about 200 Ω. This value parallels R_L to ground and reduces the net output load resistance to about 200 Ω (the parallel combination of 200 Ω and 10 kΩ). Thus, the *RC* time constant is reduced to 2 percent of its original value. The circuit of Fig. 6-15 can also be modified to accept large values of input voltages in Sec. 6-3-3 (for the circuit of Fig. 6-13).

Practical MOSFET series-shunt choppers for low-input voltages. A series-shunt chopper capable of low-level chopping is shown in Fig. 6-16. Two N-channel MOSFETs with matched capacitances are used in the circuits. The gate drives for the pair are produced by a high-current-type (also known as a current-mode type) free-running multivibrator, discussed in Sec. 4-5.

The high-current-mode multivibrator drive results in good frequency stability. In addition, the complementary outputs are not

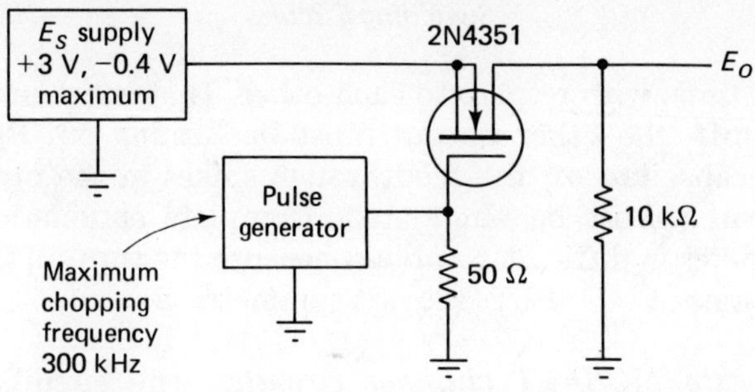

Figure 6-14. Practical series chopper using N-channel MOSFET.

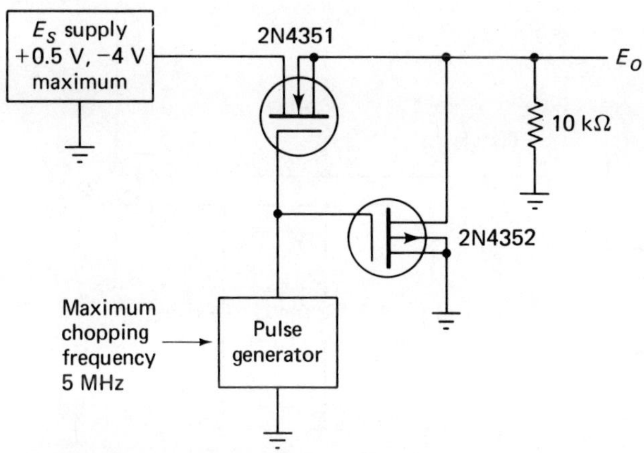

Figure 6-15. Series-shunt chopper for high-frequency applications using complementary enhancement mode MOSFETs.

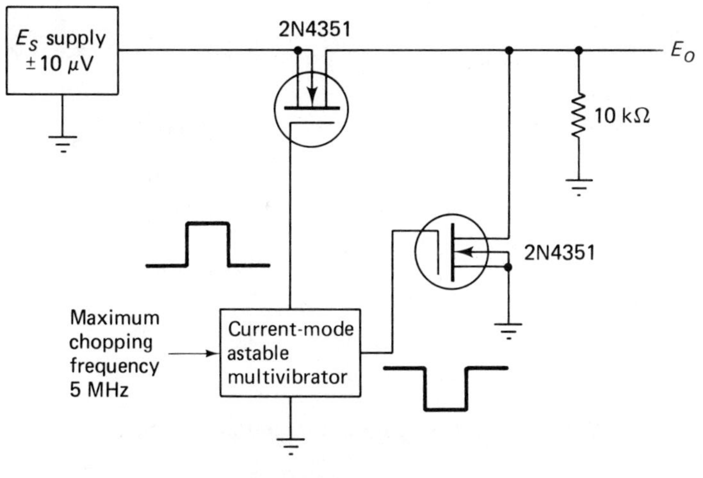

Figure 6-16. Series-shunt chopper for low-input voltages.

delayed in time, with respect to each other. That is, when one output is turning off, the other output must be turning on. By matching MOSFET capacitances, the feedthrough spikes in the output of the chopper can almost be eliminated. Complete elimination of feedthrough spikes is difficult to obtain because the turn-on and turn-off characteristics of MOSFETs are not symmetrical.

Dual-gate MOSFET chopper circuits. The circuits shown in Figs. 6-17 and 6-18 use dual-gate MOSFETs in chopper or gating cir-

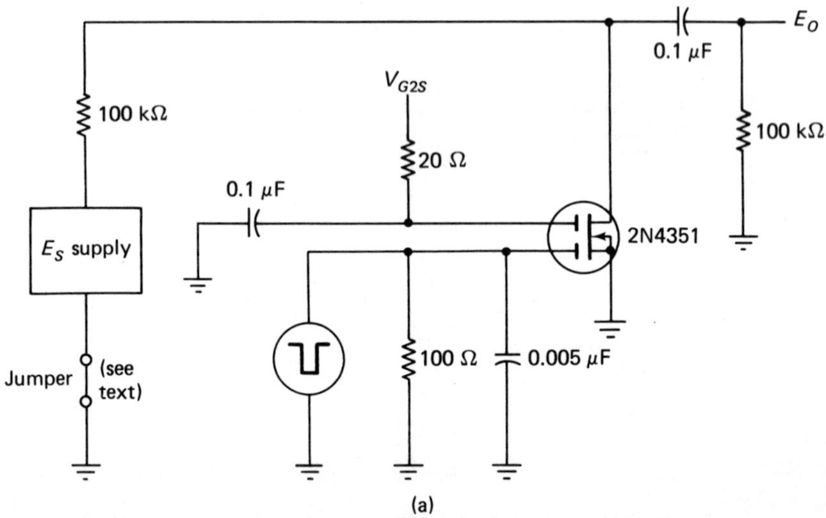

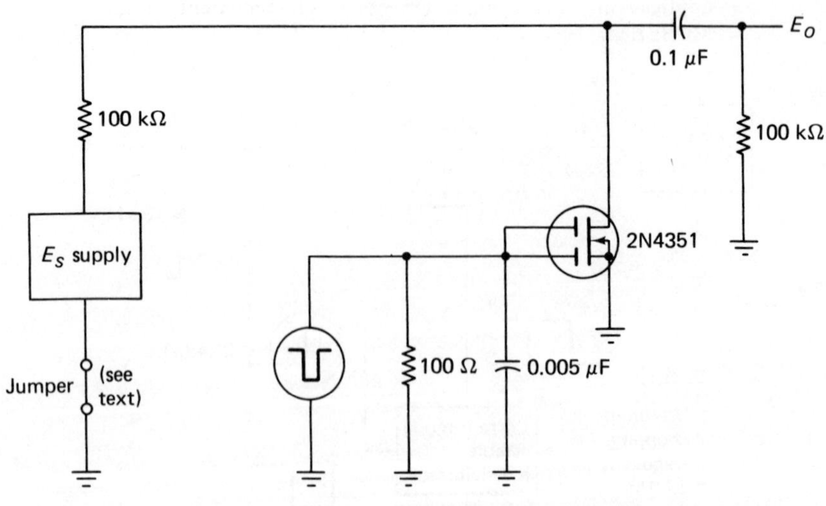

Figure 6-17. Dual-gate shunt chopper circuits.

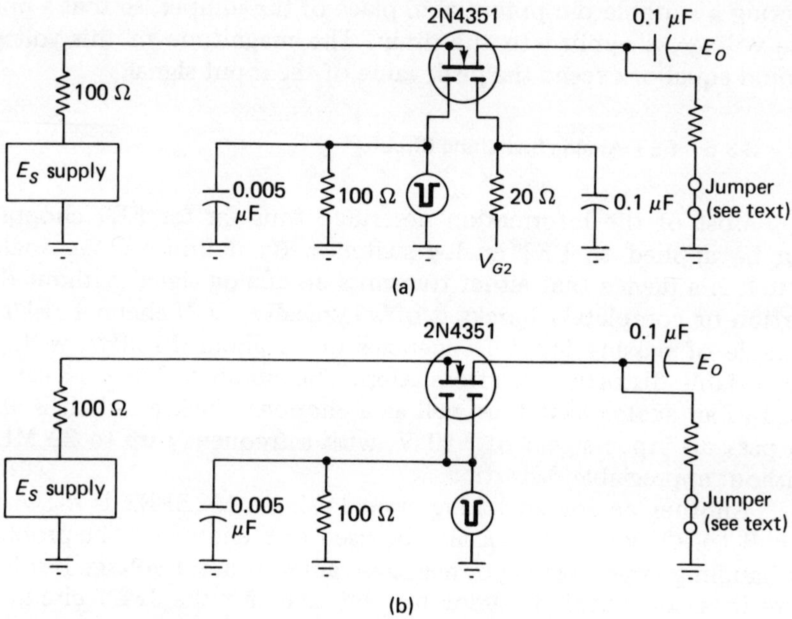

Figure 6-18. Dual-gate series chopper circuits.

cuits. In the shunt choppers of Fig. 6-17, the MOSFET is normally conductive, so the output is low. A negative gating pulse turns off the MOSFET so that approximately 50 percent of the input appears at the output terminals. The circuit of Fig. 6-17a features an additional control voltage, V_{G2}. A d-c voltage may be applied to the second gate, thereby establishing the value of desired channel ON resistance. As an alternative, the second gate can function as a coincidence gate to reduce the output to a low value. This requires a positive-going pulse applied to gate 2 simultaneously with a positive pulse to gate 1.

The circuits shown in Fig. 6-18 operate in a manner opposite to those of Fig. 6-17. That is, in the Fig. 6-18 circuits, output voltage appears in the absence of a gating signal. Consequently, a negative gating signal reduces the level of output. In the circuit of Fig. 6-18a, a negative voltage or signal at V_{G2} also reduces output level, and the circuit thus provides dual-gate control.

The circuits of Figs. 6-17 and 6-18 show a jumper connected between two terminals in the drain-to-ground-return circuits. The circuits assume a peak input level of less than 0.2 V. Should the signal exceed this value, it is possible that the parasitic "diode" between the drain and the semiconductor substrate will be driven into conduction and load the signal. This can be overcome by connecting a suitable d-c potential in place of the jumper, so that a positive voltage is applied to the drain. The magnitude of this voltage should equal or exceed the peak value of the input signal.

6-3-5 FET Analog Switching Circuits

Most of the information described thus far for FET choppers can be applied to FET analog switches. By definition, the analog switch is a device that either transmits an analog signal without distortion or completely blocks it off. Typically, an N-channel JFET is capable of passing input frequencies up to about 20 MHz, without appreciable distortion or attenuation. The circuit of Fig. 6-13 can be used as an analog switch as well as a chopper. Such a switch is able to pass an input signal of ±10 V, with a frequency up to 20 MHz, without appreciable distortion.

Another analog switching circuit using a MOSFET is shown in Fig. 6-19. (This circuit can also be used as a chopper.) The problem of handling large positive or negative values of input voltage is solved here in a somewhat different manner than for the JFET circuit of Fig. 6-13.

With the MOSFET circuit of Fig. 6-19, there is no PN junction to worry about since the gate is insulated from the rest of the FET. This eliminates the need for a diode at the gate. There are, however,

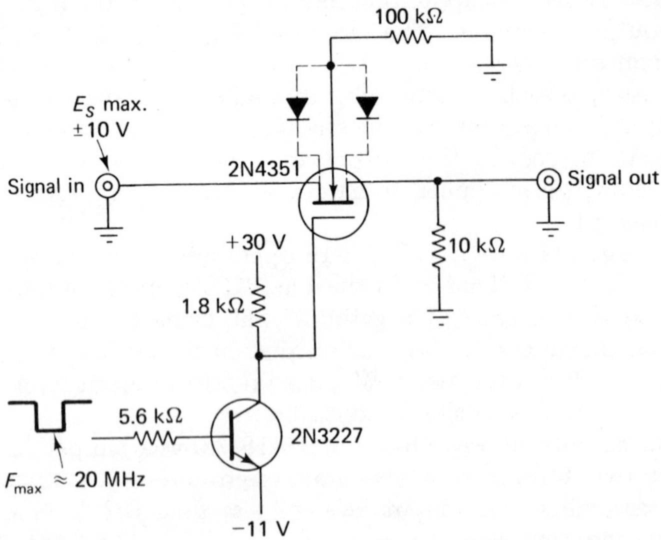

Figure 6-19. MOSFET analog switch for large input voltages.

PN junctions from substrate-to-source and substrate-to-drain. These PN junctions must not be allowed to become forward-biased. One way to prevent this is to cut off the substrate lead and leave it floating. However, since the substrate is connected to the package (metal can), this can result in noise pickup. A more practical solution is shown in Fig. 6-19. Here, the substrate-to-source junction is coupled with a diode. In turn, the substrate is coupled back to ground through a 100-kΩ resistor.

Commutator using FET analog switches. The analog switch can be used in a commutator circuit, as shown in Fig. 6-20. Each switch has a three-input AND gate in series with the gate drive. In order to turn a switch ON, a positive voltage is required at the gate. To accomplish this, the three inputs of the AND gate must be "true" (at a positive potential).

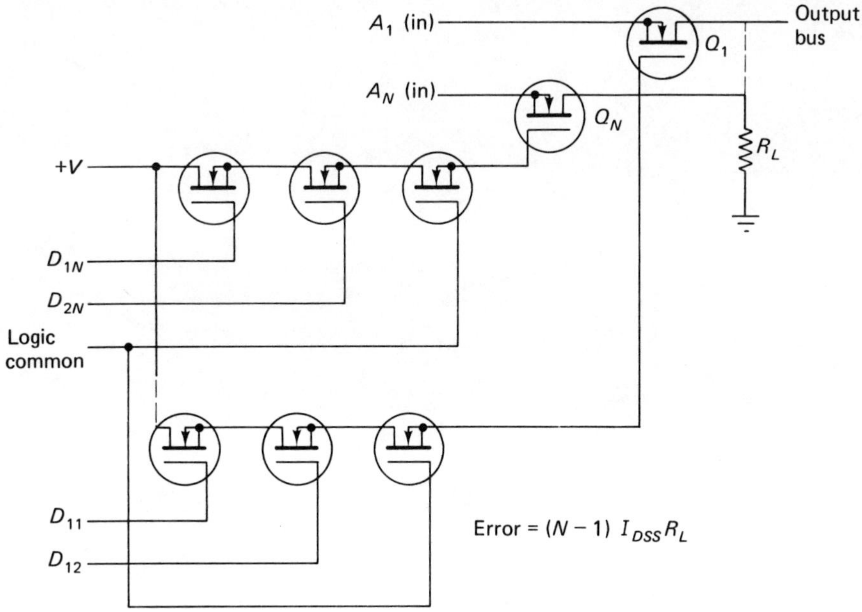

Figure 6-20. Commutator network using N-channel MOSFETS.

Assume that A_N is to be sampled. Further assume that the logic common is a clock signal, that D_{1N} is an order from the control system to sample A_N, and that D_{2N} is a ready signal from the device to be sampled. When all these conditions are true at the same time, switch Q_N is turned on and A_N is sampled. The circuit of Fig. 6-20 can be modified to accept large values of input voltage.

In this type of commutator circuitry, where only one channel is turned on at a time, an error can be introduced due to the leakage of the other FET switches. Assuming that the leakage is the same for all switches, an approximation of the error signal is given by

$$(\text{number of channels} - 1) \times \text{drain-source current} \times R_L$$

If you are not familiar with the digital electronic terms used to describe this commutator circuit (such as AND gate, clock signal, and true logic) your attention is invited to the author's best selling *Handbook of Digital Electronics* (Prentice-Hall, Inc., Englewood Cliffs, N. J., 1981).

Chapter 7

Electronic Control Circuits

This chapter is devoted entirely to electronic control circuits. Such circuits include controlled rectifiers, thyristors, SCRs, SCSs, triacs, diacs, SUSs, SBSs, LASCRs, LASCSs, opto-couplers, phase-control, motor-control, and digital-control systems.

7-1 ELECTRONIC CONTROL DEVICES

Before going into control circuit operation, let us review electronic control devices. There is an increasing use of electronic devices in both control and instrumentation systems. At one time, electron tube devices such as the diode, triode, thyratron, and ignatron were used extensively in control systems. Today, these electron tube devices have generally been replaced by solid-state electronics. For example, the electron tube diode and triode have been replaced by the solid-state diode and triode. The thyratron and ignatron have been replaced by solid-state *controlled* rectifiers or *thryistors.* The subject of electronic control devices is quite extensive and cannot be covered fully in one book, much less in this chapter. For that reason, we concentrate here on the basics of a few selected solid-state control devices.

7-1-1 Controlled Rectifiers and Thyristors

The controlled rectifier is similar to the basic diode, with one specific exception. The controlled rectifier must be "triggered" or "turned on" by an external voltage or signal. The controlled rectifier has a high forward and reverse resistance (no current flow), without the trigger. When the trigger is applied, the forward resistance drops to zero (or very low) and a high forward current flows just as with the basic diode. The reverse resistance remains high, and no reverse current flows to the controlled-rectifier a-c power in the normal manner. As long as the forward voltage is applied, the forward current continues to flow. The forward current stops, and the controlled rectifier turns off, if the forward voltage is removed.

There are a great number of controlled rectifiers in use. Many of these are actually the same type (or a slightly different version), but are manufactured under different trade names or designations. The term *thyristor* is applied to many controlled rectifiers. Technically, a thyristor defines any semiconductor switch using PNPN regenerative feedback (Sec. 7-1-3). Thyristors can be two-, three-, or four-terminal devices. Both unidirectional and bidirectional controlled rectifiers and thyristors are used in electronic control circuits.

7-1-2 Silicon- or Semiconductor-controlled Rectifier (SCR)

With some manufacturers, the letters SCR refer to *semiconductor-controlled rectifier* and can mean any type of solid-state controlled rectifier. However, SCR usually refers to *silicon-controlled rectifier.*

If four semiconductor materials, two P-type and two N-type, are arranged as shown in Fig. 7-1a, the device can be considered as three diodes arranged alternately in series, as shown in Fig. 7-1b. Such a device acts as a conventional diode rectifier in the reverse direction and as a combined electronic switch and series rectifier in the forward direction. Conduction in the forward direction can be controlled or "gated" by operation of the switch.

Figure 7-2 shows the circuit symbol, block diagram, and basic physical construction of a typical SCR. Note that there are two basic arrangements for the SCR: one with the gate terminal connected to the cathode and one with the gate terminal connected to the anode.

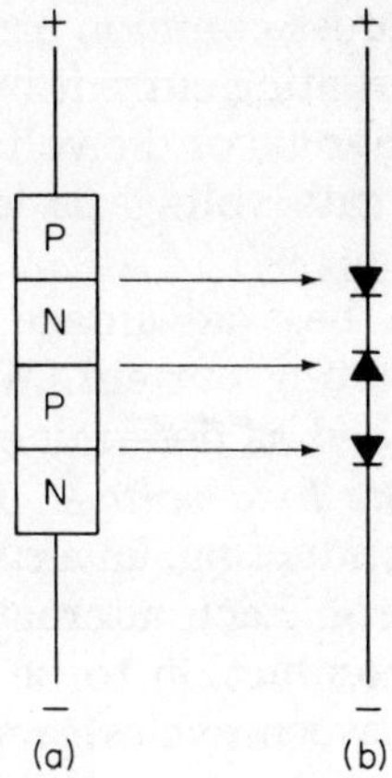

Figure 7-1. Semiconductor material arrangement and equivalent diode relationship of an SCR.

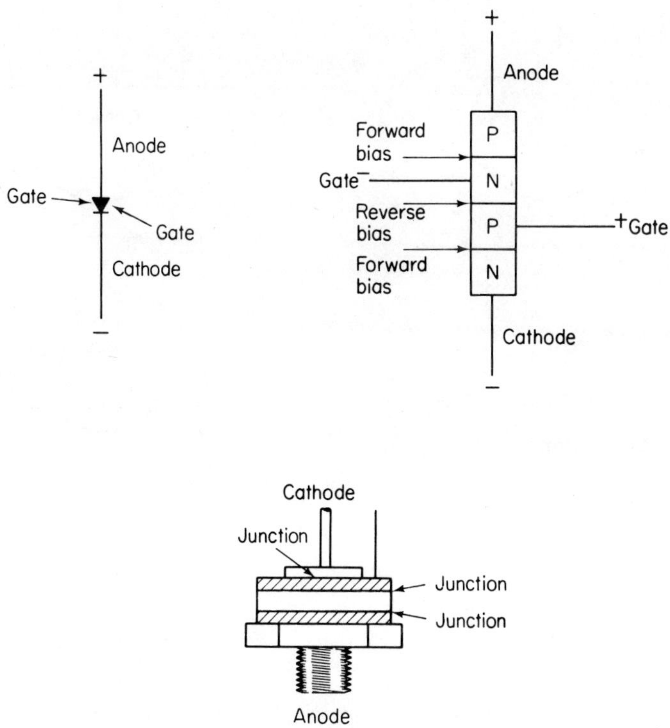

Figure 7-2. Symbol, block diagram, and basic construction of a typical SCR.

The cathode gate is the most common arrangement. SCRs are most often used to control alternating current but can be used to control a direct-current circuit. Either a-c or d-c voltage can be used as the gate signal provided that the gate voltage is large enough to trigger the SCR into the ON condition.

An SCR is used to best advantage when both the load and trigger circuits use alternating current. With a-c power, control of power is applied to the load as determined by the *relative phase of the trigger signal versus the load voltage.* Because the trigger control is lost once the SCR is conducting, an a-c voltage at the load permits the trigger to regain control. Each alternation of alternating current through the load causes conduction to be interrupted (when the a-c voltage drops to zero between cycles), regardless of the polarity of the trigger signal.

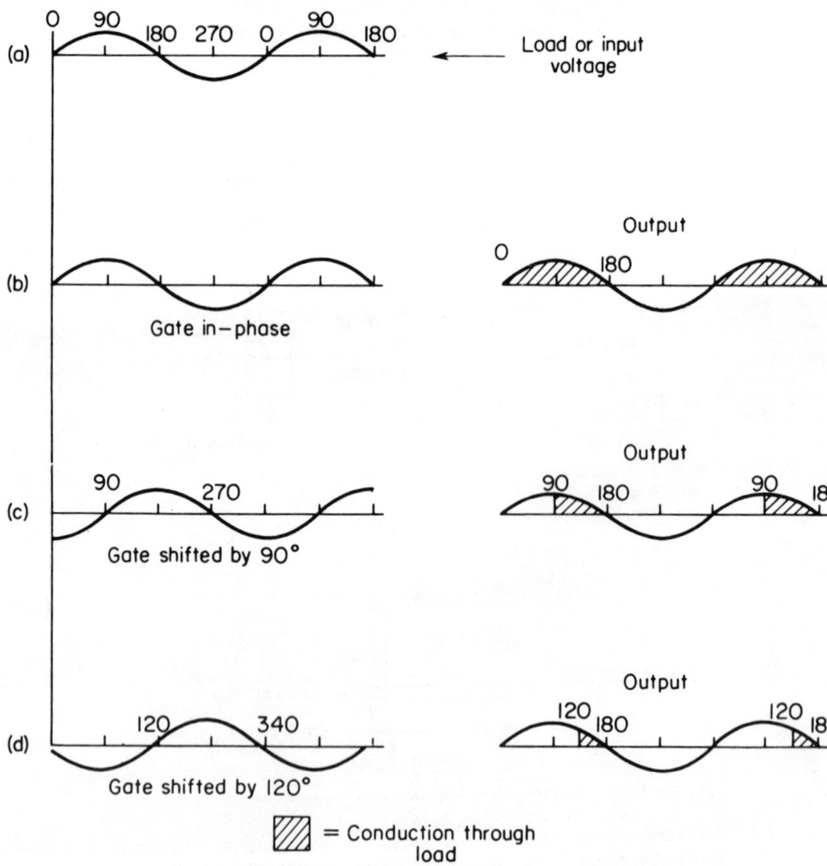

Figure 7-3. Operation of an SCR with ac at both load and gate or trigger.

Phase relationship between gate and load voltages. Figure 7-3 shows operation of an SCR with a-c voltages at the trigger circuit and load circuit. If the trigger voltage is in phase with the a-c power input signal as shown in Fig. 7-3b, the SCR conducts for each successive positive alternation at the anode. When the trigger is positive-going at the same time as the load or anode voltage, load current starts to flow as soon as the load voltage reaches a value that will cause conduction. When the trigger voltage is negative-going, the load voltage is also negative-going, and conduction stops. The SCR thus acts as a half-wave rectifier in the normal manner when the trigger is in phase with the load.

If there is a 90° phase difference between the trigger voltage and the load voltage (say, the load voltage lags the trigger voltage by 90°) as shown in Fig. 7-3c, the SCR does not start conducting until the trigger voltage swings positive, even though the load voltage is initially positive. When the load voltage drops to zero, conduction stops, even though the trigger voltage is still positive.

If the phase shift is increased between the trigger and load voltages as shown in Fig. 7-3d, conduction time is even shorter, and less power is applied to the load circuit. By shifting the phase of the trigger voltage in relation to the load voltage, it is possible to vary the power output even though the voltages are not changed in amplitude.

7-1-3 PNPN-controlled Switch (SCS)

The PNPN-controlled switch (often referred to as an SCS or silicon-controlled switch) is similar in operation to an SCR. However, the SCS is a PNPN device where all four semiconductor regions are made accessible by means of terminals. Figure 7-4 shows the circuit symbols, block diagram, and basic physical construction of a typical SCS. Note that two circuit symbols are used and that both are in common use by manufacturers. Often, the SCS is used as an SCR, with the extra gate terminal not connected.

In some control circuits, the SCS can be considered as a transistor and diode in series. Figure 7-5 shows this arrangement. If a negative load voltage is applied to terminal 4, with a positive at terminal 1, the SCS does not turn on, no matter what trigger signals are applied. However, with a positive load voltage at terminal 4 and a negative at terminal 1, the SCS is turned on by either a positive voltage at terminal 2 or a negative voltage at terminal 3.

The SCS also has the ability to turn off by means of a gate signal. The arrangement is shown in Fig. 7-6. It should be noted that the gate turn-off method applies only when the conducting current is below a certain value. The SCS is often considered as two transis-

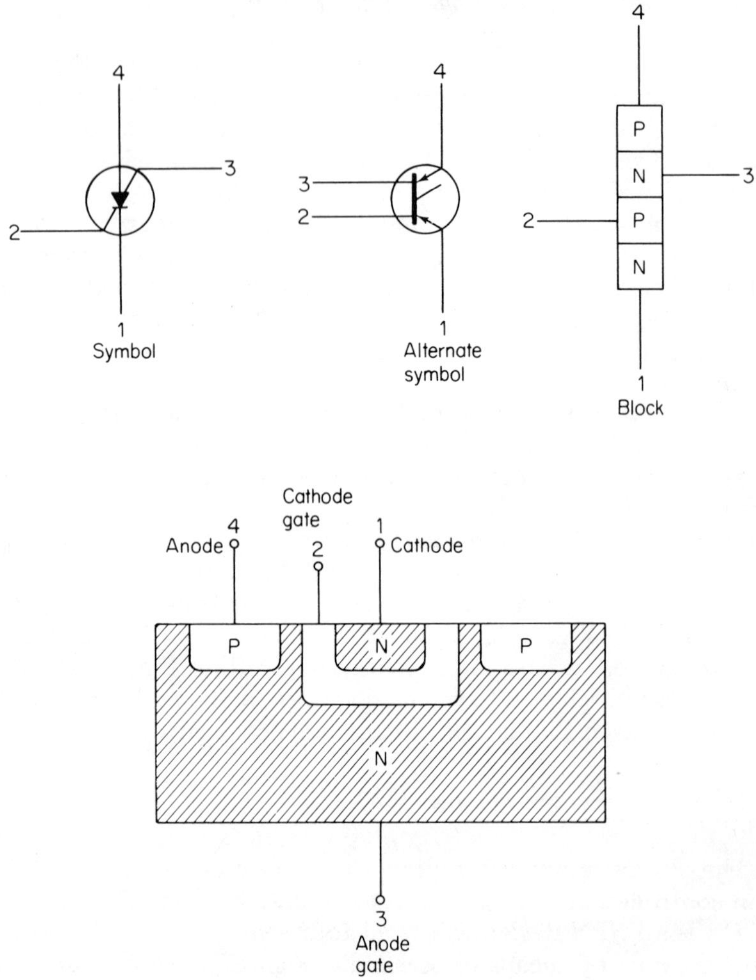

Figure 7-4. Circuit symbols, block diagram, and basic physical construction of a typical SCS.

tors (an NPN and a PNP) connected as shown in Fig. 7-7. Both transistors are connected so that the collector output of the NPN feeds into the base input of the PNP, and vice versa. If a positive trigger voltage is applied to the NPN base, the NPN is turned on and some current flows in the NPN emitter-collector circuit. Since the NPN collector feeds the PNP base, the PNP is also turned on, and the PNP collector output feeds into the NPN base, adding to the trigger voltage.

Load current then flows through the two transistors from the NPN emitter (or rectifier cathode) to the PNP emitter (or rectifier anode). Load current continues to flow even though the trigger is

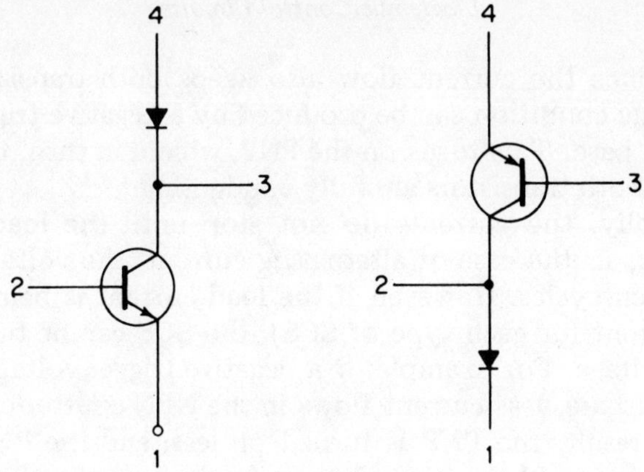

Figure 7-5. Equivalent diode-transistor representation of an SCS.

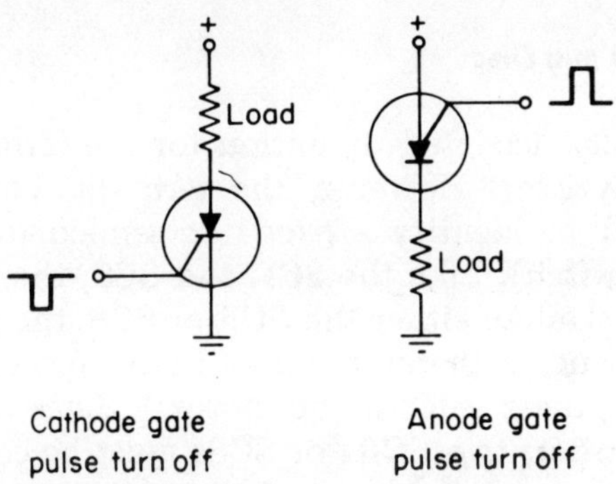

Figure 7-6. Gate turn-off arrangement for SCSs.

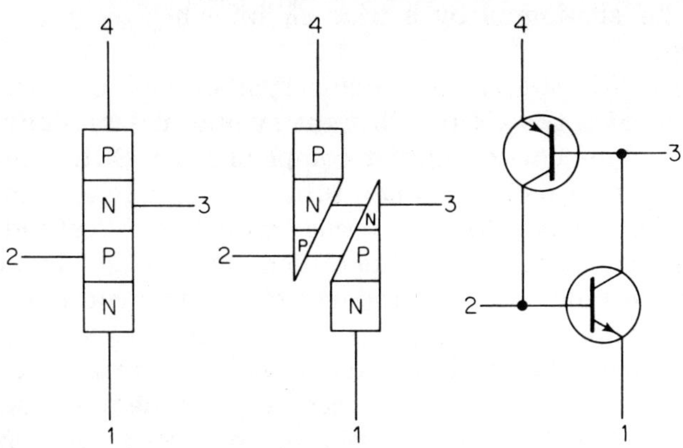

Figure 7-7. Equivalent two-transistor representation of an SCS.

removed, since the current flow also keeps both transistors turned on. The same condition can be produced by a negative trigger applied to the PNP base. This turns on the PNP, which, in turn, turns on the NPN, until both transistors are fully conducting.

Normally, the currents do not stop until the load voltage is removed or, in the case of alternating current, the voltage drops to zero between cycles. However, if the load current is below a certain level (different for each type of SCS), the SCS can be turned off by a trigger voltage. For example, if a negative trigger voltage is applied to the NPN base, less current flows in the NPN emitter-collector circuit. As a result, the PNP is turned on less, and the PNP collector output drops to aid the turn-off trigger voltage at the NPN base. The feedback process continues until the load current is completely stopped.

7-1-4 Triacs and Diacs

The generally accepted definition for the term *triac* is bidirectional triode thyristor. However, the term (as coined by General Electric) is used to identify a *tri*ode *ac* semiconductor switch (or three-terminal switch). Like the SCR and SCS, the triac is triggered by a gate signal. Unlike either the SCR or SCS, the triac conducts in *both directions* and is therefore most useful for controlling devices operated by a-c power (such as a-c motors). Since an SCR or SCS is essentially a rectifier, two SCRs or SCSs must be connected back to back (in parallel or bridge) to control alternating current. (Otherwise, the a-c is rectified into direct current.) The use of two or more SCRs or SCSs requires elaborate control circuits (in many cases). Much of this can be eliminated by a triac circuit when a-c power is to be controlled.

Figure 7-8 shows the circuit symbol and outward physical appearance of a triac. Note that the symbol is essentially one SCR symbol combined with another complementary SCR symbol. Since the triac is not a rectifier (when turned on and conducting, current flows in both directions), the terms "diode" and "cathode" do not apply. Instead, terminals are identified by number. Terminal T_1 is the reference point for measurement of voltages and currents at the gate terminals and at terminal T_2. The area between terminals T_1 and T_2 is essentially a PNPN switch in parallel with an NPNP switch.

Like the SCS or SCR, the triac can be made to conduct when a "breakdown" or "breakover" voltage is applied across terminals T_1 and T_2 and when a trigger voltage is applied. Current continues to flow in one direction until that half-cycle of the a-c voltage (across T_1 and T_2) is complete. Current then flows in the opposite direction

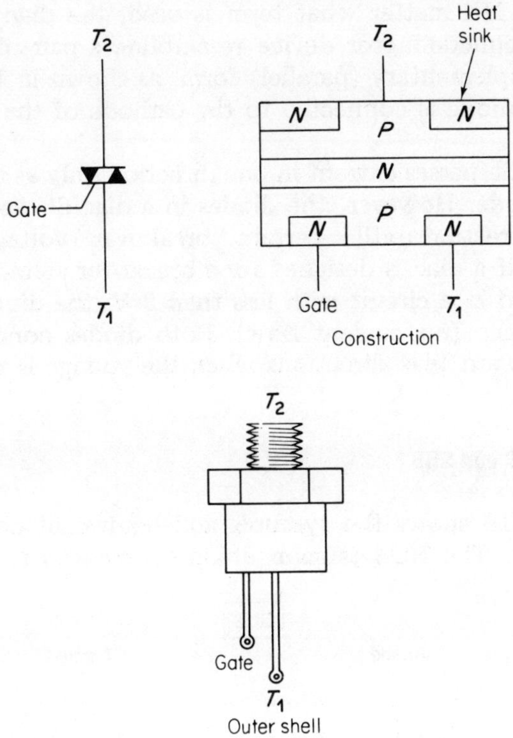

Figure 7-8. Symbol, outer shell, and construction of a typical triac.

for the next half-cycle. The triac does not conduct on either half-cycle unless a gate-trigger voltage is present during that half-cycle (unless the breakdown voltage is exceeded).

Triac trigger sources. Triacs can be triggered from many sources, as can SCRs and SCSs. One of the most common trigger sources is the diac, shown in Fig. 7-9. Note that there are two types of diacs, diode and transistor. The term *diac* (officially bidirectional diode thyristor) was coined by General Electric to identify a *di*ode *ac* semiconductor device. The diac is also known as a bidirectional

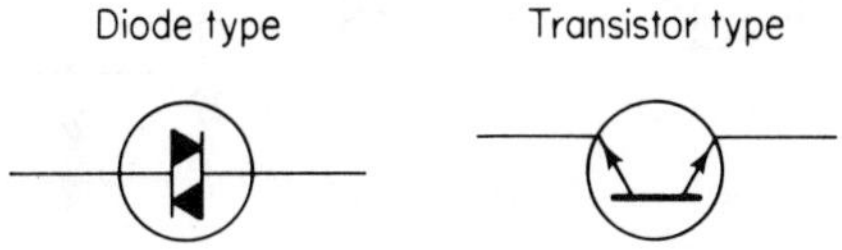

Figure 7-9. Symbols for diode-type and transistor-type diacs.

trigger diode. No matter what term is used, the diac can be considered as a semiconductor device resembling a pair of diodes connected in complementary (parallel) form, as shown in Fig. 7-9. The anode of one diode is connected to the cathode of the other diode, and vice versa.

Each diode passes current in one direction only as in the case of a common diode. However, the diodes in a diac do not conduct in the forward direction until a certain "breakover" voltage is reached. For example, if a diac is designed for a breakover voltage of 3 V and the diac is used in a circuit with less than 3 V, the diodes appear as a high resistance (no current flow). Both diodes conduct in their respective forward bias directions when the voltage is raised to any value over 3 V.

7-1-5 SUS and SBS

Figure 7-10 shows the symbols and equivalent circuits of the SUS and SBS. The SUS (silicon unilateral switch) is essentially a

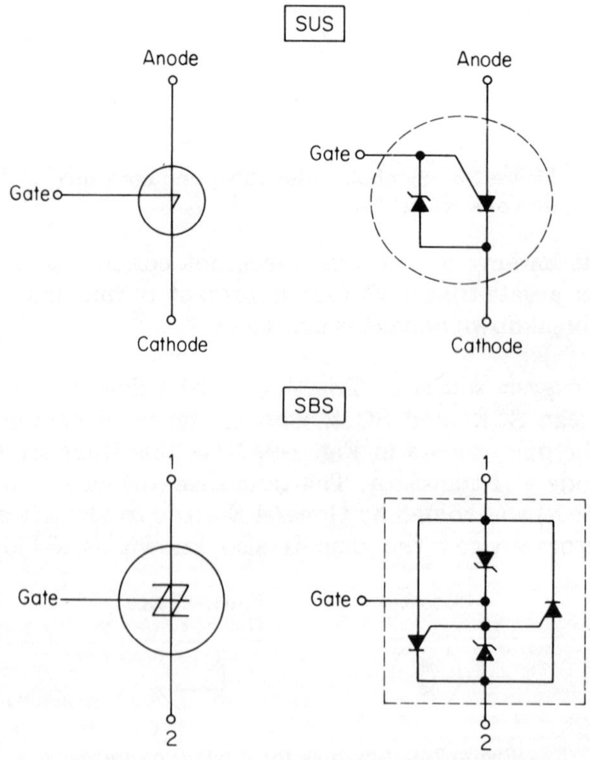

Figure 7-10. Symbols and equivalent circuits for SUS and SBS.

miniature SCR having an anode gate (instead of the usual cathode gate) and a built-in low-voltage avalanche (zener) diode between the gate and cathode. The SUS operates in a manner similar to the UJT (Chapter 8). However, the SUS switches at a fixed voltage, determined by the internal zener diode, rather than a fraction of percentage of the supply voltage. Also, the switching current of the SUS is generally higher than that of the UJT.

The SBS (silicon bilateral switch) is essentially two identical SUS structures arranged in inverse parallel. Since the SBS operates as a switch with both polarities of applied voltage, the SBS is particularly useful for triggering triacs with alternating positive and negative gate pulses.

7-2 TYPICAL ELECTRONIC CONTROL CIRCUITS

The following paragraphs describe some basic control circuits using the electronic devices described in Sec. 7-1. Keep in mind that these represent only a very small fraction of the electronic control circuits in present use.

7-2-1 Basic Phase Control

Phase control is the process of rapid ON-OFF switching which connects an a-c power supply to a load for a controlled fraction of each cycle. This is a highly efficient means of controlling the average power to loads such as lamps, heaters, and motors. Control is accomplished by governing the *phase angle* of the a-c waveforms at which the control device is triggered. The control device then conducts for the remainder of the half-cycle. Figure 7-11 shows the basic forms of a-c phase control.

The simplest form of phase control is the half-wave control shown in Fig. 7-11a. This circuit uses one SCR for control of current flow in one direction only and is used for loads that require power control from zero to one-half of full-wave maximum. The circuit is also useful where direct current is required (or permitted) for the load.

The addition of one rectifier, as shown in Fig. 7-11b, provides a fixed half-cycle of power that shifts the power control range to half-power minimum and full-power maximum. The use of two SCRs, as shown in Fig. 7-11c, controls from zero to full power, but requires two gate signals. The single gate or trigger signal is required for the zero to full-power control circuit of Fig. 7-11d. The most flexible circuit (Fig. 7-11e) uses one SCR inside a bridge rectifier and may be used for control of either a-c power or full-wave rectified d-c power.

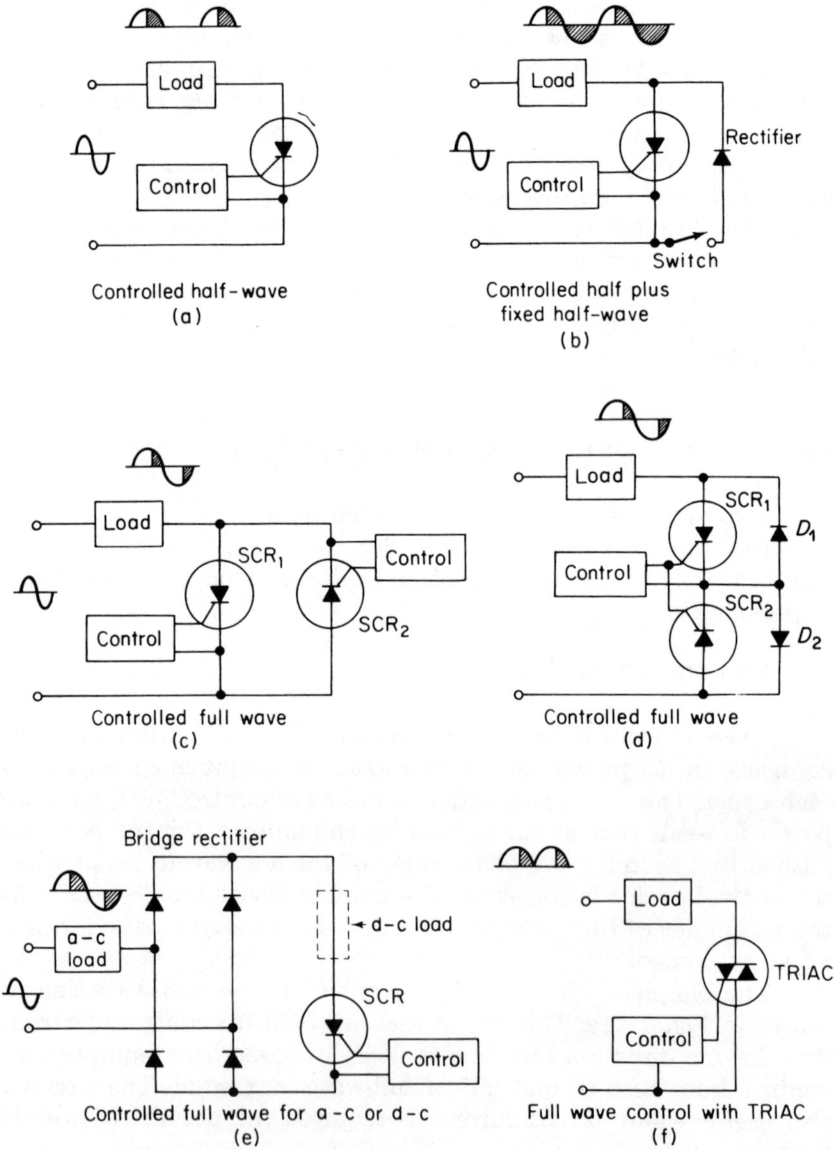

Figure 7-11. Basic forms of a-c phase control.

Generally, the most simple, efficient, and reliable circuit for controlling a-c power uses a triac, as shown in Fig. 7-11f. This circuit provides full-wave control, in both directions, using only one gate or trigger (often a diac). For that reason, the triac-diac combination has been by far the most popular a-c power control circuit.

7-2-2 Basic Trigger Circuits for Phase Control

Most phase-control trigger circuits use some form of relaxation oscillator (Chapter 4). Since relaxation oscillators are simply *timing circuits*, they must be synchronized to the a-c power supply in order to provide proper phase control. That is, their timing cycles must start simultaneously with each alternation of the a-c power. This is usually done by taking the oscillator input voltage from the a-c supply.

Figure 7-12 shows the basic half-wave and full-wave phase-control circuits. With either circuit, the load, control device, and trigger device are all operated from the a-c power supply. Capacitor *C* is charged through adjustable resistor *R* on alternations of the a-c supply. In the half-wave circuit, *C* is charged once for each half-alternation or half-cycle. In the full-wave circuit, *C* is charged on both alternations of each cycle. When *C* is charged to a certain voltage, the trigger device turns on and provides a trigger signal to the control device. This switches the control device on for the remainder of the half-cycle, and power is applied through the load.

The time required to charge *C* to the trigger point is set by adjustment of *R*. In turn, this determines the portion of the half-

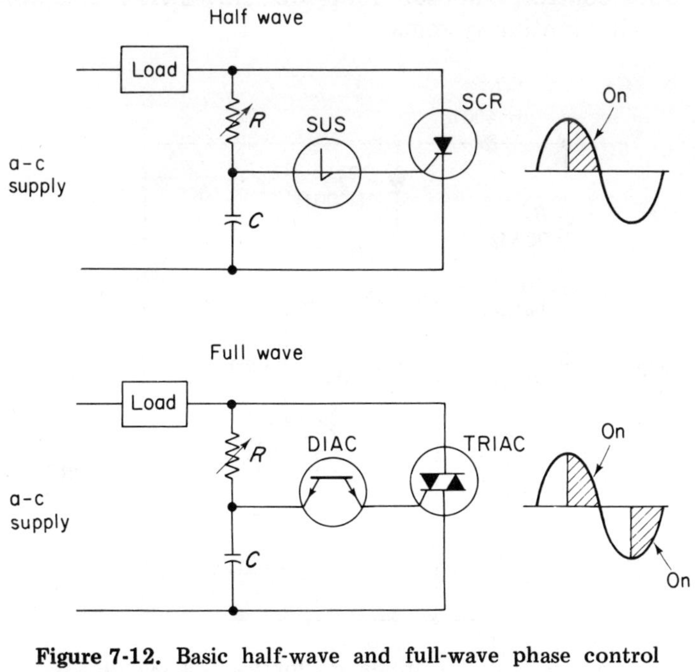

Figure 7-12. Basic half-wave and full-wave phase control circuits.

cycle in which the control device is on and the amount of power applied to the load. For example, assume that the value of *R* is adjusted to minimum resistance. This causes *C* to charge quickly and switch both the trigger and control devices on early in the half-cycle. Thus, maximum power is applied to the load. With *R* at maximum resistance, *C* charges slowly, and the devices switch on late in the half-cycle, providing minimum power to the load. Note that both trigger and control devices are switched off when the a-c power goes to zero between each half-cycle.

7-2-3 Static Switching Circuits

Electronic control devices are not limited to phase control. In fact, almost any static switching function otherwise performed by various mechanical and electromechanical switches can be done with electronic control circuits. The following paragraphs describe two examples.

Electronic proximity switches. Figure 7-13 shows a classic electronic proximity switch. This circuit is typically used in elevators (both for door safety controls and floor selector buttons), supermarket door control, bank-safe monitors, flow-switch actuators, and conveyor-belt counting systems.

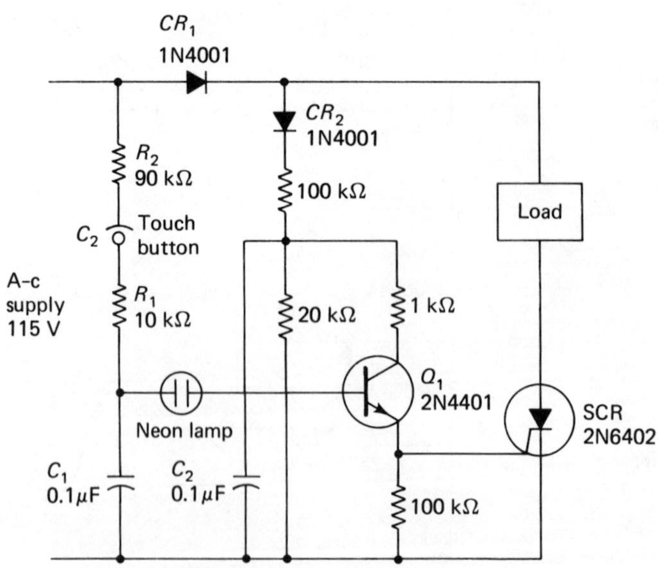

Figure 7-13. Electronic proximity switch circuit.

Capacitor C_1, resistors R_1/R_2, and the sensor plate "capacitor" C_2 form a voltage divider across the a-c supply. The voltage across C_1 depends on the C_1/C_2 ratio and the supply voltage. The capacitance of C_2, in turn, depends on the proximity to the sensor plate of any reasonably conductive and grounded object (metals, human body, etc.). As soon as the voltage across C_1 exceeds the breakdown voltage of the neon lamp, capacitors C_1 and C_2 discharge through the base and emitter of transistor Q_1. This discharge current is amplified by Q_1, and turns the SCR on to energize the load.

The load remains energized as long as the sensor plate or "button" is touched. When the button is no longer touched, the load is deenergized. A latching action can be provided by replacing CR_1 with a short circuit (so that the SCR remains triggered on both half-cycles). The circuit must then be reset by an auxiliary contact in series with the SCR.

Electronic heater control. The familiar mercury-in-glass technique is often used to sense and measure temperature. The same

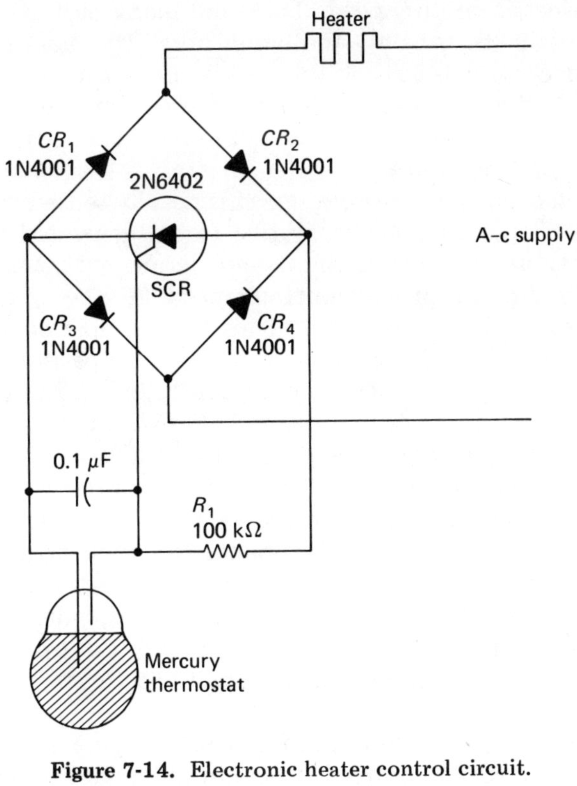

Figure 7-14. Electronic heater control circuit.

technique can be used as the basic elements of a thermostat to control operation of an electrical heater. One drawback to the mercury-in-glass system is very low current handling capability, typically well below 1 mA. This current is not sufficient to control an electrical heater (to switch the heater on and off). This problem can be overcome by means of the circuit shown in Fig. 7-14. Here, an SCR is used to amplify the current of a mercury thermostat and to carry the heavy current required by the heater.

When the thermostat is open (low temperature), capacitor C is in the circuit and charges on each half-cycle of the a-c power. This triggers the SCR on each half-cycle and delivers power to the heater. When the thermostat is closed (high temperature), capacitor C is shorted out of the circuit and cannot charge. The SCR can no longer trigger, and the heater shuts off.

7-2-4 Electronic Motor Control Circuits

One of the most common electronic control circuits is for phase control of electric motor speed. There are many motor control circuits, so we will discuss only the basics here. The main problem in motor speed control is sensing the motor speed and then providing feedback to offset changes in speed. If speed increases (due to a change in load or in supply voltage), the feedback signal slows the motor down, and vice versa.

In d-c and universal motors, speed is usually sensed by monitoring the CEMF (counter electromotive force) generated during the SCR on-time. In the case of shunt and permanent-magnet motors, the CEMF is directly proportional to speed. If speed increases, the CEMF increases, and vice versa. Thus, by comparing the CEMF against an adjustable voltage, and using this comparison to control the SCR on-time, motor speed can be controlled within close tolerances. In the case of series motors, the CEMF is almost proportional to speed, so the same basic principle can be used.

In a-c induction motors, speed is often sensed by means of a *tachometer generator*. If speed increases, tachometer output voltage increases, and vice versa. This tachometer feedback signal is used to control SCR on-time. If you are not familiar with the basics of electrical motors and their use in controls systems, your attention is invited to the author's best selling *Handbook of Controls and Instrumentation* (Prentice-Hall, Inc., Englewood Cliffs, N. J., 1980).

D-c and universal motor speed control. Figure 7-15 shows the basic circuit for speed control of d-c and universal motors using an

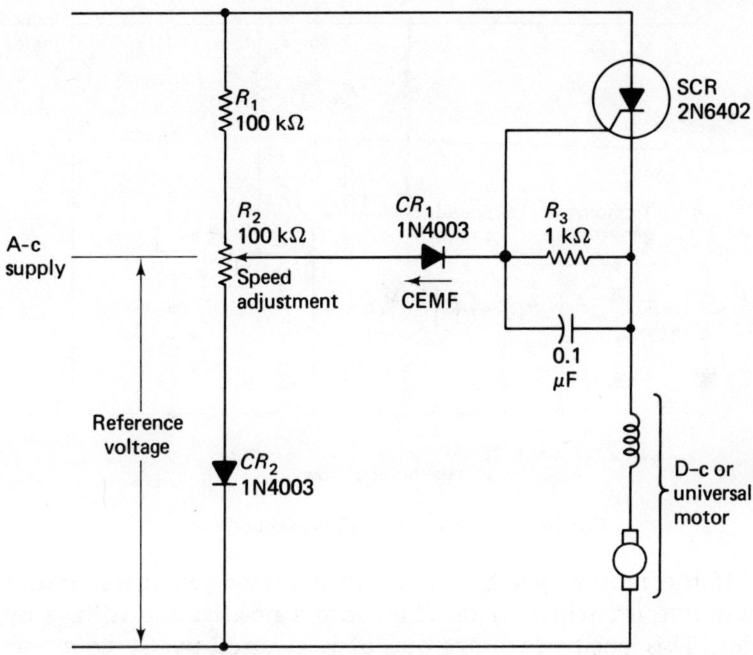

Figure 7-15. Basic d-c and universal motor speed control circuit.

SCR. This phase-control circuit operates by comparing the motor CEMF voltage against a reference voltage (which is rectified by CR_2 and set by adjustment of R_2).

If the CEMF is greater than the reference voltage (the motor is going faster than the selected speed), CR_1 is reverse-biased (the cathode is more positive than the anode), the SCR is not triggered, and no power is applied to the motor. When the motor slows down, the CEMF voltage drops and is less than the reference voltage, CR_1 is forward-biased (anode more positive), the SCR is triggered, and power is applied to the motor. The speed at which this occurs may be varied by changing the magnitude of the reference voltage (by adjustment of R_2).

Induction motor control. Figure 7-16 shows the basic circuit for speed control of a-c motors using a tachometer generator geared to the motor. The output of the tachometer generator (or feedback signal) is rectified by diodes D_1 through D_4, and applied through resistor R_1 to amplifier transistor Q_1. This feedback signal is inverted by Q_1 and is used to charge capacitors C_1 through Q_2. The charge on C_1 determines the SCR on-time.

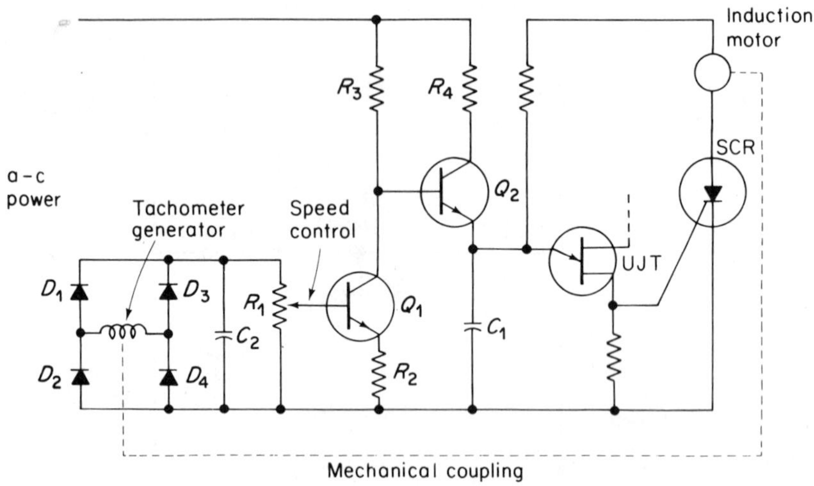

Figure 7-16. Induction motor speed control.

If the motor speeds up, the tachometer generator produces a greater output, which is rectified into a positive d-c voltage by the diodes. This positive voltage output is inverted by Q_1 to a negative voltage that charges C_1 to a lower value. The SCR then triggers at a later point and the motor slows down. The charge on C_1 (and thus the motor speed) can also be set by adjustment of R_1.

7-3 LIGHT-ACTIVATED CONTROL DEVICES

There are many types of devices available for converting light or radiant energy into electrical power. When this principle is used in electronic control devices, they are often referred to as *light-activated semiconductors* (LAS). When light strikes silicon, there is an increase in current flow. In PNPN control devices, such as the SCR and SCS, this current increase acts as a trigger to turn on the device. The most common LAS devices are the LASCR (light-activated SCR), LASCS (light-activated SCS), and the opto-coupler (or light-activated switch). In addition, basic *photocells, phototransistors,* and *photo-Darlingtons* are used in light-activated control circuits.

7-3-1 Basic Photocell Theory

There are two types of photocells, *photovoltaic* and *photoconductive.* The symbols for photocells are shown in Fig. 7-17. The photovoltaic cells produce an output voltage and current in the

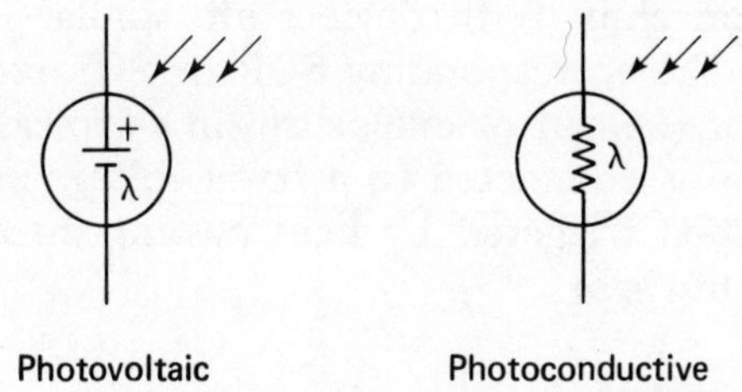

Figure 7-17. Basic photocell symbols.

presence of light. For this reason, photovoltaic cells are often called *solar batteries* or *solar cells*. The *photoconductive cells* are often called *light-sensitive resistors* because they function as a resistor and do not generate an output. Instead, photoconductive cells act as a resistance which varies in the presence of light, thus changing the amount of current being conducted through the circuit.

7-3-2 LASCR and LASCS

Figure 7-18 shows the symbols and construction for the LASCR and LASCS. Note that the arrows indicate light or radiant energy

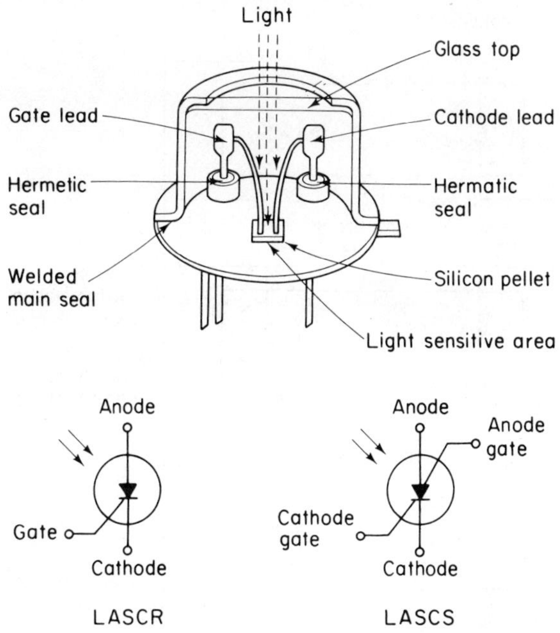

Figure 7-18. Symbols and construction for LASCR and LASCS.

striking the silicon chip. Both devices are similar in operation and characteristics to the corresponding SCR or SCS except for the glass window on top of the can or enclosure. In a typical circuit, the gate or trigger terminal is connected to a fixed voltage in the circuit. The SCR or SCS is then triggered by light passing through the window onto the silicon chip area.

7-3-3 Opto-couplers

Figure 7-19 shows a typical light-activated switch or opto-coupler. When a photocell is exposed to a very strong light, the internal resistance drops to zero (or near zero). This has the same effect as closing the contacts of a switch. When all light is removed from the photocell, the internal resistance increases to several million ohms. This has the same effect as opening the switch contacts.

In an opto-coupler (also known as an optoelectronic coupler), both the photocell and a light source are sealed in a lightproof enclosure. When the light source is off, the photocell resistance is very high and the "switch" is "open." When the light source is on, the light strikes the photocell and causes the resistance to drop to zero. This "closes" the "switch." Such a switch is completely noise-free

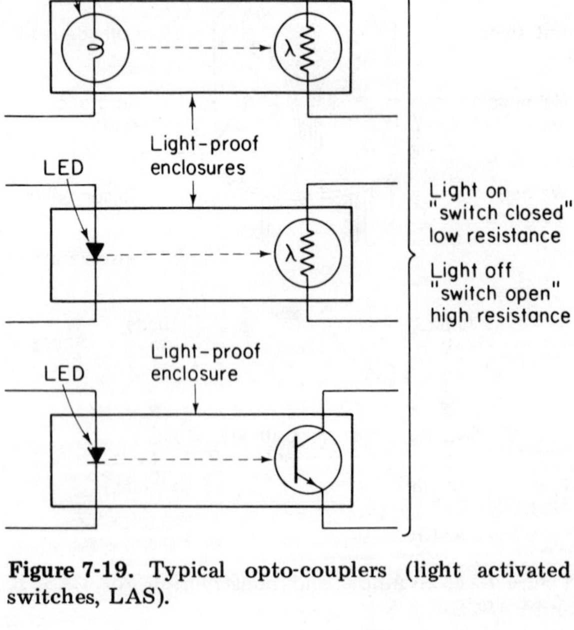

Figure 7-19. Typical opto-couplers (light activated switches, LAS).

and spark-free, which are important factors in many control circuits. The light source can be an incandescent lamp. However, most present-day opto-couplers use a light-emitting diode (LED) similar to the LEDs used for digital readouts or displays. In many present-day opto-couplers the detector is a phototransistor or photo-Darlington rather than a basic photocell.

Opto-couplers are designed for applications requiring electrical isolation, and have input/output isolation voltages as great as 2500 V, with isolation resistances being typically in the 1 million-megohm range. Owing to the high isolation properties of opto-couplers and a typical bandwidth capability from dc to about 300 kHz (in some cases), opto-couplers can be used where it is required to isolate higher voltage and power loads from low-level circuits (such as digital circuits).

The coupling from input to load is simplified as the opto-coupler can replace a d-c-coupled circuit (with the associated general requirement for level translation) and a pulse transformer or coupling transformer (with their limited bandwidths). Isolation is complete between the input and the load in that the power supplies and their circuit grounds can be completely independent of each other.

7-3-4 Typical Photocell Circuits

Figure 7-20 shows two basic photocell circuits. In Fig. 7-20a, the output of a photovoltaic cell is used to control operation of a relay through a basic transistor amplifier circuit. The transistor is PNP, so the photovoltaic cell output forward-biases the base-emitter circuit. If an NPN transistor is used, the cell terminals must be reversed. When light is applied to the cell, the base-emitter circuit is forward-biased and emitter-collector current flows. If the light is sufficient, the forward bias causes enough collector current to actuate the relay. As light is reduced, the forward bias also decreases, lowering the collector current and deactuating the relay.

This basic circuit can be used to provide automatic control of house lights or an electric sign. At nightfall, light striking the photocell is reduced so that the relay deactuates. The relay contacts can be arranged so that deactuation turns on the lights. At dawn, as light striking the cell is increased, the relay is actuated, removing power to the lights.

The circuit of Fig. 7-20b is the photoconductive version. Here, the amount of base-emitter forward bias is controlled (rather than produced) by the light striking the cell. However, operation of the circuit is the same.

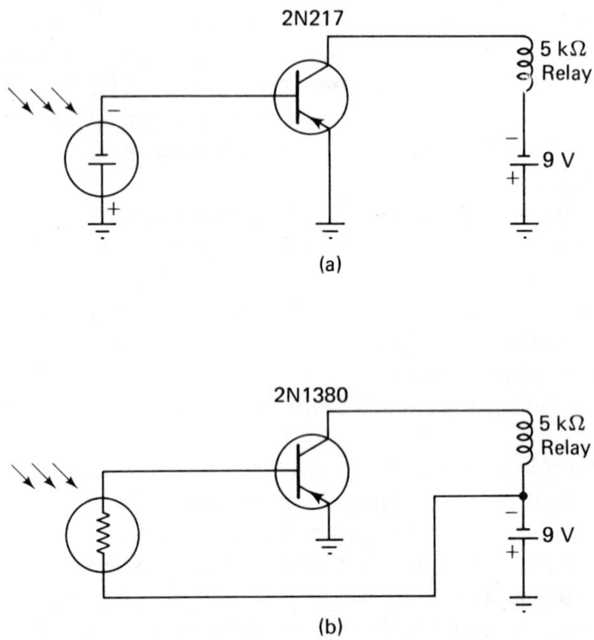

Figure 7-20. Basic photocell circuits.

7-3-5 Typical Light-activated Power Control Circuits

Figure 7-21 shows two basic light-activated power control circuits. In the case of the *light-presence detector* (Fig. 7-21a), power to the load is switched on only when light is applied. Capacitor C starts to charge only when light strikes phototransistor Q_1. The rate at which C charges is set by resistor R_1. When C has charged sufficiently, the UJT fires and triggers the SCR, applying power to the load. When light is removed from Q_1, capacitor C is disconnected from the power source and does not charge. Thus, the UJT and SCR do not trigger and power is removed from the load. A typical application from such a circuit is to activate warning or advertising signs at night by the headlights of approaching vehicles.

In the case of the *light-absence detector* (Fig. 7-21b), power to the load is switched on when light is interrupted. As long as there is light on the LASCS, capacitor C is shorted and does not charge. When the light striking the LASCS is removed, capacitor C charges and the SUS anode goes positive on the next positive cycle of the power supply. This triggers the SUS and SCR, applying power to the load. When light is again applied to the LASCS, capacitor C is shorted

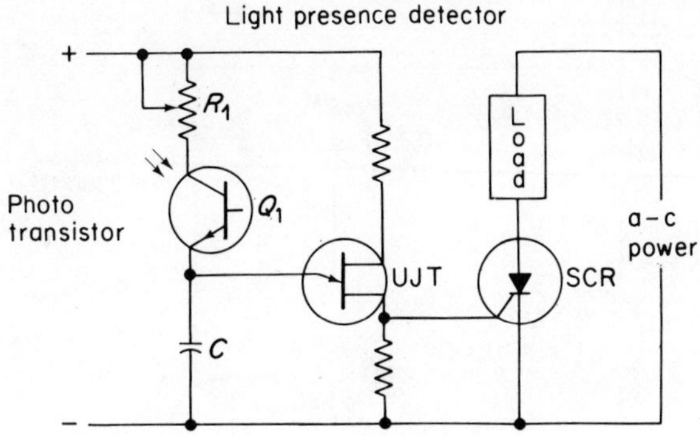

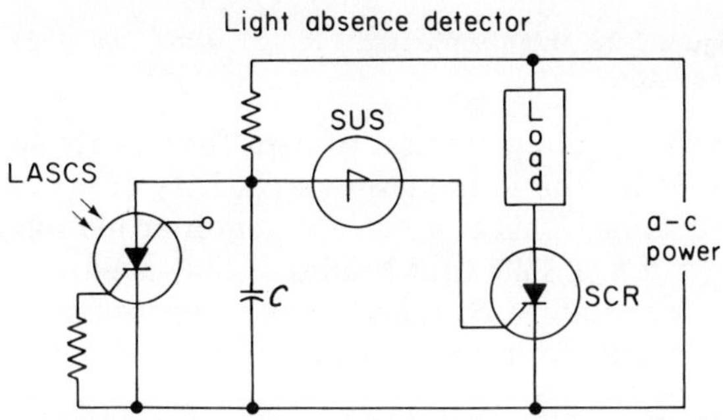

Figure 7-21. Light-actuated control circuits.

and prevented from charging. Thus, the SUS and SCE do not trigger and power is removed from the load. A typical application for such a circuit is to switch on house lights in the evening or whenever it starts to get dark (such as on an overcast day).

7-3-6 Typical Light-operated Relay Circuits Using a Phototransistor

Figure 7-22 shows a circuit in which the presence of light causes a relay to operate. When light is applied to phototransistor Q_1, transistor Q_2 conducts and operates relay K_1. The relay used in this circuit draws about 5 mA when Q_2 is in saturation. The gain for Q_2 is rated at 55 minimum. A base current of 0.5 mA applied to Q_2 will thus provide more than enough current to operate the relay when Q_2 saturates. In theory, when 0.5 mA is applied to Q_2, the Q_2 current through the relay is about 27 mA minimum. Transistor Q_1 provides the 0.5-mA base current if sufficient light is applied.

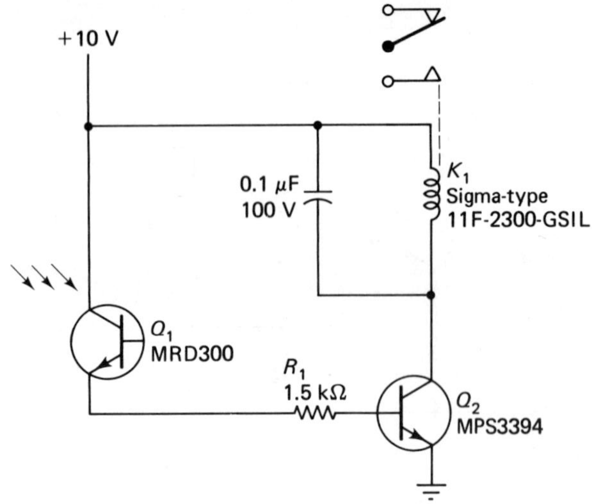

Figure 7-22. Light-operated relay using a phototransistor.

To find the required amount of light for Q_1, divide the required Q_2 base current by the illumination sensitivity of Q_1, which is rated as 4 μA footcandle. This results in a required illumination of 125 footcandles (4 μA $\times$ 125 footcandles = 500 μA or 0.5 mA). This light level can be supplied by a flashlight or any other equivalent light source. (About 200 footcandles can be expected from a flashlight with two D-cells.)

7-4 DIGITAL ELECTRONIC CONTROL CIRCUITS

Many modern electronic control systems involve the use of digital circuits. Since digital devices operate at voltage and signal levels entirely different from those used in power control systems, *interfacing* circuits are usually required between the digital device and the control system. The following paragraphs describe some of the most common interface circuits. If you are not familiar with the digital terms and devices described here, your attention is invited to the author's best selling *Handbook of Digital Electronics* (Prentice-Hall, Inc., Englewood Cliffs, N. J., 1981).

7-4-1 Thyristor-controlled A-C Loads

Figures 7-23 through 7-26 show how CMOS and HTL digital circuits can be interfaced to higher-power a-c loads using thyristors as the a-c control element. (CMOS is complementary metal oxide semiconductor; HTL is high threshold logic.)

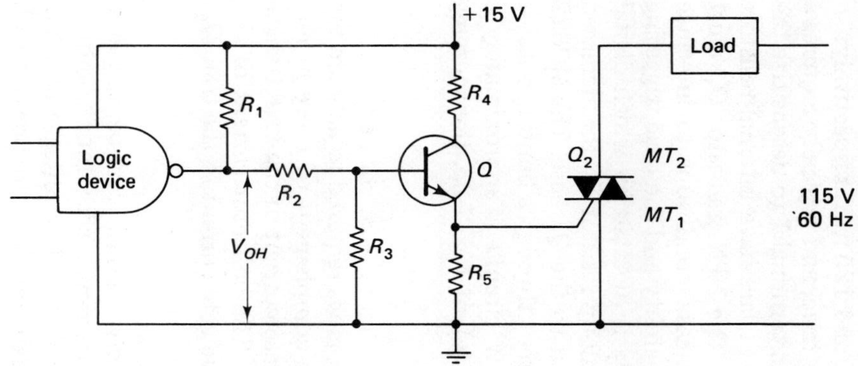

Condition	Logic device	Load	Q_1	Q_2	R_1	R_2	R_3	R_4	R_5	V_{OH}
a	MHTL MC672	300 W	2N4401	MAC10-4	*	2.2 kΩ	10 kΩ	150 2 W	*	9 V
b	MHTL MC670	300 W	2N4401	MAC10-4	5.6 kΩ	†	*	150 2 W	*	1.7 V
c	McMOS MC14011	25 W	2N4401	MAC92A-4	*	12 kΩ	10 kΩ	560 1 W	1 kΩ	14.6 V
d	McMOS MC14011	300 W	2N4401	2N6346	*	2.2 kΩ	10 kΩ	110 4 W	*	12.8 V

*Infinity, open circuit.
†Zero, short circuit.

Figure 7-23. Interfacing between logic and a-c loads, activation high.

Triac control, logic output high. Figure 7-23 shows the interface circuits for high-level activation of both MHTL and McMOS digital gates with triac-controlled a-c loads. (MHTL and McMOS are Motorola designations for HTL and MOS, respectively.) The active logic 1-level (high-level) output of the digital device turns on the effective common-emitter transistor Q_1, the emitter current of which (as determined by collector resistor R_4) triggers the triac. The required trigger current for the three triacs used, MACIO-4, MAC92A-4, and 2N6345, are approximately 90, 25, and 125 mA, respectively.

The MAC92A-4 is a sensitive gate triac, and typically requires a 1-kΩ resistor between the gate and main terminal 1 to desensitize the triac to noise transients. The other two triacs have internal resistors (built-in) and require no desensitizing. The NPN transistor Q_1 must have adequate current gain for worst-case conditions to interface between the digital device sourcing capability and the triac triggering requirement. The input network of Q_1 is dictated by which digital device is used. The MC670 requires a pull-up resistor R_1 for increased current sourcing and is directly coupled to the Q_1 base. The MC672 and MC14011 can source adequate current through a limiting resistor R_2. The sourcing current in the 300-W circuits is approximately 3 mA. For the 25-W circuit, condition (c) of Fig. 7-23, the sourcing current is about 1 mA.

Triac control, logic output low. Figure 7-24 shows the interface circuits when an active low-level logic is required to activate the load. These circuits are essentially the complement of those in Fig. 7-23, in that the logic zero sinks the base current of the PNP transistor Q_1 and drives Q_1 into saturation. The collector current of Q_1, determined by R_3, supplies the positive gate current to the triac Q_2, turning on both Q_2 and the load.

Triac control with negative supply. Triacs can be triggered with both positive and negative gate currents, with four combinations of gate current and main terminal 2 (MT_2) voltages possible. The most sensitive mode is when the gate and MT_2 are positive with respect to MT_1 (this is known as quadrant 1). The least sensitive condition is when the gate is positive and MT_2 is negative (known as quadrant 4).

When gate sensitivity for logic current drive capability is a criterion for switching an a-c load, it is recommended to use negative gate trigger current with its associated nominal sensitivity (known as quadrants 2 and 3). The circuits of Fig. 7-25a and b show both high and low CMOS logic activating a 25-W load. To achieve negative gate current (current flowing out of the gate) requires that the logic and

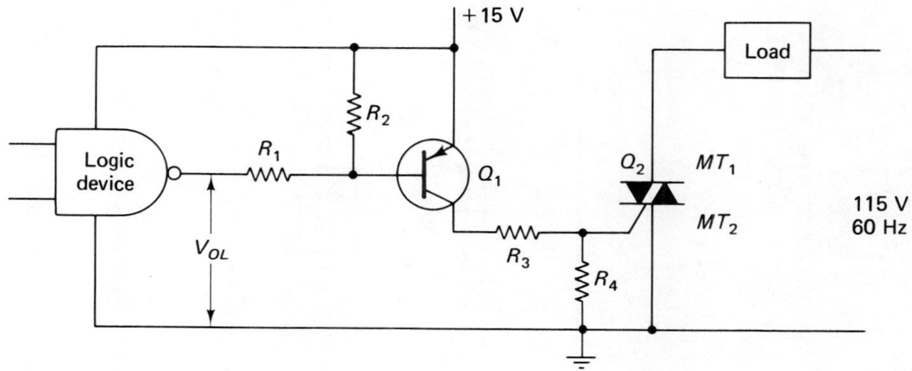

Condition	Logic device	Load	Q_1	Q_2	R_1	R_2	R_3	R_4	V_{OL}
a	MHTL MC672	300 W	MPS3638A	MAC10-4	2.2 kΩ	10 kΩ	150 2 W	*	1.1 V
b	MHTL MC670	300 W	MPS3636A	MAC10-4	2.2 kΩ	*	150 2 W	*	0.25 V
c	McMOS MC14011	25 W	MPS3638A	MAC92A-4	12 kΩ	10 kΩ	560 1 W	1 kΩ	0.15 V
d	McMOS MC14011	300 W	2N4403	2N6346	3.9 kΩ	10 kΩ	110 4 W	*	0.50 V

*Infinity, open circuit.

Figure 7-24. Interfacing between logic and a-c loads, activation low.

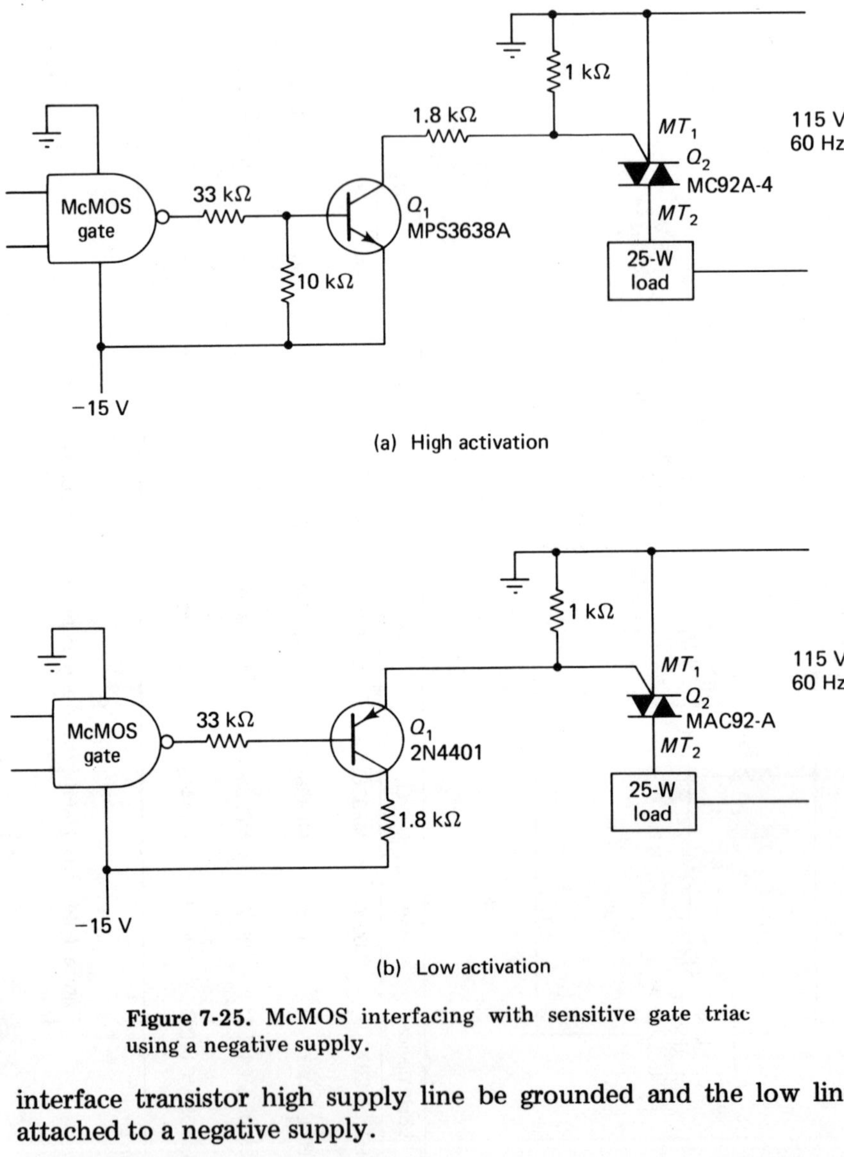

Figure 7-25. McMOS interfacing with sensitive gate triac using a negative supply.

interface transistor high supply line be grounded and the low line attached to a negative supply.

Triac control using capacitor storage. The relatively large power-supply drain of the Fig. 7-25 circuit can be excessive in some applications. The circuit of Fig. 7-26, using capacitor storage techniques, overcomes this problem. Storage capacitor C_2 completely charges to V_{CC} (+15 V) in five R_3C_2 time constants. Shortly after C_2 is charged, Q_1 is fired by the positive-going, differentiated pulse derived from the input square wave. Capacitor C_2 is discharged

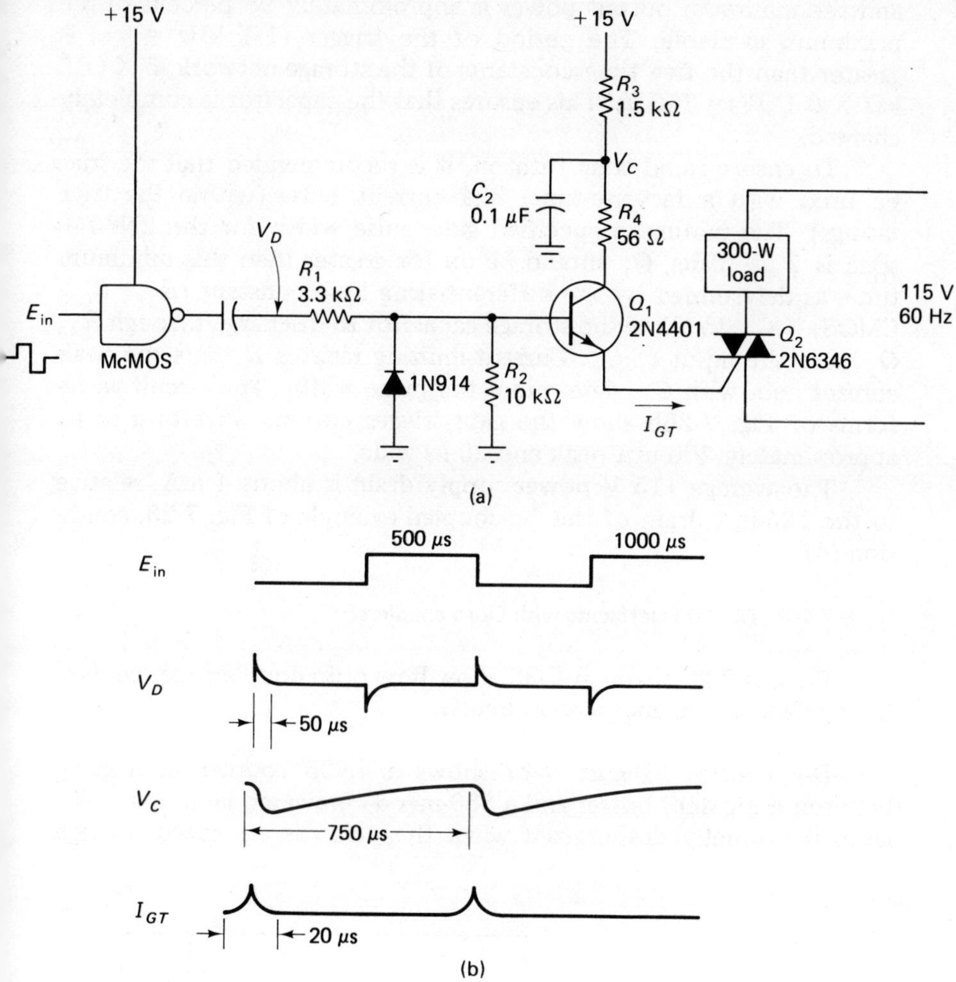

Figure 7-26. McMOS driving triac using capacitor storage.

through R_4 and Q_1, causing triac Q_2 to fire and energize the a-c load.

To get maximum power to the load, the triac should be fired early in its conduction angle. As an example, if the triac is fired at a conduction angle of 18°, greater than 99 percent of the power is delivered; at 30°, approximately 97 percent. When the input square-wave trigger is 1 kHz, the greatest change in condition angle on a cycle-to-cycle basis for a 60-Hz load is approximately

$$\text{conduction angle} = \frac{360^\circ}{1\ \text{kHz}/60\text{Hz}} \quad \text{or} \quad 21.6^\circ$$

and the minimum output power is approximately 98 percent of the maximum available. The period of the trigger (1/1 kHz = ms) is greater than the five time constants of the storage network, 5 × (1.5 kΩ × 0.1 μF) = 750 μs. This ensures that the capacitor is completely charged.

To ensure rapid triac turn on, it is recommended that the triac be fired with a fast-rise-time, high-current pulse (within the triac ratings). The minimum specified gate pulse width for the 2N6346 triac is 2 μs. Thus, Q_1 should be on for greater than this minimum time, as determined by the differentiating time constant ($R_1 + R_{ON}$ CMOS) × C_1, to allow the storage capacitor to discharge through R_4, Q_1, and the input of Q_2. Current-limiting resistor R_4 sets the peak current and, with C_2, determines the pulse width. The circuit waveforms of Fig. 7-26b show the gate trigger current waveform to be approximately 200 mA peak and 20 μs wide.

The average +15 V power supply drain is about 4 mA, relative to the 125-mA drain of the d-c-coupled example of Fig. 7-23, condition (d).

7-4-2 Digital Interfacing with Opto-couplers

Figures 7-27 through 7-30 show how opto-couplers can be used to interface digital and power circuits.

D-c control. Figure 7-27 shows a 4N25 coupler interfacing between a McMOS buffer and a 350-mA d-c-powered lamp load. The lamp is normally deenergized when the 4N25 is energized. A high

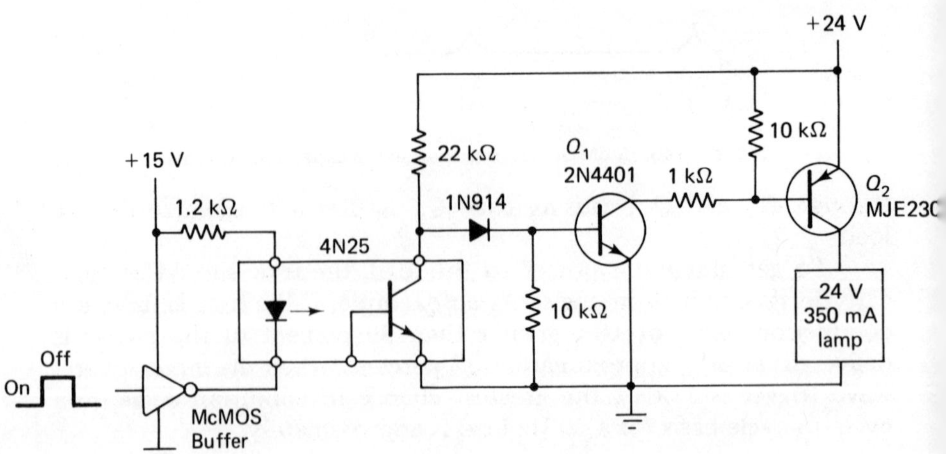

Figure 7-27. D-c control using a 4N25 coupler.

level on the input of the McMOS inverter energizes the 4N25, clamping the base-emitter junction of Q_1 off. Transistor Q_1 removes drive to Q_2, thus deenergizing the load. When a logic 0 is applied to the 4N25, the clamp to Q_1 is removed and Q_2 is driven into conduction through the load, thus energizing the load. With the values shown, about 10 mA of input coupler current can control a completely isolated 350-mA load.

Half-wave a-c control. Figure 7-28 shows a circuit for isolated, half-wave a-c control using the 4N26 coupler. The detector is used as a static series switch in the gate circuit of the 2N5064 SCR. When the logic input goes high, the coupler is energized, allowing the 2N5064 SCR principal current to flow. This current turns on the power SCR 2N6402 and the load for that positive half-cycle of the a-c line voltage. Conversly, when the logic goes low, the coupler is deenergized, removing the 2N5064 gate current. The load current ceases at the next zero excursion of the a-c source. Thus, with only about 5 mA of isolated d-c control current, a half-wave load of up to 16 A can be controlled using the 2N6402.

Full-wave normally off control. Figure 7-29 shows a circuit for isolated full-wave control of an a-c load using the 4N26 and the 2N6071B sensitive-gate triac. For full-wave operation of the 2N5064, a bridge rectifier (MDA920-1) is required to convert the bidirectional

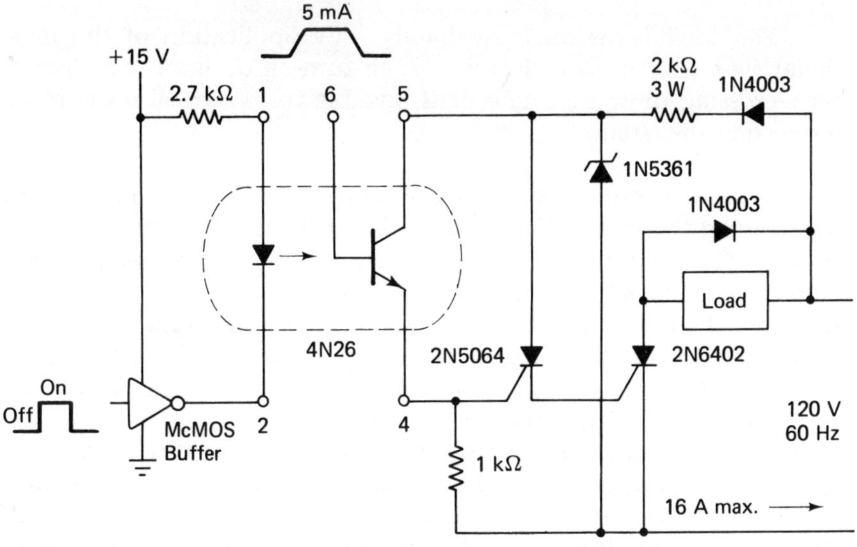

Figure 7-28. Half-wave a-c control using 4N26 coupler.

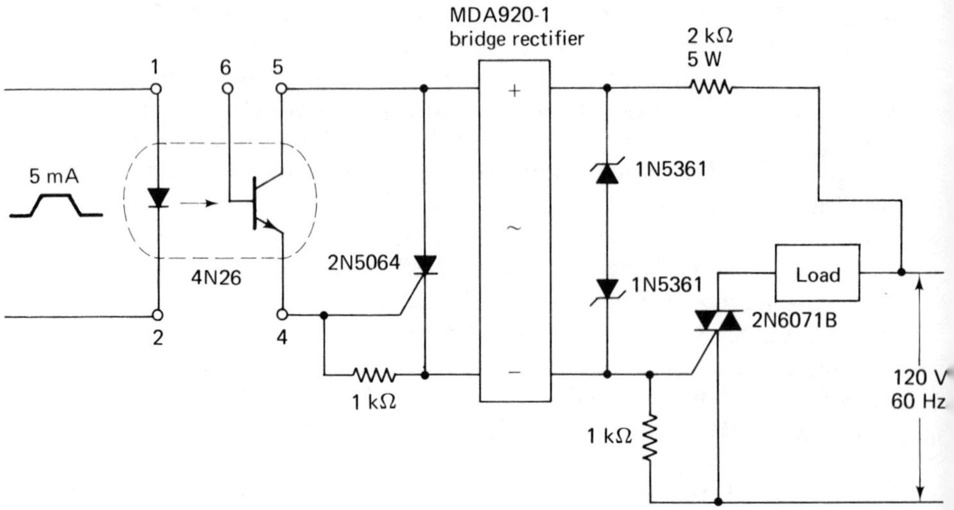

Figure 7-29. Full-wave a-c control using 4N26 coupler and 2N6071B sensitive-gate triac.

a-c line voltage into a unidirectional (full-wave rectified) voltage. The SCR is placed across the bridge d-c output. When the SCR conducts, a patch for the triac gate trigger current is completed through the bridge and the SCR. Thus, a high-level logic input energizes the opto-coupler, which triggers the SCR and provides gate current for the triac.

The load turns on immediately after application of the logic signal (less propagation delays). When turn-on occurs other than at zero crossing of the a-c line, it is possible for electrical noise to be induced in the circuit.

Full-wave zero-crossing control. Figure 7-30 shows a circuit for overcoming the induced electrical noise problem. The circuit consists of triac Q_1, with the trigger circuit R_1, C_1, S_1 and a clamp circuit consisting of the 4N26, 2N5064 SCR and bridge rectifier. To energize the load, the logic input goes low, the coupler is deenergized, and the clamp is removed from across the trigger capacitor C_1, allowing C_1 to charge through timing resistor R_1. When the voltage across C_1 reaches the triggering voltage of S_1 (approximately 8 V), S_1 fires, allowing C_1 to dump the charge into the gate of the triac. This turns on both the triac and the load. Capacitor C_1 is chosen small enough to fire the triac early in the conduction cycle (near zero crossing), thus minimizing electrical noise and maximizing power delivered to the load.

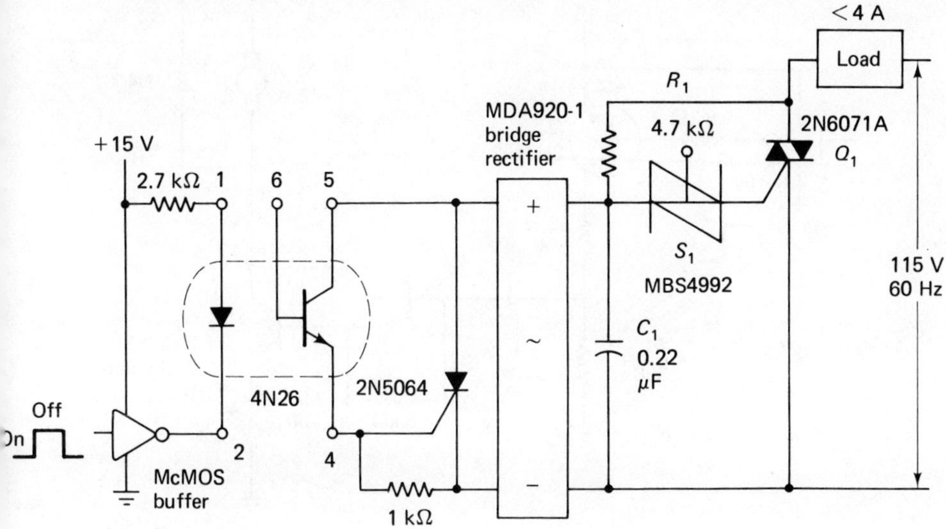

Figure 7-30. Full-wave zero-crossing control interface circuit.

To deenergize the load, the logic high input energizes the coupler, which, with the SCR and bridge rectifier, clamps the trigger capacitor, thus inhibiting the triac from firing.

7-4-3 Power-transistor-controlled D-C Loads

Industrial digital devices can interface with any power load level, assuming that the circuit has the required number of stable power gain stages. Previous examples described low d-c power circuits and thyristor-controlled a-c circuits. By cascading additional power-gain stages to the low-level examples, higher-power loads can be readily driven. The circuits of Fig. 7-31 through 7-33 describe the interfacing between industrial digital devices, using both low- and high-level activation and a 24-V 18-A d-c motor.

Low-level activation. Figure 7-31 shows how low-level logic output activation of the motor circuit can be accomplished with McMOS hex/buffers. When the output is low, the buffer sinks the approximate 8-mA base current for Q_1. Transistor Q_1 then furnishes base current for the Darlington Q_2.

High-level activation. Figure 7-32 shows high-level activation of the motor circuit. The circuit of Fig. 7-32 minimizes current drain and is essentially a complement of the Fig. 7-31 circuit. In the circuit

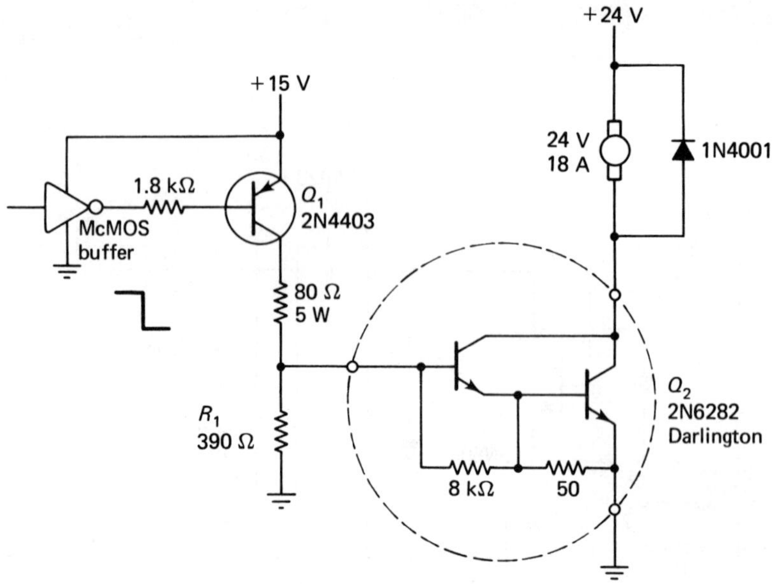

Figure 7-31. McMOS low-level activation of motor load.

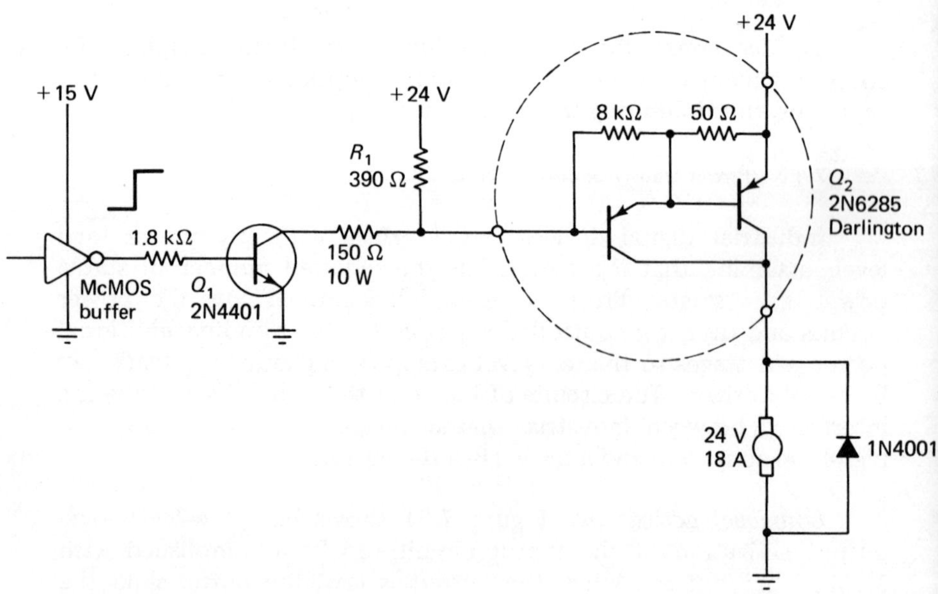

Figure 7-32. McMOS high-level activation of motor load.

of Fig. 7-32, the buffer sources the 8-mA base current to Q_1, which, in turn, sinks the 150-mA base current to Q_2. This current is furnished by the +24-V motor supply, thus reducing the drain on the digital supply by about 150 mA. For high-temperature operation (where junction temperatures exceed 100°C), the addition of R_1 across the Darlington input is recommended. This additional leakage current shunt ensures that Q_2 will not go into thermal runaway condition.

Clamped Darlington motor control. Figure 7-33 shows another motor control circuit. The approach here is to supply drive current through a pull-up resistor, with motor control obtained by means of a transistor clamp across the input of the Darlington. (The previous circuits supplied Darlington base current through a switched transistor.) The clamp driver (Fig. 7-33a) is a passive pull-up gate. The clamp (Fig. 7-33b) is a high-current driver.

The circuit of Fig. 7-33 uses two discrete transistors for the Darlington. The MJE205 (a 5-A plastic-case transistor) drives the 2N5301 (a 30-A TO-3-case transistor). An input current of about 100 mA is required by the Darlington. When the shunt clamp is deenergized, input current flows, the Darlington saturates, and the motor is powered.

The clamp, when energized (motor deenergized), must be able to sink the input current and maintain a saturation voltage less than the on-voltage of the Darlington. Both circuits of Fig. 7-33 meet these requirements. Diode D_1 is used for high-temperature operation to further ensure proper clamping.

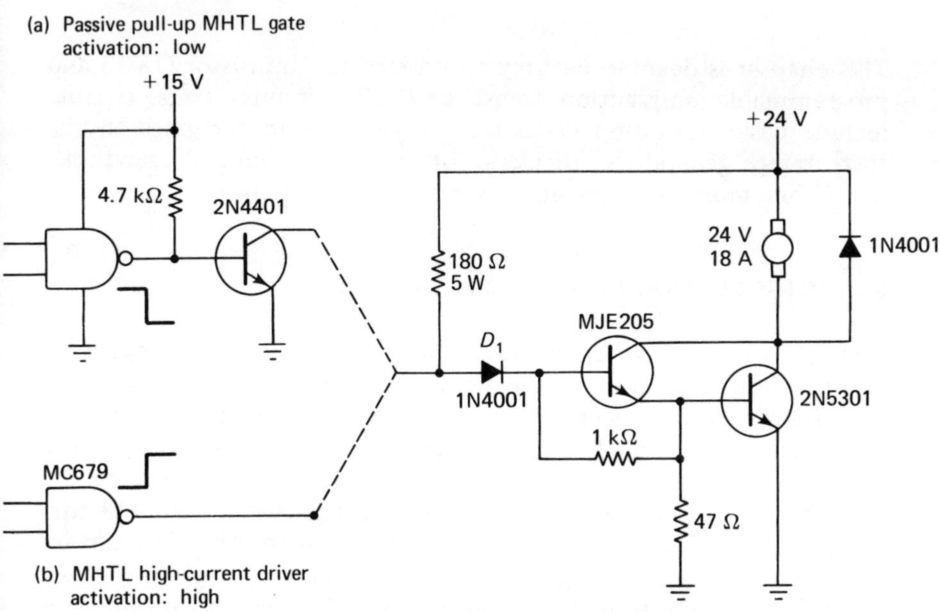

Figure 7-33. Clamped Darlington motor control circuit.

Chapter 8

Unijunction and Programmable Unijunction Transistor Circuits

This chapter is devoted entirely to unijunction transistor (UJT) and programmable unijunction transistor (PUT) circuits. These circuits include basic relaxation oscillators, trigger circuits for gated thyristors, ramp generators, dividers, timers, phase control, sawtooth oscillators, multivibrators, and regenerative amplifiers.

8-1 INTRODUCTION TO UJT AND PUT DEVICES

The UJT operates on an entirely different principle from that of the two-junction transistor or the field-effect transistor. The UJT is a *negative-resistance* device. That is, under the proper conditions, the input voltage or signal can be decreased; yet the output or load current increases. Once the UJT is turned on, it will not turn off until the circuit is broken or the input voltage is removed. For this reason, the UJT makes an excellent trigger source. The UJT can be biased just below the firing point. When a small trigger voltage (either intermittent or constant) is applied, the UJT fires and produces a large output voltage pulse or signal that remains on until the circuit is broken (by switching off the base voltage).

The PUT is a four-layer device similar to an SCR (Chapter 7) except that the anode gate (rather than the cathode gate) is brought out. The PUT is normally used in conventional UJT circuits.

The characteristics of both UJT and PUT devices are similar, but the triggering voltage of the PUT is programmable and can be set by an external resistive voltage-divider network. The PUT is faster and more sensitive than the UJT. The PUT finds limited application as a phase-control element and is most often used in long-duration timer circuits. In general, the PUT is more versatile and is a more economical device than the UJT and can replace the UJT in many applications.

Although the basic function of the UJT is that of a switch, a relaxation oscillator (Chapter 4) is the primary building block in most UJT circuits. This basic relaxation oscillator provides phase control of thyristors such as SCRs and triacs. For that reason, much UJT theory is based on the relationship to relaxation oscillator principles.

8-1-1 Basic UJT Theory

Figure 8-1 shows the standard UJT symbol with appropriate terms for current and voltage. As shown, the UJT is a three-terminal device. The three terminals are emitter (E), base 1 (B_1), and base 2 (B_2). Note that the base 1 and base 2 leads are shown at right angles to the base because they are nonrectifying (not a junction, as in two-junction transistors and JFETs). The emitter connection is represented by an arrow, however, because it is a rectifying PN junction and the arrow is slanted to indicate the emitting properties of the junction. The arrowhead is pointing toward the base, showing a P-type emitter and an N-type base.

Figure 8-1 also shows a static emitter-characteristic curve for a single value of voltage between base 1 and base 2 (V_{B1B2}). The emitter curve is not drawn to scale (to show the different operating regions in more detail). The region to the left of the peak point is called the *cutoff region.* The emitter junction is reverse-biased in most of the cutoff region but is slightly forward-biased at the peak point. The region between the peak point and the valley point, where the emitter junction is forward-biased and conductivity modulation takes place, is called the *negative-resistance* region. The region to the right of the valley point, where the emitter current is limited only by the saturation resistance, is called the *saturation region*. The curve

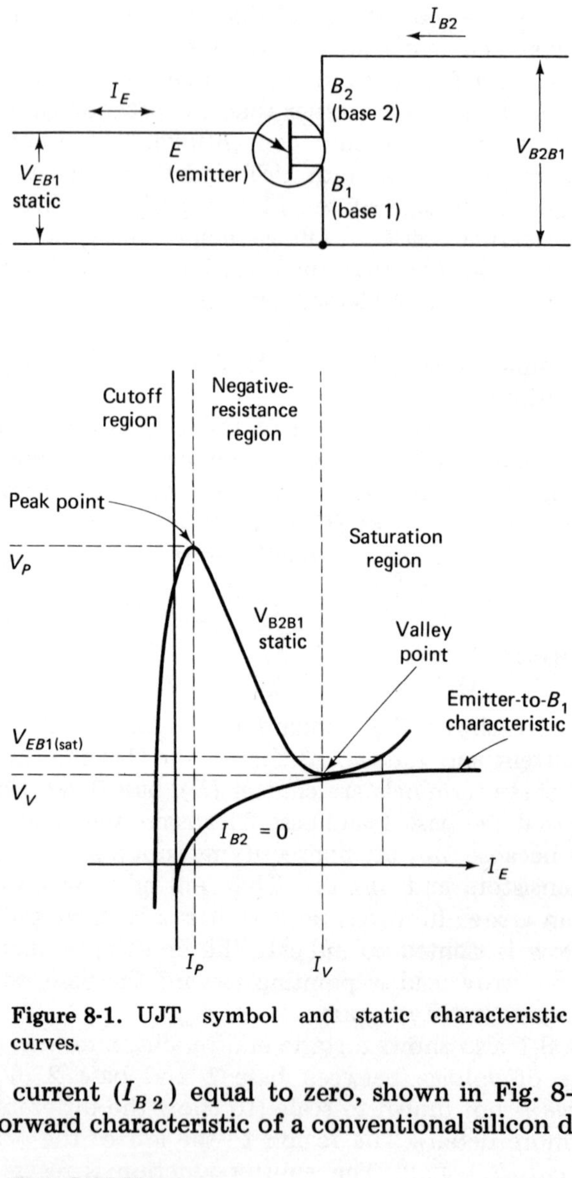

Figure 8-1. UJT symbol and static characteristic curves.

for base 2 current (I_{B2}) equal to zero, shown in Fig. 8-1, is essentially the forward characteristic of a conventional silicon diode.

8-1-2 Basic PUT Theory

Figure 8-2 shows the symbol and a transistor equivalent circuit for a PUT. As shown, the PUT has three terminals: an anode (A), a gate (G), and a cathode (K). As seen from the equivalent circuit, the PUT is actually an anode-gated SCR. This means that if the gate is made negative with respect to the anode, the PUT switches from a

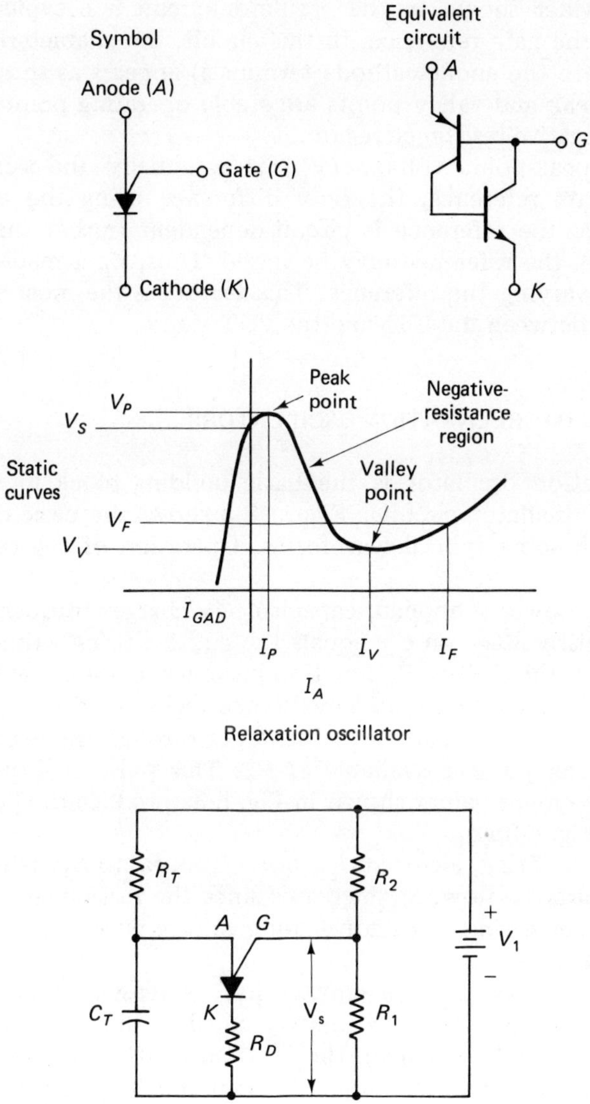

Figure 8-2. Symbol, equivalent circuit, static curves, and basic relaxation oscillator circuit for PUT.

blocking state to an ON state. Since the PUT is normally used as a UJT, the UJT terminology is often used to describe PUT parameters. However, in order to operate a PUT as a UJT, it is necessary to maintain an external reference voltage at the PUT gate terminal.

Figure 8-2 also shows the PUT connected in a typical relaxation oscillator circuit, as well as typical characteristics for the PUT. The voltage divider shown in the oscillator circuit is a typical way of obtaining the gate reference. In this circuit, the characteristic curve (looking into the anode-cathode terminals) appears as shown in Fig. 8-2. The peak and valley points are stable operating points at either end of a negative-resistance region.

The peak-point voltage (V_P) is essentially the same as the external gate reference, the only difference being the gate diode drop. Since the reference is circuit-dependent (rather than device-dependent), the reference may be varied. Thus, V_P is made programmable by varying the reference. This feature is the most significant difference between the UJT and the PUT.

8-2 BASIC UJT RELAXATION OSCILLATORS

The relaxation oscillator is the basic building block in most UJT timer and oscillator circuits. Figure 8-3 shows the basic circuit, together with some typical waveforms. Operation of the circuit is as follows.

When power is applied, capacitor C_E charges through the resistor R_E until voltage on C_E equals the emitter firing voltage V_P. At this voltage, the emitter B_1 junction becomes forward-biased and the emitter goes into the negative-resistance region of the characteristic curve (Fig. 8-1). Capacitor C_E discharges through the emitter and a positive-going pulse is available at B_1. This pulse is shown in Fig. 8-3b. (The circuit values shown in Fig. 8-3a produce the pulse values shown in Fig. 8-3b-d.)

Prior to firing, a current I_{B2} flows from B_2 to B_1. When emitter current starts to flow, I_{B2} increases since the resistance from B_2 to ground is decreasing. A negative-going voltage pulse appears at B_2, as shown in Fig. 8-3c.

When voltage at the emitter has decreased to peak emitter voltage V_P (a voltage approximately equal to the valley voltage V_V when R_1 is purely resistive), the UJT turns off if R_E meets certain conditions. C_E starts to charge up again and the cycle repeats. The waveform that appears at the emitter is shown in Fig. 8-3d.

8-2-1 Design Considerations for Basic UJT Oscillators

The period of oscillation (and thus the frequency) of the UJT oscillator is determined primarily by the values of R_E and C_E, as shown by the equations of Fig. 8-3. It is most practical to start design by selecting a trial value for R_E rather than for C_E because R_E must meet certain conditions for the oscillator to operate. If R_E

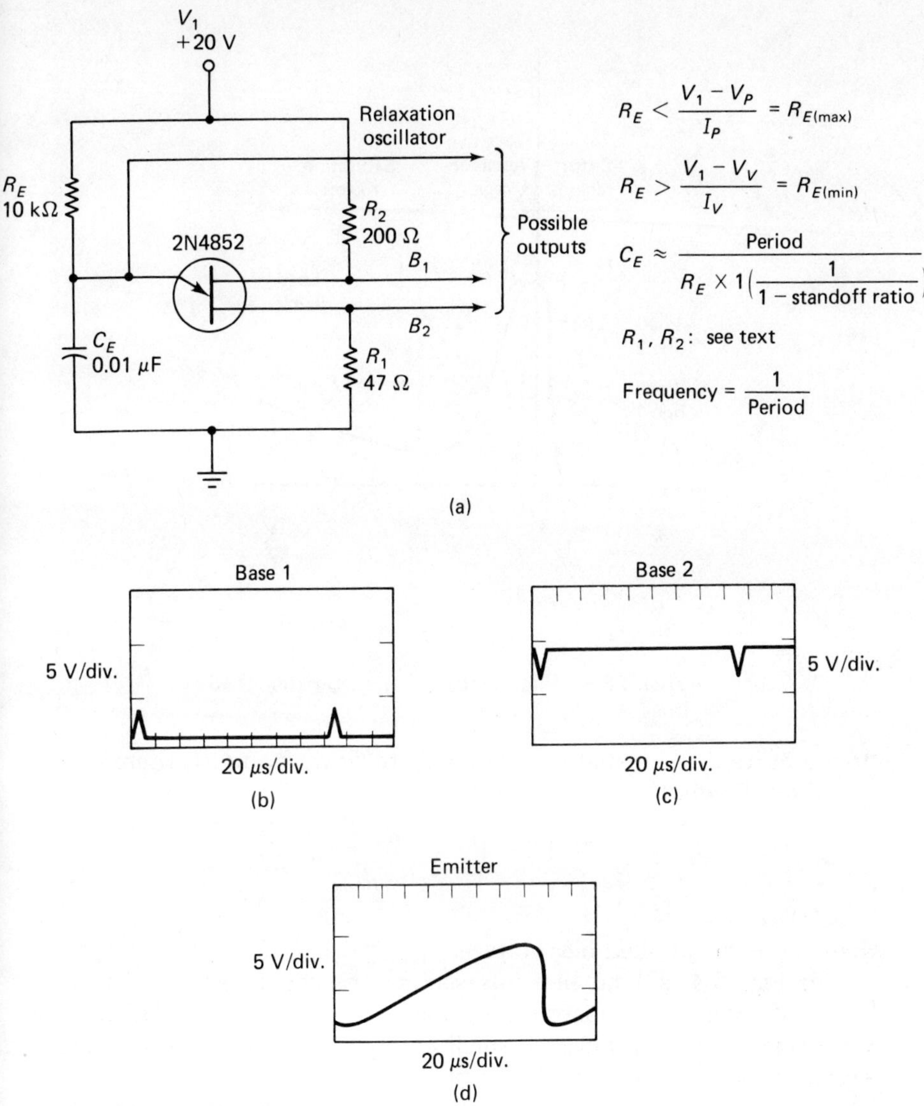

Figure 8-3. Basic UJT relaxation oscillator.

is too large, the UJT never fires; if R_E is too small, the UJT never turns off (until power is removed). These conditions can best be explained by means of the emitter-characteristic curve in Fig. 8-4. (Again, this curve is not drawn to scale in order to show more detail.)

Selecting emitter resistor R_E. As discussed, the emitter capacitor C_E charges until the emitter voltage is equal to V_P. At this point on the characteristic curve, peak-point emitter current I_P flows and, in order to fire the UJT, the value of R_E must be small enough to

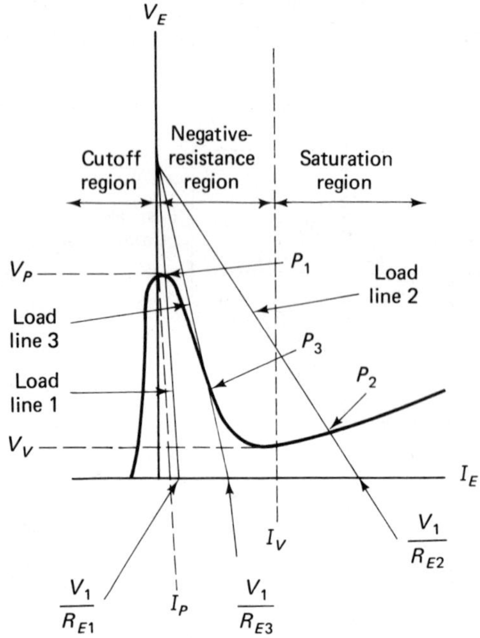

Figure 8-4. UJT emitter characteristic load lines.

allow a current somewhat larger than I_P to flow. R_E must, therefore, meet the following requirements:

$$R_E < \frac{V_1 - V_P}{I_P} = R_{E\,(\mathrm{max})}$$

where V_1 is the applied bias voltage.

In Fig. 8-4, a load line intersecting the characteristic curve in the cutoff region, as illustrated by load line 1, keeps the UJT from ever firing. Thus, R_E must be small enough to keep the UJT in the negative-resistance region.

By keeping R_E smaller than $R_{E\,(\mathrm{max})}$, the UJT turns on and C_E discharges through the emitter. If R_E is too small, however, an emitter current larger than the valley current I_V flows, but the UJT will not turn off. Under these conditions, the UJT operates in the saturation region, as illustrated by load line 2 in Fig. 8-4. The minimum R_E that can be used to assure oscillation is set by the following condition:

$$R_E > \frac{V_1 - V_V}{I_V} = R_{E\,(\mathrm{min})}$$

where V_1 is the applied bias voltage.

An emitter resistance R_E selected to meet the requirements of Fig. 8-3 for $R_{E\,(min)}$ and $R_{E\,(max)}$ results in a load line that intersects the characteristic curve somewhere in the negative resistance region. This is illustrated by load line 3 in Fig. 8-4. In a practical UJT circuit, however, the emitter-voltage variation in the neighborhood of the valley point is small. Thus, to assure turnoff, the value of R_E should be two or three times larger than $R_{E\,(min)}$.

Selecting base 2 resistor R_2. The primary function of R_2 (Fig. 8-3) is to provide temperature compensation. Practically all UJT characteristics are temperature-dependent, some more than others. The interbase resistance and emitter reverse current increase; whereas the peak and valley voltage (and current), the UJT standoff ratio, and the junction diode drop decrease with increasing temperature. The term *standoff ratio* is a characteristic of the UJT and is found by the equation

$$\text{standoff ratio} = \frac{V_P - V_D}{V_{B2B1}}$$

where V_D is the forward voltage drop of the emitter junction. Generally, the oscillator frequency is the main factor affected by these temperature variations. Output voltage can also be affected, however, and is of some concern in many applications.

Since both V_D and standoff ratio decrease with temperature, V_P also decreases. This is, of course, a very undesirable condition in many applications, particularly in timers and oscillator circuits. The change in V_P can be compensated for by adding a resistor R_2 in series with the base 2 terminal. If R_2 is selected properly, V_P can be made to vary less than 1 percent over a 50°C temperature variation. The value of R_2 is best selected by using data-sheet curves. As an alternative, an approximate or first-trial value for R_2 can be found by the equation

$$R_2 = 0.015 \times V_1 \times \text{interbase resistance} \times \text{standoff ratio}$$

Selecting base 1 resistor R_1. The primary function of R_1 (Fig. 8-3) is to provide an output load for the oscillator. In some applications, R_1 is included to provide a path for the interbase current. Such a path is necessary in some circuits to prevent current from flowing through the device being driven by the UJT oscillator. Typically, R_1 is less than 100 Ω, but it can be as high as 2 or 3 kΩ. The value of R_1 is most often selected on the basis of output voltage, but can be selected on the basis of a current path for the external circuit.

When R_1 is selected on the basis of output voltage, a minimum output voltage is usually specified. For example, if the UJT oscillator is to trigger an SCR or other thyristor device and the SCR requires a minimum of 3 V for the trigger, the peak drop across R_1 must be 3 V.

When R_1 is selected on the basis of a current path for an external device, a maximum voltage drop is usually specified. For example, if the UJT oscillator is to trigger an SCR (from an output pulse across R_1) and the fixed SCR voltage must not exceed 50 mV (between firing cycles), the drop across R_1 must not exceed 50 mV. Assuming a 2.5-mA interbase current (and the 50-mV maximum voltage drop, the value of R_1 is 20 Ω.

Some UJT data sheets include curves for calculating the best value of R_1 to meet various circuit conditions. These curves should be used when available, keeping in mind the old electronics rule: "when all else fails, follow instructions."

Selecting emitter capacitor C_E. The value of capacitor C_E is determined by the desired period (or frequency) of operation. The equation for C_E shown in Fig. 8-3 is *very* approximate since it does not take into account the on- and off-times of the UJT or other characteristics of the circuit, such as source voltage V_1, voltage drop V_D, valley voltage V_V, standoff ratio, or interbase voltage V_{B2B1}. However, the period equation of Fig. 8-3 is accurate enough for a first trial value.

8-3 UJT TRIGGER CIRCUITS FOR GATED THYRISTORS

UJTs provide a simple, convenient means for obtaining a thyristor trigger pulse that is synchronized to the a-c line at a controlled phase angle (Chapter 7). This phase control system, using UJTs and thyristors, is one of the most common means of controlling the flow of power to electric motors, lamps, and heaters.

With a-c voltage applied to such circuits, the thyristor (SCR, triac, etc.) remains in its OFF state for the first portion of each half-cycle of the power-line voltage. Then, at a time (phase angle) determined by the control circuit, the thyristor switches on for the remainder of the half-cycle. By controlling the phase angle at which the thyristor is switched on, the *relative power* in the load may be controlled.

UJT trigger circuits are based on the simple relaxation oscillator circuit of Fig. 8-5. Note that this circuit is essentially the same as the basic UJT oscillator of Fig. 8-3, with two major exceptions. In the Fig. 8-5 circuit, the source voltage V_S is "controllable" in that it can be turned on and off by switch S_1. Second, the base 1 output is applied to the gate of a thyristor, which, in turn, is in series with a load.

The R_T and C_T in Fig. 8-5 form the timing network that determines the time between the application of voltage to the control

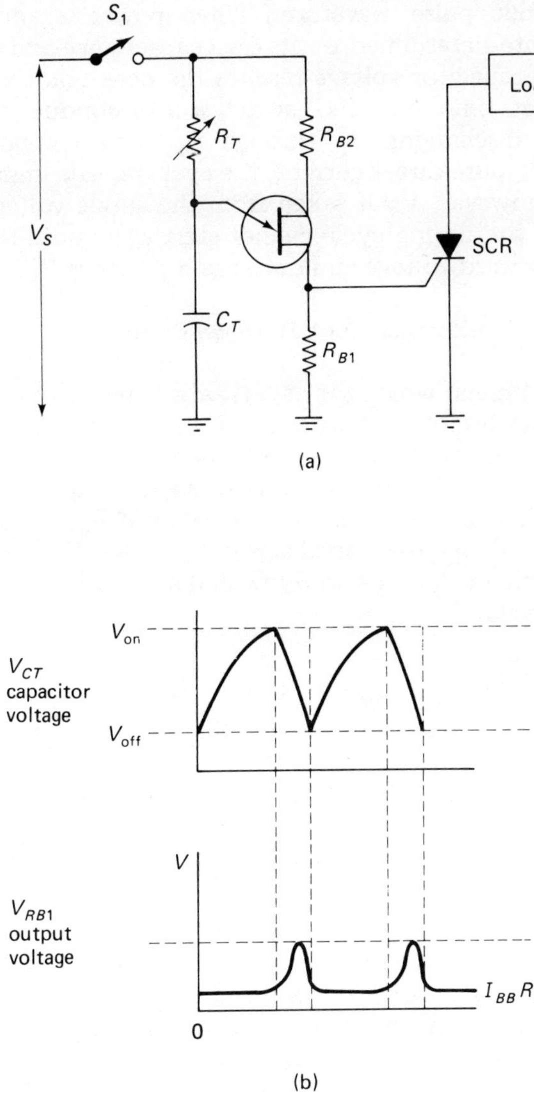

Figure 8-5. Basic UJT trigger circuit.

circuit (represented by the closing of S_1) and initiation of the pulse to the thyristor. In the circuit of Fig. 8-5, with V_S pure direct current, the UJT oscillator is free-running. R_T and C_T determine the frequency of oscillation. The peak of the output pulse voltage is clipped by the forward conduction voltage of the gate-to-cathode diode in the thyristor.

Operation of the circuit may best be understood by referring to the capacitor voltage waveforms shown in Fig. 8-5b, together with the base 1 output pulse waveform. When power is applied, C_T charges, at the rate determined by its own capacitance and the value of R_T, until the capacitor voltage reaches the peak-point voltage of the UJT. At that time, the UJT switches into conduction in the normal manner, discharging C_T through R_{B1} and the gate of the thyristor. With V_S pure direct current, the cycle repeats immediately. In many cases, however, V_S is taken from the anode voltage of the thyristor so that the timing cycle cannot start again until the thyristor is blocking forward voltage and once again provides V_S.

8-3-1 Design Considerations for UJT Trigger Circuits

During the time in which the capacitor is being charged, current flows through the interbase resistance (R_{BB}) of the UJT. R_{B1} is included in the circuit to provide a path for this current so that the current does not flow through the thyristor gate. This can cause an undesirable turn-on of the thyristor. The value of R_{B1} is selected so that a maximum voltage developed across R_{B1} is less than 0.2 V. For a typical UJT with an R_{BB} of 4 to 9 kΩ, and a typical operating voltage of 20 V, the value of R_{B1} is

$$R_{B1} = \frac{0.2R_{BB\,(\text{min})}}{V_S} = \frac{0.2 \times 4\ \text{k}\Omega}{20} = 40\ \Omega$$

As discussed in Sec. 8-2-1, some UJT data sheets specify a recommended value of base 1 resistance for a specific type of thyristor. This recommended value should always be used as the first trial value.

Base 2 resistance R_{B2} is necessary only if some degree of temperature compensation is required. The value of R_{B2} is found by using the equation or data-sheet discussed in Sec. 8-2-1. In the case of a simple power control circuit, particularly in the feedback systems described in following paragraphs of this section, the base 2 resistance may be left out. In such systems, the base 2 of the UJT is connected directly to the positive side of V_S.

It is often necessary to synchronize the timing of the output pulses to power-line voltage zero-crossing points. One simple method to accomplish such synchronization is shown in Fig. 8-6. Zener diode D_1 clips the rectified supply voltage, resulting in a V_S as shown in Fig. 8-6b. Since V_{B2B1} (and the peak voltage of the UJT) drops to zero each time the line voltage crosses zero, C_T is discharged at the end of every half-cycle and begins each half-cycle in the discharged state.

With the circuit of Fig. 8-6, even if the UJT has not triggered during one half-cycle, the capacitor begins the next half-cycle discharged so that the phase angle at which the pulse occurs is directly controlled for each cycle (by the values of C_T and R_T). The zener diode also provides voltage stabilization for the timing circuit, giving the same pulse phase angle regardless of normal line-voltage fluctuations.

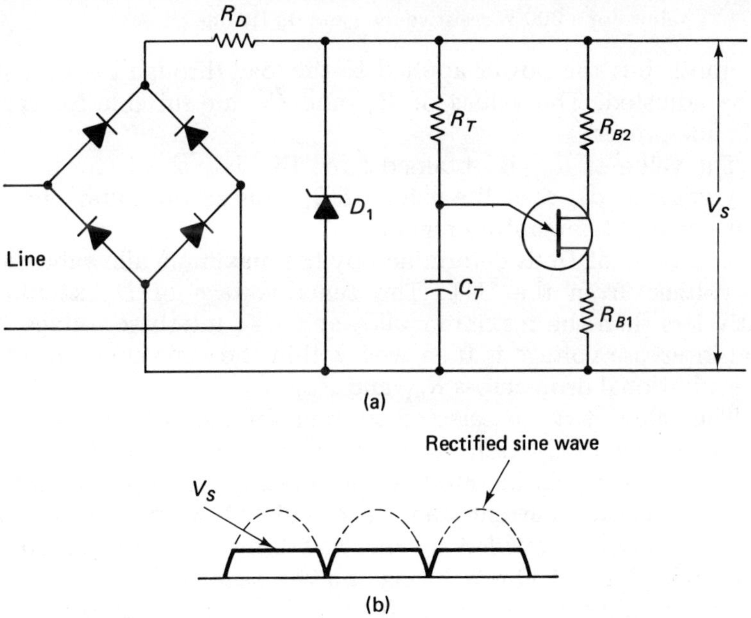

Figure 8-6. Basic control circuit for synchronizing a UJT trigger.

8-3-2 Basic Half-Wave UJT Trigger Circuit

The most elementary application of the UJT trigger is a half-wave circuit shown in Fig. 8-7. The values of R_T and C_T set the frequency of operation. Note that R_T is made variable so that the phase

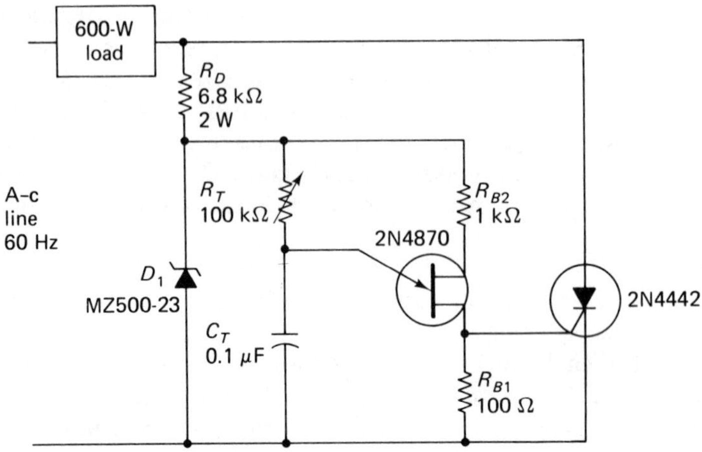

Figure 8-7. Half-wave UJT control circuit with typical values for a 600-W resistive load and 60-Hz line.

angle (and thus the power applied to the load through the thyristor) can be adjusted. The values of R_T and C_T are suitable for typical 60-Hz line power.

The value of R_{B1} is obtained from the data sheet (for the type of thyristor used), as is the value of R_{B2} (as an optimum value for the anticipated temperature range).

The value of D_1 is determined by the maximum allowable interbase voltage from the UJT. The zener voltage of D_1 should be slightly less than the maximum allowable UJT interbase voltage. The actual interbase voltage is then well within the allowable range (due to the additional drop across R_{B1} and R_{B2}).

The value of R_D is selected to limit the current through D_1 so that the diode dissipation capability is not exceeded. Since D_1 is conducting the regulating mode only during positive half-cycles, the allowable positive current can be calculated when the allowable diode dissipation is divided by one-half the zener voltage. For example, if D_1 has a 150-mW rating and the zener voltage is 24 V, the allowable positive current is 12.5 mA ($24 \times 0.5 = 12$; $0.150/12 = 0.0125$). If the zener data sheet shows a specific current, usually listed as I_{ZT} or some similar term, use that figure for I_{positive}.

Once the positive half-cycle current is determined, the value of R_D is found by subtracting 0.7 times the zener voltage from the root-mean-square (rms) line voltage and dividing the result by the positive current. For example, assume that the zener voltage is 24 V, line voltage is 115 V, and I_{positive} is 15 mA. R_D is then $115 - (0.7 \times 24)/0.015 = 6.8$ kΩ (or the nearest standard value).

Although only the positive half-cycles need be considered in calculating the resistance value, the power rating of R_D must be calculated on the basis of full-wave conduction. (D_1 is conducting on the negative half-cycle acting as a shunt rectifier, as well as providing V_S on the positive half-cycle.) Using the example of 6.8 kΩ and 1.5 mA, the minimum power rating of R_D is $(I^2R) \times 15^{-3} \times 15^{-3} = 225^{-6}$; $6800 \times 225 \times 10^{-6} = 1.53$ W; use a 2-W value for R_D to provide safe power dissipation.

8-3-3 Providing Half-Wave Control with Switching for Full-Wave Operation

In the circuit of Fig. 8-7, the thyristor is acting both as a power control device and a rectifier, providing variable power to the load during the positive half-cycle and no power to the load during the negative half-cycle. The circuit of Fig. 8-7 is designed to be a two-terminal control that can be inserted in place of a switch.

If full-wave power is desired at the upper extremes of this control, a switch can be added that will short-circuit the SCR when R_T is turned to its maximum power position. The switch may be placed in parallel with the SCR if the load is resistive. If the load is inductive, the load must be transferred from the SCR to the direct line, as shown in Fig. 8-8.

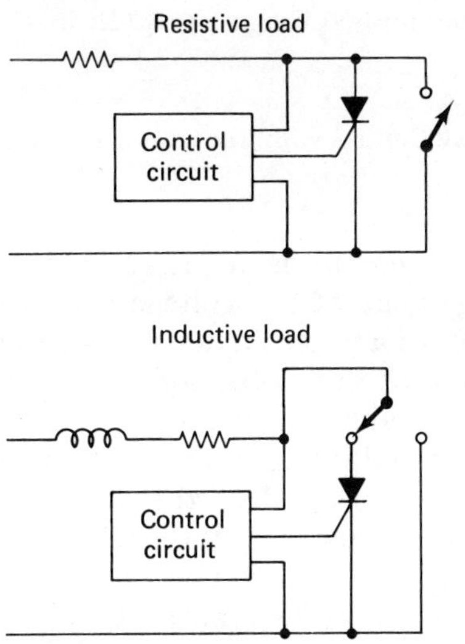

Figure 8-8. Half-wave controls with switching for full-wave operation.

8-3-4 Basic Full-Wave Control with UJT Trigger

Full-wave control may be realized by the addition of a bridge rectifier or a pulse transformer and by changing the thyristor from an SCR to a triac, as shown in Fig. 8-9. In this circuit, R_{B1} is not necessary because the pulse transformer isolates the thyristor from the steady-state UJT current.

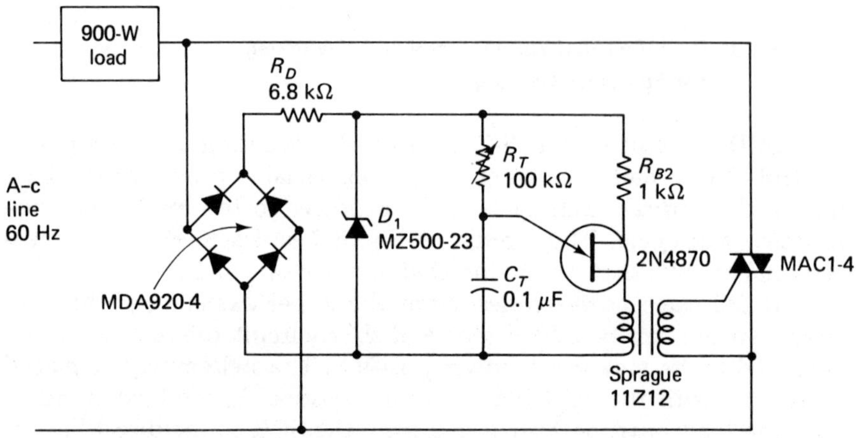

Figure 8-9. Simple full-wave UJT trigger circuit with typical values for 900-W load and 60-Hz line.

8-3-5 Full-Wave Control with Constant UJT Trigger Output

Occasionally, a circuit is required that will provide constant output voltage regardless of line-voltage changes. This can be accomplished by addition of a potentiometer in series with R_D as shown in Fig. 8-10. The value of this potentiometer P_1 should be 10 percent (or less) of the R_D value. In operation, P_1 is adjusted to provide reasonably constant output over the desired range of line voltage. As line voltage increases, so does the voltage on the wiper of P_1, increasing interbase voltage (and the peak voltage of the UJT). The increased peak-voltage point results in C_T charging to a higher voltage and thus taking more time to trigger. The additional delay reduces the thyristor conduction angle and maintains the average voltage at a reasonably constant value.

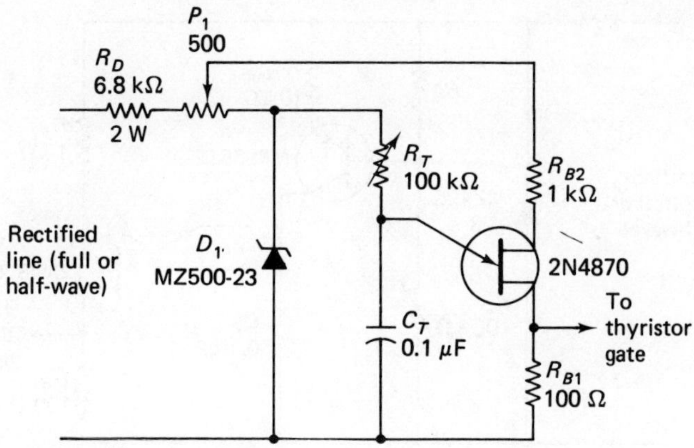

Figure 8-10. UJT trigger circuit with line voltage compensation.

8-3-6 UJT Trigger Circuits with Feedback

The circuits described thus far have been manual controls. That is, the power output is controlled by a potentiometer R_T turned by hand. Simple feedback circuits may be constructed by replacing R_T with heat- or light-dependent sensing resistors. However, such circuits have no means of adjusting the operating levels. By adding a transistor to the basic UJT trigger circuit, it is possible to provide both manual and automatic (feedback) control.

Figure 8-11 shows a feedback control using a sensing resistor R_S for feedback. The sensing resistor may respond to any one of many stimuli, such as heat, light, pressure, moisture, or magnetic field. R_S is the sensing resistor and R_C is the manual control resistor that establishes the desired operating point. Transistor Q_1 is connected as an emitter follower such that an increase in the resistance of R_S decreases the voltage on the base of Q_1, causing more current to flow.

Current through Q_1 causes voltage to charge C_T, triggering the UJT at some phase angle. As R_S becomes larger, more current flows into the capacitor; the voltage builds up faster, causing the UJT to trigger at a smaller phase angle; and more power is applied to the load. Thus, this circuit is for a sensing resistor that decreases in response to excessive power in the load. If the sensing resistor increases with load power, then R_S and R_C should be interchanged.

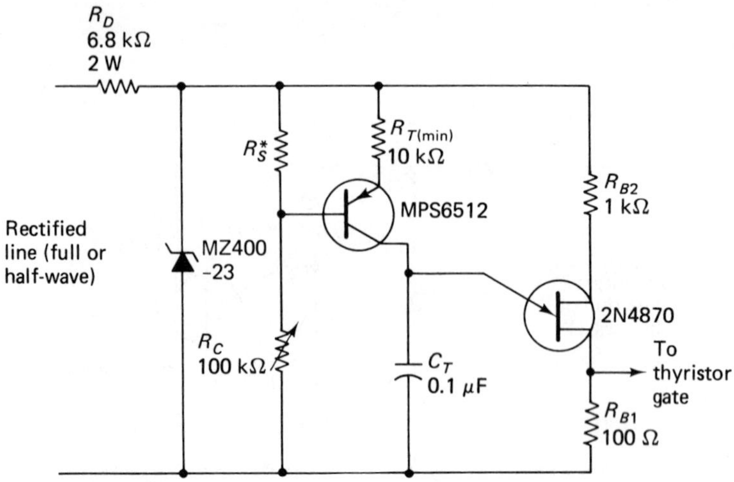

*R_S should be selected to be about 3 to 5 kΩ at the desired output level.

Figure 8-11. UJT trigger with feedback control.

8-4 UJT SAWTOOTH OSCILLATORS

The UJT has certain characteristics that make it well suited for use as a sawtooth waveform source. The voltage waveform at the emitter of the UJT in the basic relaxation oscillator is a fair approximation of a sawtooth waveform. However, if the emitter output of a UJT oscillator is connected directly to a load (either inductive or resistive), the circuit may fail to oscillate. Even if oscillation continues, the waveform will probably be distorted.

The most practical method to couple the emitter output of a UJT oscillator to a load is by means of a direct-coupled emitter follower. Such a circuit is shown in Fig. 8-12. Note that simple direct coupling is made possible by the fact that typical minimum voltage at a UJT emitter $V_{E\,(\min)}$ is about 1.2 V. If $V_{E\,(\min)}$ is less than the normal base-to-emitter drop of the transistor, then the waveform across the load R_L is clipped. However, a typical base-emitter drop of a silicon transistor is about 0.7 V (or considerably less than 1.2 V).

8-4-1 Design Considerations

The primary consideration for the circuit of Fig. 8-12 is the loading effect of the emitter follower. A small amount of loading will shift oscillator frequency. A large amount of loading stops oscillation. To minimize the effects of loading on the frequency, the

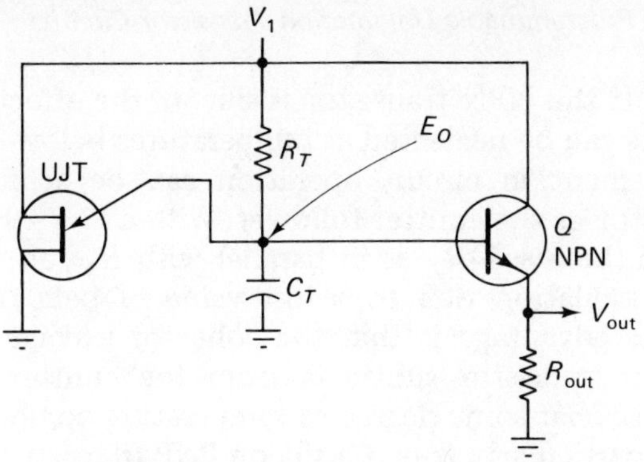

$$\frac{(\text{beta} + 1)R_L}{R_T + (\text{beta} + 1)R_L} > \text{maximum standoff ratio}$$

$$R_T \approx \text{load impedance} \approx 0.1 \text{ to } 0.2 \times (\text{beta} + 1)R_L$$

$$C_T \approx \frac{\text{Period}}{R_T + 1\left(\dfrac{1}{1 - \text{standoff ratio}}\right)}$$

$$V_{out} \approx E_0 \times 0.9 \approx V_1 \times 0.3 \qquad R_{out} = \text{desired output impedance}$$

$$V_1 \approx \frac{V_{out}\ (\text{desired})}{0.3}$$

Figure 8-12. Basic UJT sawtooth oscillator.

values of R_L and the NPN transistor Q_1 beta should be as large as possible. If the value of beta or R_L is too small, the circuit will not oscillate. To ensure oscillation, beta and R_L must satisfy the condition in relation to the maximum standoff ratio shown in Fig. 8-12.

The effects of temperature must also be considered. Two important temperature effects are involved in the use of the emitter-follower stage. First, the variation in beta with temperature changes the loading and thus affects the frequency of oscillation. To minimize this temperature effect, the (beta + 1)R_L factor shown in Fig. 8-12 should be much greater than the value of R_T. The values of C_T and R_T are chosen to provide the desired sawtooth frequency, as shown by the equations of Fig. 8-12. However, design should start with R_T, by making R_T a value between 0.1 and 0.2 times (beta + 1)R_L. If this results in an impractical value for C_T, increase the value of R_T only as necessary to reduce the size of C_T to practical limits.

The second temperature effect results from the collector leakage current of the two-junction transistor Q_1. Note that this current adds to the emitter leakage current of the UJT. Both leakage currents tend to increase the frequency with increasing temperature. The effect of leakage currents on frequency can be minimized by using a large value for C_T. If the NPN transistor is silicon, the effects of the two leakage currents can be neglected at temperatures below 100°C.

Some improvement in circuit operation can be achieved by using a PNP transistor as the emitter follower. With a PNP, the effective load resistance (beta + 1)R_L is in parallel with R_T, so that the possibility of nonoscillation due to a low value of beta or R_L is eliminated. Another advantage is that the collector leakage current of the two-junction transistor subtracts from the emitter leakage current of the UJT so that some degree of temperature compensation is obtained. This is particularly true if a silicon PNP transistor is used.

8-4-2 Design Example of UJT Sawtooth Oscillators

Assume that the circuit of Fig. 8-12 is to provide a sawtooth output of approximately 5 V minimum into a 1-kΩ load. The available power source is 20 V, and a transistor with a beta of 50 is to be used. The UJT has a maximum standoff ratio of 0.7.

The value of R_L is 1 kΩ to match the 1-kΩ load. With (beta + 1)R_L at 51 kΩ, the value of R_T should be between 5.1 and 10.2 kΩ. Assume a value of 5.1 kΩ. Substituting these values for comparison against the maximum standoff ratio, we have

$$\frac{(50+1) \times 1000}{5100 + [(50+1) \times 1000]} = \frac{51{,}000}{59{,}000} \approx 0.9$$

Since 0.9 is greater than the maximum standoff ratio of 0.7, oscillation should be sustained with no difficulty.

With an available source voltage V_1 of 20 V, the V_O output voltage should be approximately 6 V (20 × 0.3), which is greater than the required 5-V output. With 6 V across an R_L of 1 kΩ, there is 6 mA of current through the emitter follower. Of course, the emitter follower must be capable of dissipating this power plus whatever current is passing through the load. With R_T at 5.1 kΩ, select the value of C_T to produce the desired frequency (or period) of operation (as shown by the equations of Fig. 8-12).

8-4-3 Improving Linearity of UJT Sawtooth Oscillators

For many applications, the sawtooth linearity obtained with the basic UJT relaxation oscillator is inadequate. Those UJTs with the lowest standoff ratio produce the most linear sawtooth waveforms. However, 10 percent is about the best linearity that can be obtained from the basic circuit, even with low standoff ratio UJTs. A number of simple circuit techniques can be used to improve linearity of the sawtooth waveform from a UJT oscillator. Some practical circuits are described in the following paragraphs.

High voltage. Figure 8-13 shows the direct approach of using a higher supply voltage for charging the timing capacitor. This is an inexpensive method of improving linearity if the high voltage supply is available. The circuit of Fig. 8-13 has some disadvantage in that frequency is not as stable as with a single power supply.

Charging inductance. Figure 8-14 shows the use of a choke to maintain a constant current for charging the timing capacitor. This circuit requires a time constant for the charging circuit which is

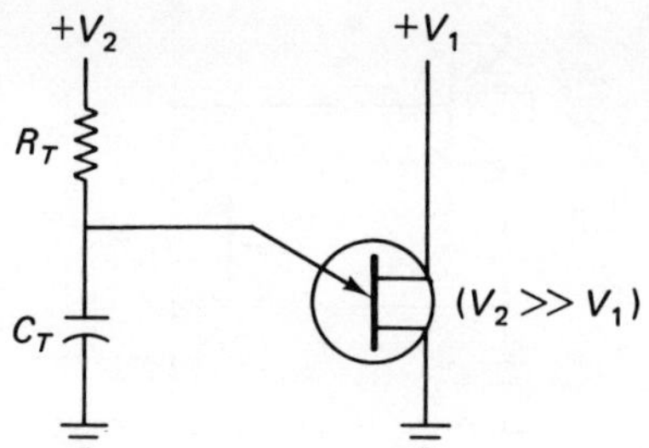

Figure 8-13. Use a high voltage supply for charging of UJT sawtooth oscillator timing capacitor.

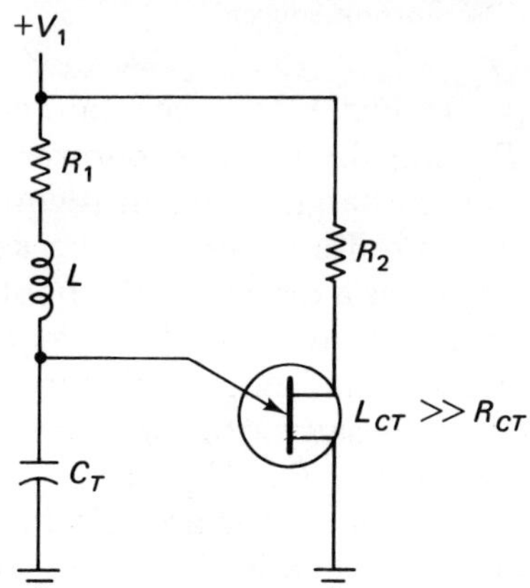

Figure 8-14. Use of charging inductance for maintaining constant UJT sawtooth oscillator charging current.

much greater than the period of oscillation, as shown by the equation. This condition usually produces an impractical size of L at oscillator frequencies of less than 1 kHz.

Collector characteristics. Figure 8-15 shows the use of the high output impedance of a common-base transistor to maintain a constant charging current for the capacitor. The values of R_T and C_T remain the same as for the basic circuit.

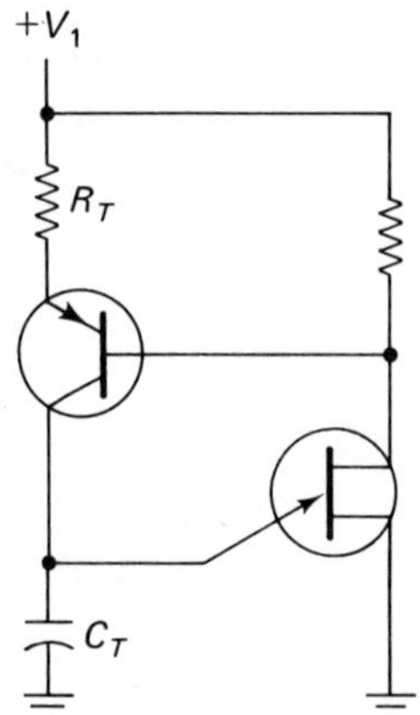

Figure 8-15. Use of collector characteristics for constant UJT sawtooth oscillator current source.

Zener bootstrap. In Fig. 8-16, a constant voltage is maintained across R_3 by zener D_1 and the emitter-follower transistor amplifier stage, so that the capacitor charging current is constant over the complete cycle. This circuit is quite economical because it makes double use of the transistor, both as a driver for the bootstrap circuit and as an output amplifier stage. Note that R_4 is returned to a negative voltage. If R_4 is grounded, the current flowing through D_1 causes clipping at the bottom of the sawtooth waveform.

The values shown for C_1, R_1, R_2, and R_3 provide a sawtooth output at about 2 kHz. These values are selected on the same basis as for the basic circuit. The value is D_1 is 6 V, which is typical for a supply voltage in the range 20 to 25 V. The value of R_4 is chosen to match a given load impedance. However, a change in R_4 value

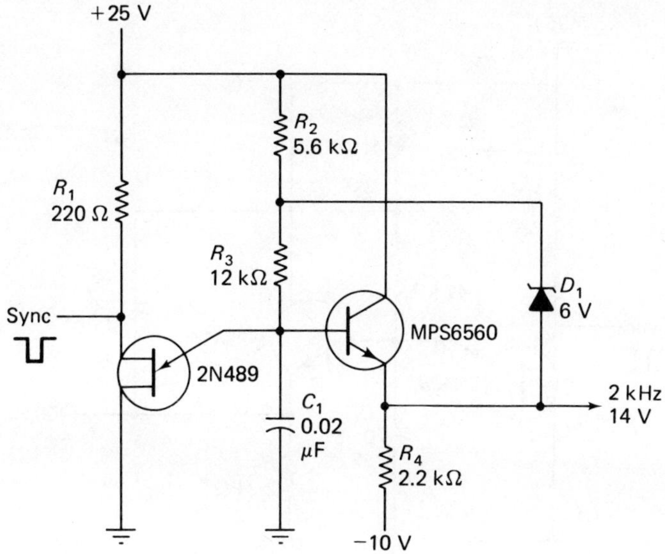

Figure 8-16. Bootstrap UJT sawtooth oscillator circuit with zener diode feedback.

changes the output voltage, all other factors remaining the same. Note that the frequency of the Fig. 8-16 circuit is somewhat dependent upon the supply voltage.

Capacitor bootstrap. In Fig. 8-17, the circuit uses a capacitor C_2 in place of the zener diode. This variation permits the negative supply to be eliminated, thus making frequency largely independent of the supply voltage. In each of the circuits shown in Figs. 8-13 through 8-17, the linearity is limited by the loading effect of the output stage, so that it will not be possible to increase linearity beyond a certain value. This value is set by (beta + 1)R_L.

RC integrator. The circuit of Fig. 8-18 shows a method of compensating for both the loading of the output stage and the variable charging current of the timing capacitor. Resistor R_3 and capacitor C_2 act as an integrating network of the waveform. By varying the values of R_3 and C_2, the output waveform can be made concave upward, concave downward, or linear. In a practical application, the circuit is assembled in experimental form, using a potentiometer for R_3. The output is monitored on an oscilloscope, and the value of R_3 is adjusted to provide the most linear waveform.

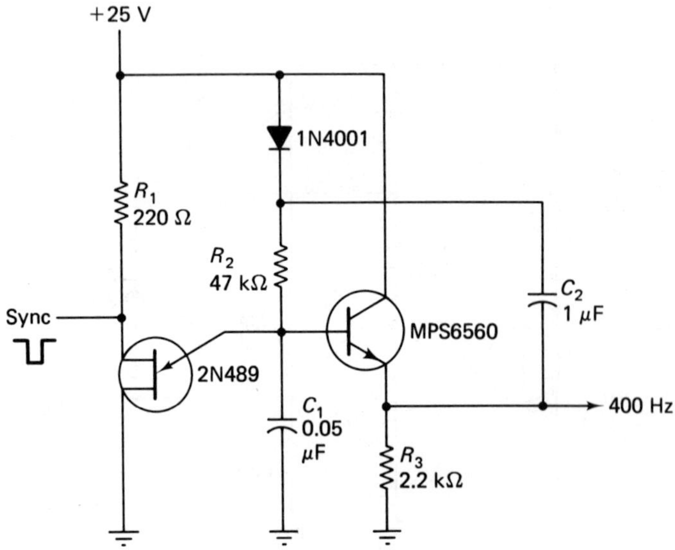

Figure 8-17. Bootstrap UJT sawtooth oscillator circuit with capacitor feedback.

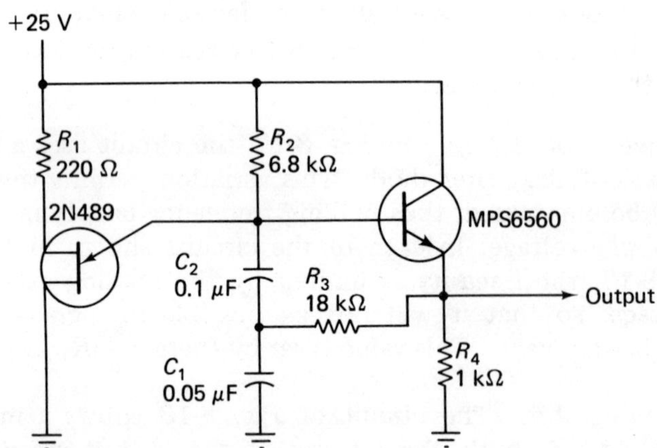

Figure 8-18. Second order UJT sawtooth oscillator compensation circuit with RC integrator.

8-5 UJT MULTIVIBRATORS

It is possible to use a single UJT as a multivibrator (Chapter 4). The basic UJT multivibrator circuit is shown in Fig. 8-19. Note that the circuit is essentially the same as for the basic relaxation oscillator,

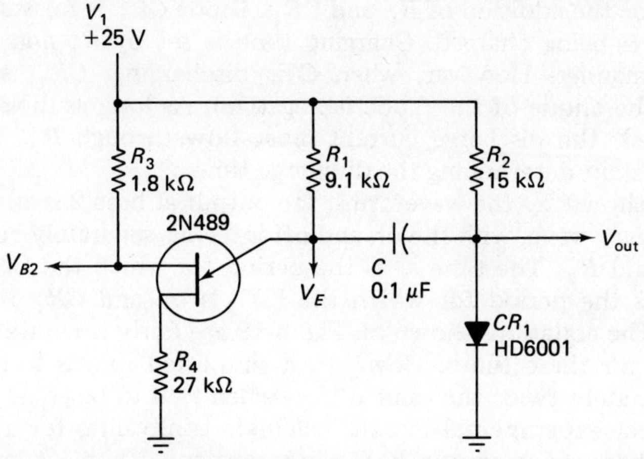

$$t_1 \approx R_1 C \times 1 \left(\frac{V_1 - V_E}{V_1 - V_P}\right)$$

$$t_2 \approx R_2 C \times 1 \left(\frac{V_1 + V_P + V_E}{V_1}\right)$$

When $t_1 = t_2$, then $R_2 \approx 2R_1$

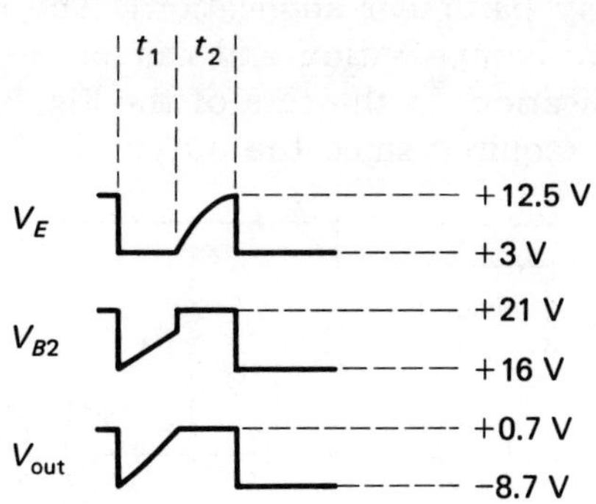

$$V_E \approx \text{emitter voltage measured at } I_E \approx \frac{V_1 (R_1 + R_2)}{R_1 R_2}$$

Figure 8-19. UJT multivibrator circuit.

except for the addition of R_2 and CR_1. Diode CR_1 is forward-biased when C is being charged. Charging time is set by R_1 and C in the normal manner. However, when C is discharging, CR_1 is reverse-biased (the anode of the diode is negative). As long as the capacitor is charged, the discharge current must flow through R_2, with the R_2C constant determining the discharge time.

As shown by the waveforms, the output at base 2 is an approximate square wave, with the on and off intervals separately controlled by R_1 and R_2. The time t_1 is the period for which the UJT is off, and t_2 is the period for which the UJT is on and CR_1 is reverse-biased. The equations shown on Fig. 8-19 are fairly accurate approximations for these times. However, a simplified rule is to make R_2 approximately twice the value of R_1, when t_1 is to be equal to t_2. In a practical experimental circuit, calculate trial values for a given t_1 on the basis of a simple R_1C time constant. Then use a variable resistance for R_2, starting with a value twice that of R_1.

The basic UJT multivibrator may be coupled to a conventional transistor by means of a circuit such as shown in Fig. 8-20. In this circuit, the emitter-to-base diode of the transistor takes the place of the diode in the circuit of Fig. 8-19. The circuit of Fig. 8-20 has the advantage that the load section is completely isolated from the timing section. However, all the timing values remain the same.

Note that in the circuit of Fig. 8-20, both base 1 and base 2 resistors are omitted. The base 1 resistor is not required in either circuit, although the addition of a base 1 resistor to either circuit minimizes excessive emitter current when the UJT fires (if that is a problem with any particular application). The base 2 resistor provides temperature compensation and can be used or omitted to fit a particular application. In the case of the Fig. 8-19 circuit, the base 2 resistor R_3 is required since the output is developed across this

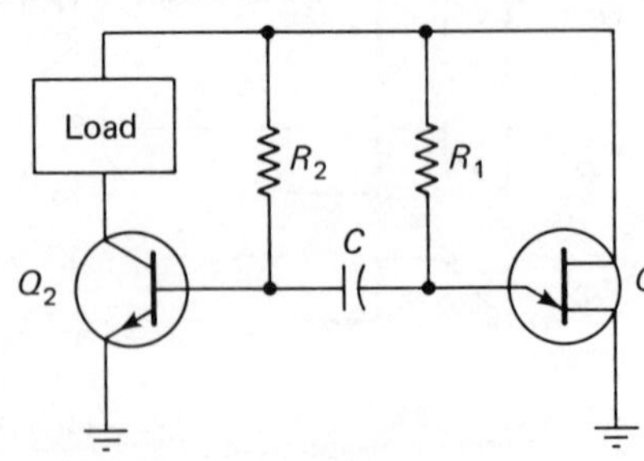

Figure 8-20. UJT multivibrator coupled to load through an NPN transistor.

resistor. In the Fig. 8-20 circuit, output is taken from transistor Q_2, and no temperature compensation is provided.

8-6 UJT REGENERATIVE AMPLIFIERS

The basic UJT relaxation oscillator can also be adapted to form a regenerative pulse amplifier. Such an amplifier circuit is shown in Fig. 8-21. Note that this circuit is essentially the basic relaxation oscillator with two resistors added. Resistor R_L provides a load for a trigger input or a pulse output. Resistor R_3 forms a voltage divider with resistor R_E. The voltage at the junction of R_3 and R_E is set so

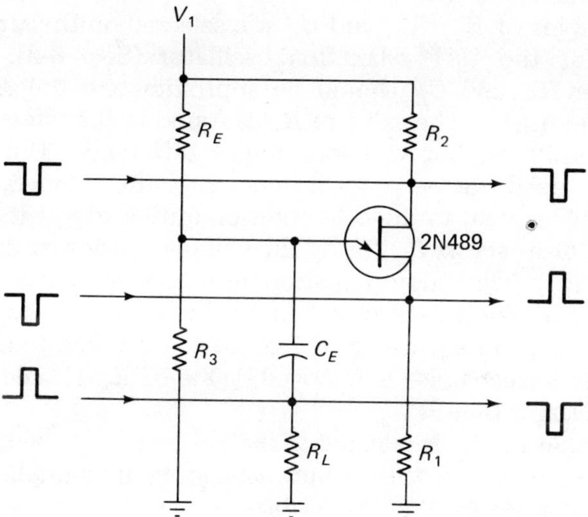

$$R_L \approx Z \text{ of trigger, or } Z \text{ of output load}$$

$$R_3 \approx \left(\frac{R_E}{V_1 - V_{bais}}\right) \times V_{bias}$$

$$R_E >> \frac{V_1 - V_V}{I_V} \qquad V_{bias} = V_P - \text{one-half of trigger}$$

$$C_E \approx \frac{\text{desired period}}{R_E \times 1\left(\frac{1}{1 - \text{standoff ratio}}\right)}$$

Figure 8-21. Basic UJT regenerative amplifier circuit.

that the UJT does not fire except in the presence of a trigger. That is, R_3 and R_E have a ratio such that the emitter voltage does not exceed the peak-point (firing) voltage for the OFF state.

As shown in Fig. 8-21, inputs can be applied to any of three points (base 1, base 2, or emitter). Similarly, outputs can be taken from any of these points. In operation, the trigger pulse is applied to one of the inputs, and a corresponding trigger output (in greatly amplified form) is taken from another of the points. For example, a trigger of 0.1 V can be applied to the emitter, with a 7-V output taken from base 1.

8-6-1 Design Considerations

The values of R_1, R_2, and C_E are selected on the same basis as discussed for the UJT relaxation oscillator (Sec. 8-2). The time constant for R_E and C_E should be approximately the same as the period of the trigger. The value of R_E is found in the same way as for the basic oscillator, but with one minor difference. The maximum value of R_E need not be of great concern in the regenerative amplifier. The UJT is held in the off-condition until a trigger is applied by the voltage drop across R_3 (or by the voltage division at the junction of R_E and R_3). Thus, no calculation need be made for $R_{E\,(\max)}$. The minimum value for R_E is found in the same way as for the basic oscillator, using the equations of Fig. 8-21. However, to assure turn-off after the trigger pulse is removed, value of R_E should be two or three times larger than $R_{E\,(\min)}$.

The value of R_L is selected to match the load impedance of the input trigger source, or the output, depending upon application. The value of R_3 is set by the desired bias voltage. The ratio of R_E and R_3 sets the fixed bias that must be overcome by the input trigger. V_{bias} is set at V_P (peak voltage of the UJT), less one-half of the input trigger.

8-7 TYPICAL PUT CIRCUITS

The following paragraphs describe a few typical PUT circuits. These circuits described by no means represent all the many applications for PUTs. Instead, the circuits given here are chosen because of their universal nature. In general, a PUT can be used almost anywhere a UJT is required. As discussed in the beginning of this chapter, however, PUTs are best suited for timing applications rather than for phase control. For that reason, we shall concentrate on timing and timing-control circuits.

8-7-1 Voltage-controlled Ramp Generators

The PUT provides a simple approach to a voltage-controlled ramp generator (VCRG), as shown in Fig. 8-22. The current source formed by Q_1 in conjunction with capacitor C_1 sets the duration time of the ramp. As the positive d-c voltage at the gate is changed, the peak-point firing voltage of the PUT is changed. In turn, the ramp-time duration is changed. For example, if the peak-point firing voltage is increased, the duration is increased, and vice versa.

Figure 8-22 also shows a plot of voltage versus ramp duration time for a 4700-pF and a 10,000-pF timing capacitor. As shown, it is possible to have a change in frequency (or period) with either a change in C_1 value or a change in gate voltage.

For example, it is possible to have a 3-ms duration ramp with a gate voltage of 8 V and a C_1 of 10,000 pF. The same 3-ms ramp can be obtained with a C_1 value of 4700 pF and a gate voltage of 14 V. From a practical design standpoint, the gate voltage must be kept within the limits specified by the data sheet (5 to 20 V, using the Motorola MPS6516 specified in Fig. 8-22). However, the C_1 value can be changed over a wide range to achieve a given ramp duration. A 10:1 or even a 100:1 time change can be achieved by a corresponding change in C_1 value (using the remaining values shown in Fig. 8-22).

8-7-2 Low-Frequency Dividers

The circuit of Fig. 8-23 is a frequency divider, with the ratio of capacitors C_1 and C_2 determining the division. Operation of the circuit is as follows.

Assume that C_1 equals C_2 (for a division of 2, as shown by the table of Fig. 8-23). Also assume that C_1 and C_2 are discharged and that a pulse is applied to the base of Q_1. Transistor Q_1 turns on and off for each positive pulse. When Q_1 turns off, both C_1 and C_2 charge to 10 V through R_3. On the next pulse to the base of Q_1, C_1 is again discharged (at the end of the pulse) but C_2 remains charged to 10 V (due to the arrangement of D_1 and D_2). On the second pulse, C_2 charges to the peak-point firing voltage of the PUT, causing the PUT to fire. This discharges C_2 and allows C_1 to charge to the line voltage. As soon as C_2 discharges and C_1 charges, the PUT turns off. The next cycle begins with another pulse on the base of Q_1, which again discharges C_1. The input and output frequency can be approximated by the equation of Fig. 8-23.

For an input frequency in the 10-kHz range, pulses of about 3 V, and a C_1 of 10,000 pF, the table of Fig. 8-23 shows the values

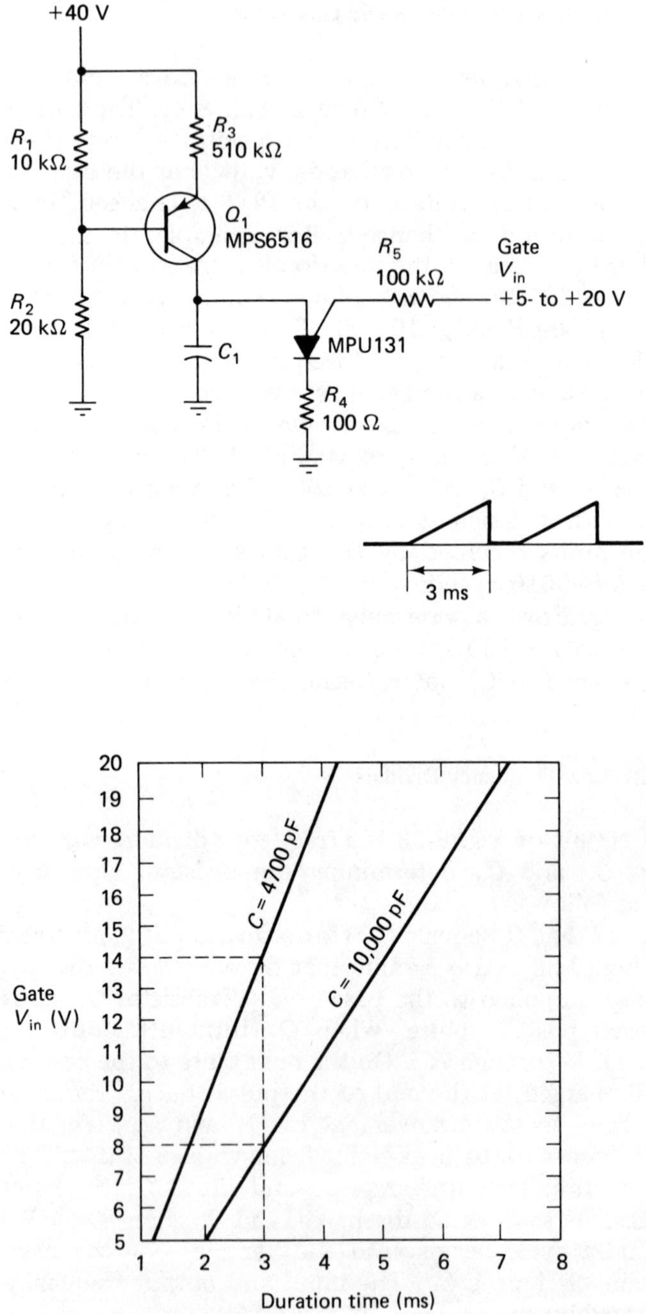

Figure 8-22. Voltage-controlled ramp generator using a PUT.

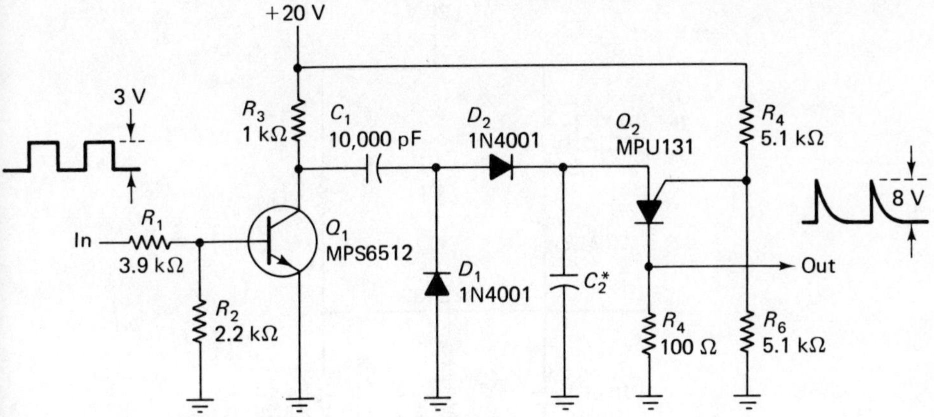

*C_2 (pF)	Division
10,000	2
20,000	3
30,000	4
40,000	5
50,000	6
60,000	7
70,000	8
80,000	9
90,000	10
100,000	11

$$F_{in} \approx \frac{C_1 + C_2}{C_1} F_{out}$$

Figure 8-23. Low-frequency divider using a PUT.

for C_2 to obtain divisions for each whole number between 2 and 11. For example, with C_1 at 10,000 pF and C_2 at 70,000 pF, the circuit divides by 8. With an input of 10 kHz, the output is 1250 Hz. This division range can be changed by using the programmable feature of the PUT and changing the voltage on the gate by changing the ratio $R_6/(R_6 + R_5)$. Decreasing the ratio with a given C_1 and C_2 decreases the division range; increasing the ratio increases the division range. The circuit shown in Fig. 8-23 is relatively insensitive to the amplitude, pulse width, and rise and fall times of the incoming pulses.

8-7-3 Long-Duration Timers

A long-duration timer circuit that can provide a time delay of up to 20 minutes is shown in Fig. 8-24. The circuit is a standard relaxation oscillator with a JFET current source in which resistor

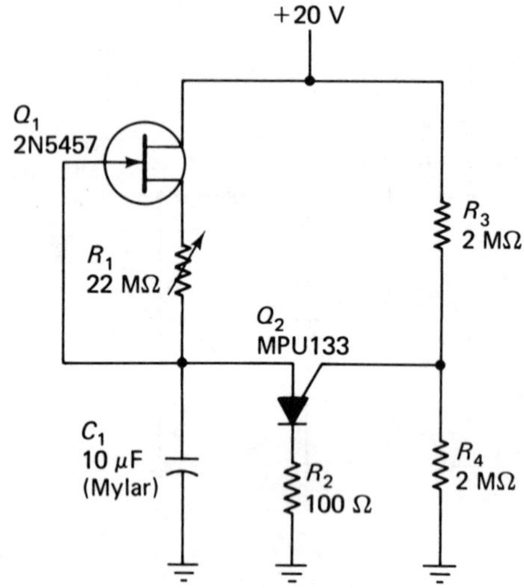

Figure 8-24. A 20-minute, long duration timer using a PUT.

R_1 is used to provide reverse bias on the gate to source of the JFET. This turns the JFET off and increases the charging time of C_1. Capacitor C_1 should be a low-leakage capacitor such as a Mylar type.

Maximum delay time of the circuit is limited by the peak-point firing current (I_P) needed to fire Q_2. For charging currents below I_P there is not enough current available from the current source to fire Q_2, causing the circuit to lock up. Thus, a PUT works better than a UJT in a long-duration timer because the PUT has a lower peak firing current. Also, because of the programmable feature of the

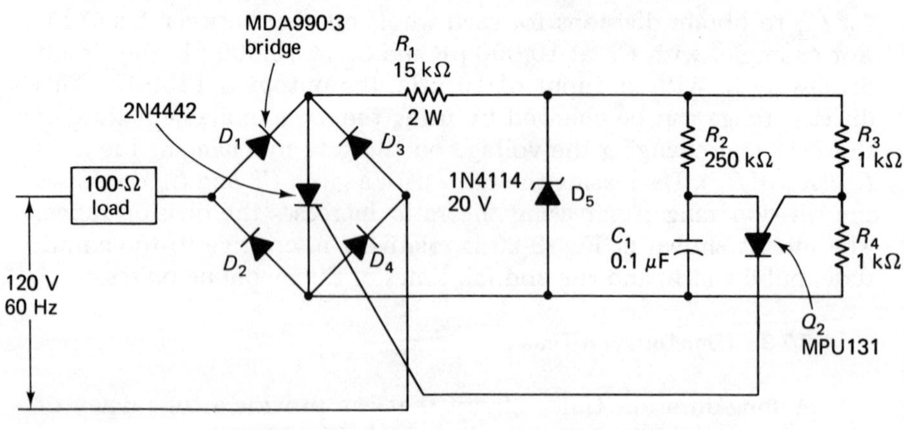

Figure 8-25. SCR phase control using a PUT.

PUT, I_P can be made very small by making R_G (the equivalent parallel resistance of R_3 and R_4) large (1 MΩ), as shown in Fig. 8-24.

8-7-4 Phase Control

Figure 8-25 shows a circuit using a PUT for phase control of an SCR. The relaxation oscillator formed by Q_2 provides conduction control of SCR Q_1 from 1 to 7.8 ms (or from 21 to 168° of conduction angle). This constitutes control over 97 percent of the power available to the load.

Only one SCR is needed to provide phase control of both the positive and the negative portions of the sine wave by putting the SCR across the bridge (diodes D_1 through D_4).

Chapter 9

Operational-Amplifier Circuits

This chapter is devoted entirely to operational amplifier (op-amp) circuits. The designation "operational amplifier" was originally applied to a series of high-performance d-c amplifiers used in analog computers. These amplifiers were used to perform mathematical operations applicable to analog computation (summing, scaling, subtraction, integration, etc.). The op-amp of today is found in IC form and is generally useful as a replacement for any low-frequency amplifier application. For that reason, we concentrate on those circuits that use the IC op-amp as the major component.

It is assumed that you are familiar with op-amp basics, including how to connect power supplies and phase-compensation components to op-amps and how to interpret op-amp data sheets. If these subjects are not familiar, or if you want further information on any phase of op-amps, your attention is invited to the author's best selling *Manual for Operational Amplifier Users* (Reston Publishing Company, Inc., Reston, Va., 1977).

Here, we concentrate on how to select external components so that the IC op-amp can perform a given circuit function. Before doing so, let us very quickly review the basic IC op-amp.

9-1 BASIC IC OP-AMPS

IC op-amps generally use several differential stages in cascade to provide both common-mode rejection and high gain. Thus, IC op-amps require both positive and negative power supplies. The circuit diagram and the equivalent circuit (or symbol) as they appear on the data sheet of a typical IC op-amp are shown in Fig. 9-1. Since a differential amplifier has two inputs, it provides phase inversion for degenerative feedback and can be connected to provide in-phase or out-of-phase amplification.

In a typical application, the op-amp output is fed back to the input through a resistance or impedance. In almost all cases, the output is fed back to the negative or inverting input (terminal 2 in Fig. 9-1) so as to produce degenerative feedback (to provide the desired gain and frequency response). As in any amplifier, the signal shifts in phase as it passes from input (terminal 2) to output (terminal 6). This phase shift is dependent on frequency. When the phase shift approaches 180°, it adds to or cancels out) the 180° feedback phase shift. Thus, the feedback is in phase with the input (or nearly so) and causes the amplifier to oscillate. This condition of phase shift with increased frequency limits the bandwidth of an op-amp. The condition can be compensated for by the addition of a phase-shift network (usually an *RC* circuit, but sometimes a capacitor), connected to terminals 1, 5, or 8 in Fig. 9.1.

9-1-1 Typical IC Op-Amp Operation

The Motorola IC op-amp shown in Fig. 9-1 is a three-stage amplifier with the first stage of a differential-in, differential-out amplifier designed for high gain, high common-mode rejection, and input overvoltage protection. The input diodes prevent damage to the circuit should the input terminals be accidentally connected to the power-supply leads (or other undesired high-voltage source). The second stage is a differential-in, single-end-out amplifier with low gain and high common-mode rejection. Common-mode feedback is used from the second stage back to the first stage to further aid in the control of a common-mode input signal. Using these two differential amplifier stages and a common-mode feedback, a typical common-mode rejection of 100 dB is obtained. The third stage is a single-ended amplifier that provides high gain, voltage translation to a ground reference, output current-drive capabilities, and output short-circuit protection.

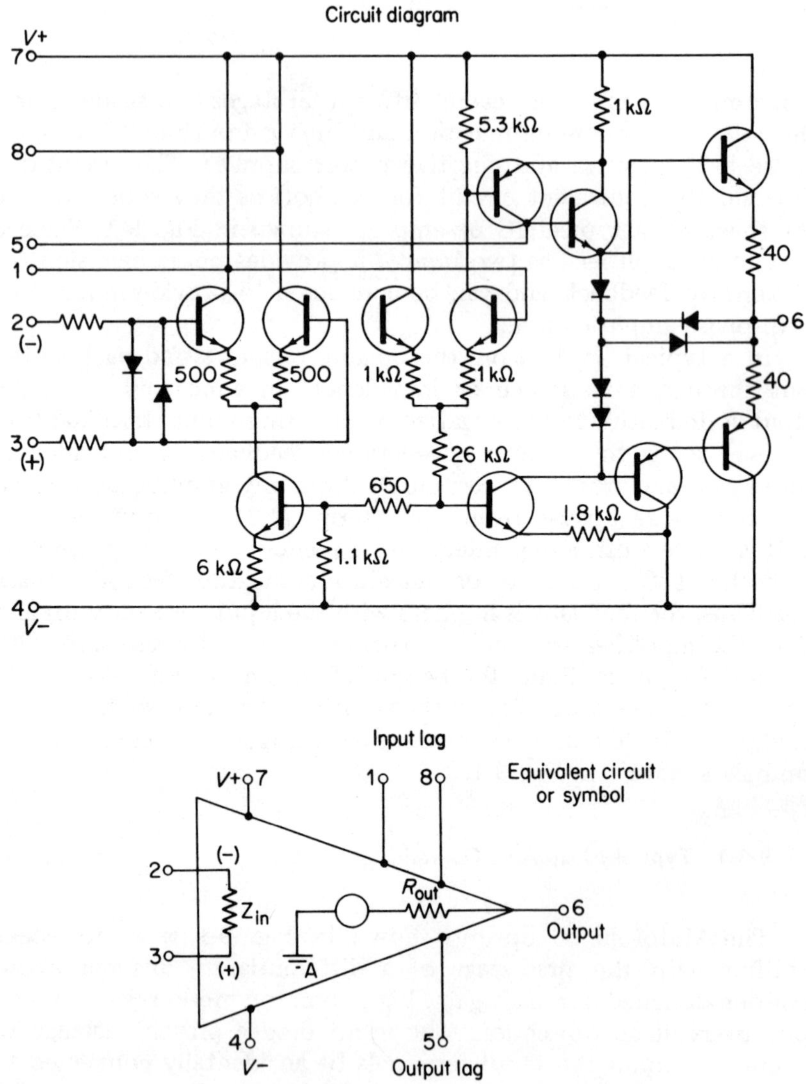

Figure 9-1. Typical IC op-amp circuit and symbol.

9-2 BASIC IC OP-AMP SYSTEM CIRCUITS

Figure 9-2 is the working schematic of a closed-loop op-amp system, complete with external circuit components. The following paragraphs provide a specific design example for the circuit.

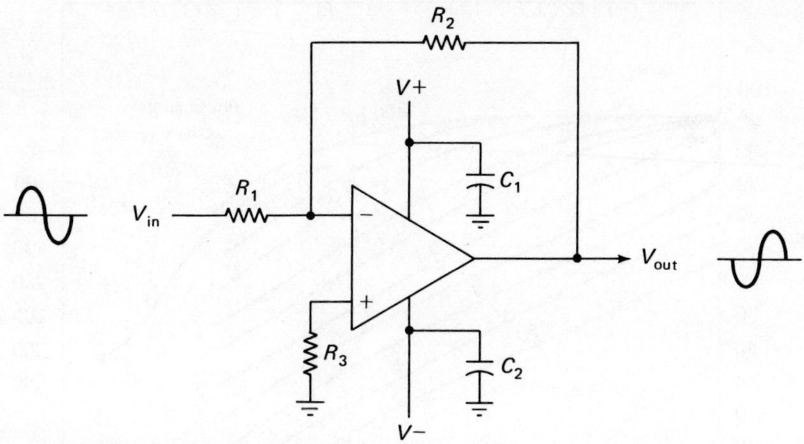

$$\text{Voltage gain} = \frac{V_{out}}{V_{in}} = \frac{R_2}{R_1} \qquad C_1 = C_2 = 0.1 - 0.001\ \mu\text{F}$$

$$R_3 = \frac{R_1 R_2}{R_1 + R_2}$$

$$\text{Peak output voltage capability} = \frac{\text{slew rate (in 1 s)}}{6.28 \times \text{maximum frequency (Hz)}}$$

Figure 9-2. Basic IC op-amp system circuit.

9-2-1 IC Op-Amp Characteristics

Assume that the IC op-amp shown in Fig. 9-2 has the following characteristics:

Supply voltage: +15 V and −15 V nominal, ±19 V maximum
Total device dissipation: 750 mW, derating 8 mW/°C
Temperature range: 0 to +70°C
Input-offset voltage: 3 mV typical
Input-offset current: 10 nA nominal, 30 nA maximum
Input-bias current: 100 nA nominal, 200 nA maximum
Input-offset-voltage sensitivity: 0.2 mV/V
Device dissipation: 300 mW maximum
Open-loop voltage gain: as shown in Fig. 9-3
Slew rate: 4 V/μs at a gain of 1; 6 V μs; 33 V/μs at a gain of 100
Open-loop bandwidth: as shown in Fig. 9-3
Common-mode rejection: 94 dB

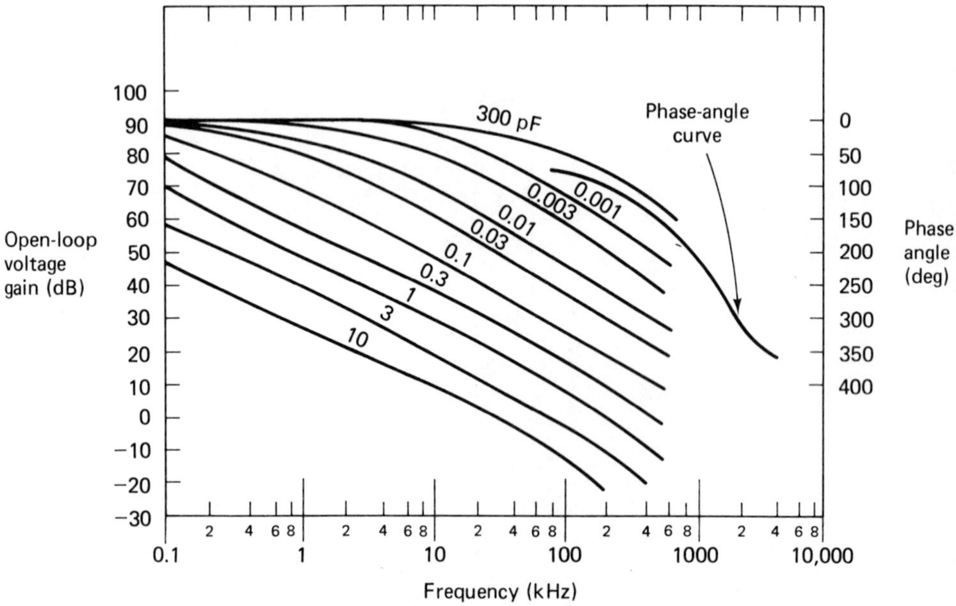

Figure 9-3. Typical phase-compensation characteristics of an IC op-amp.

Output-voltage swing: 23 V (peak to peak) typical

Input impedance: 1 MΩ

Output impedance: 300 Ω

Input-voltage range: −13 V, +10 V

Output power: 250 mW typical

9-2-2 Design Example

Assume that the circuit is to provide a voltage gain of 100 (40 dB), the input signal is 80 mV rms, the input source impedance is not specified, the output load impedance is 500 Ω, the ambient temperature is 25°C, the frequency range is dc up to 300 kHz, and the power supply is subject to 20 percent variation.

Frequency-gain relationship. Before attempting to calculate any circuit values, make certain that the op-amp can produce the desired voltage gain at the maximum frequency. This can be done by reference to a graph similar to Fig. 9-3. Note that the maximum frequency of 300 kHz intersects the phase curve at about 135°. This allows a phase margin of 45° (180 − 135 = 45), which is considered quite conservative. Any phase margin greater than about 25° is usually safe. Note that the maximum frequency of 300 kHz intersects curves above the 40-dB open-loop voltage gain level. Thus, the op-amp should be able to produce more than 40 dB of open-loop gain.

Supply voltage. The positive and negative supply voltages should both be 15 V, since this is the nominal value listed. Most IC op-amp data sheets list certain characteristics as "maximum" (temperature range, total dissipation, maximum supply voltage, maximum input signal, etc.) and then list the remaining characteristics as "typical" with a given "nominal" supply voltage. In no event can the supply voltage exceed the 19 V maximum. Since the available supply voltage (15 V) is subject to a 20 percent variation, or 18 V maximum, the supply voltage is within the 19-V limit.

Decoupling or bypass capacitors. The values of C_1 and C_2 should be found on the data sheet. In the absence of a value, use 0.1 μF for any frequency up to 10 MHz. If this value produces a response problem at any frequency (high or low), try a value between 0.001 and 0.1 μF.

Closed-loop resistances. The value of R_2 should be 100 times the value of R_1 to obtain the desired gain of 100. This value of R_1 should be selected so that the voltage drop across R_1 (with the nominal input-bias current) is comparable to the input signal (never larger than the input signal). A 50-Ω value for R_1 produces a 10-μV drop with the maximum 200-nA input-bias current. Such a 10-μV drop is less than 10 percent of the 80-mV input signal. Thus, the fixed drop across R_1 should have no appreciable effect on the input signal. With a 50-Ω value for R_1, the value of R_2 must be 5 kΩ (50 $\times$ 10 gain = 5000).

Offset minimizing resistance. The value of R_3 can be found using the equation of Fig. 9-2 once the values of R_1 and R_2 have been established. Note that the value of R_3 works out to about 49 Ω, using the Fig. 9-2 equation. A simple trial value for R_3 is always slightly less than the R_1 value. The final value of R_3 should be such that the no-signal voltages at each input are equal.

Comparison of circuit characteristics. Once the values of the external circuit components have been selected, the characteristics of the op-amp and the closed-loop circuit should be checked against the requirements of the design example. The following is a summary of the comparison.

Gain versus phase compensation. The closed-loop gain (gain with feedback through R_2) should always be less than the open-loop gain (gain with no feedback). As a guideline, the open-loop gain should be at least 20 dB greater than closed-loop gain. Figure 9-3 shows that open-loop gain up to about 66 dB is possible with a proper phase-compensation capacitor. Figure 9-3 also shows that a capacitance of 1000 pF produces an open-loop gain of slightly less than 60 dB at 3000 kHz, whereas a 300-pF capacitance produces a 66-dB open-loop gain. To assure an open-loop gain of 60 dB, use a phase-compensation capacitance value of about 700 pF.

With a 60-dB open-loop gain and values of 50 and 5000 Ω, respectively, for R_1 and R_2, the closed-loop circuit should have a flat frequency response of 40 dB (gain of 100) from zero up to 300 kHz. A rolloff starts at frequencies above 300 kHz. Thus, the closed-loop gain is well within tolerance.

Input voltage. The peak input voltage must not exceed the rated maximum input signal. In this case, the rated maximum is +10 V and −13 V, whereas the input signal is 80 mV rms, or approximately 112-mV peak (80 × 1.4). This is well below the +10-V maximum limit. When the rated maximum input signal is an uneven positive and negative value, always use the lowest value for total swing of the input signal. In this case, the input swings from +112 mV to −112 mV, far below +10 V. An input signal that started from zero could swing as much as +10 V and −10 V without damaging the op-amp. An input signal that started from −2 V can swing as much as ±11 V.

Output voltage. The peak-to-peak output voltage must not exceed the rated maximum output voltage swing (with the required input signal and selected amount of gain). In this case, the rated output voltage swing is 23 V (peak to peak), whereas the actual output is approximately 22.4 V (80 mV rms input × a gain of 100 = 8000 mV output; 8000 mV × 2.8 = 22.4 V peak to peak). Thus, the anticipated maximum output voltage is within the rated capability.

However, the actual output voltage is dependent upon slew rate, which, in turn, depends on compensation capacitance. As shown in the characteristics, the slew rate is given as 33 V/μs (or 33,000,000 V in 1 s) when gain is 100. The data sheet does not usually show a relationship between slew rate and compensation capacitance. However, since slew rate is always maximum with the lowest value of compensation capacitance, it can be assumed that the slew rate will be 33 V/μs with a capacitance of 700 pF (which is near the lowest recommended value of 300 pF, Fig. 9-3).

Using the equation of Fig. 9-2, it is possible to calculate the output voltage capability of the op-amp. With a maximum operating frequency of 300 kHz and an assumed slew rate of 33 V/μs, the peak-output-voltage capability is 33,000,000/(6.28 × 300,000), or approximately 17 V. Thus, the peak-to-peak output voltage capability is 34 V, well above the anticipated 22.4 V.

Output power. The output power of an op-amp is usually computed on the basis of rms output power (rather than peak or peak to peak) and output load. In this example, the output voltage is 8 V rms (80 mV × 100 gain = 8000 mV = 8 V). The load resistance or impedance is 500 Ω, as stated in the design assumptions. Thus, the output power is $(8)^2/500 = 0.128$ W = 128 mW.

A 128-mW output is well below the 250-mW typical output power of the op-amp. Also, 128-mW output plus a device dissipation of 300 mW is 428 mW, well below the rated 750-mW total device dissipation. Thus, the op-amp should be capable of delivering full-power output to the load.

Note that power output ratings usually are at some given temperature, 25°C in this case. Assume that the temperature is raised to 50°C. The total device dissipation must be derated by 8 mW/°C, or a total of 200 mW for the 25°C increase in temperature. This derates or reduces the 750-mW total device dissipation to 550 mW. However, the 550 mW is still well above the anticipated 428 mW.

9-3 OP-AMP SUMMING AMPLIFIERS

Figure 9-4 shows an IC op-amp used as a summing amplifier (also known as an analog adder). Summation of a number of voltages can be accomplished using this circuit. (Voltage summation was one of the original uses for op-amps in computer work.) All of the design considerations for the basic IC op-amp described in Sec. 9-2 apply to the summing amplifier, except as follows.

When the open-loop gain is high, the circuit functions with a minimum of error to sum a number of voltages. One circuit input is provided for each voltage to be summed. The single circuit output is the sum of the various input voltages (a total of four in this case) multiplied by any circuit gain. Generally, gain is set so that the output is at some given voltage value when all inputs are at their maximum value. In other cases, the resistance values are selected for unity gain.

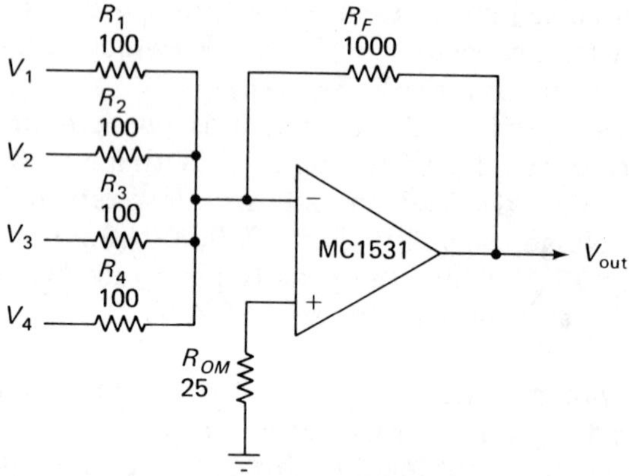

$$V_{out} = \frac{R_F}{R_1} V_1 + \frac{R_F}{R_2} V_2 + \frac{R_F}{R_3} V_3 + \frac{R_F}{R_4} V_4$$

When $R_1 = R_2 = R_3 = R_4 = R_F$

then $V_{out} = V_1 + V_2 + V_3 + V_4$

$$R_{OM} = \frac{1}{\frac{1}{R_1} + \frac{1}{R_2} + \frac{1}{R_3} + \frac{1}{R_4} + \frac{1}{R_F}}$$

Figure 9-4. Summing amplifier circuit.

The circuit values shown in Fig. 9-4 are based on the following conditions. Each of the four inputs varies from 2 to 50 mV rms. The output must be a nominal 1 V rms with full input on all four channels but must not exceed 2 V rms at any time. Resistors R_1 through R_4 are all of the value to simplify design and are selected so that the voltage drop (with nominal input-bias current) is comparable with the minimum input signal. Assume a 5-μA input-bias current for the IC. This results in 1.25 μA through each resistor. A 100-Ω value for R_1 through R_4 produces a 125-μV drop, which is less than 10 percent of the 2-mV minimum input signal.

The total (or maximum possible) signal voltage at the IC input is 200 mV (4 $\times$ 50 mV). Thus, the value of R_F should be between 500 and 1000 Ω to achieve a nominal 1 V and a maximum 2 V (200 mV $\times$ 5 = 1 V; 200 mV $\times$ 10 = 2 V). The 1000-Ω value is selected for R_F. The 25-Ω value of R_{OM} is found using the equation of Fig. 9-4.

9-4 OP-AMP INTEGRATION AMPLIFIERS

Figures 9-5 and 9-6 show IC op-amps used as integration amplifiers (also called integrators). Integration of various signals (often square waves) can be accomplished using these circuits. The output voltage

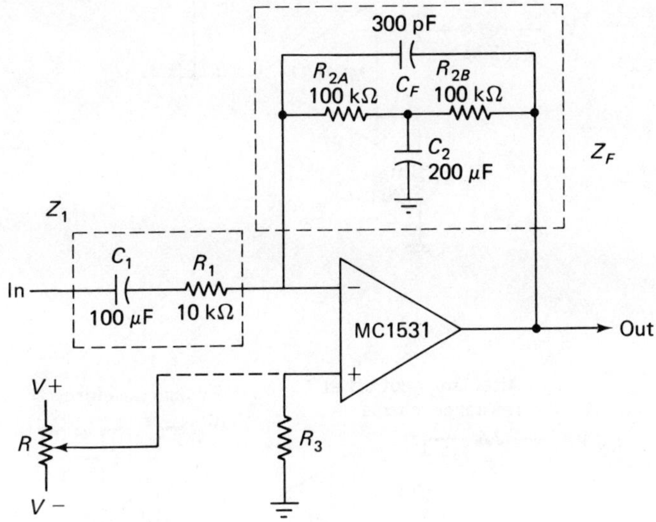

$R_2A = R_2B = R_1 \times 10$

R_3 = see text

R = see text

Freq. A = low freq. (0.016 Hz)

Freq. B = one decade higher than freq. A (0.16 Hz)

Freq. C = high freq. (5 kHz)

$$C_2 \approx \frac{\left(\dfrac{2}{6.28 \times \text{Freq. } A}\right)}{R_2A} = 200\ \mu\text{F}$$

$$C_F \approx \frac{\left(\dfrac{1}{\text{Gain} \times \text{Freq. } C \times 6.28}\right)}{R_1} = 300\ \text{pF}$$

$$C_1 \approx \frac{\left(\dfrac{1}{6.28 \times \text{Freq. } B}\right)}{R_1} = 100\ \mu\text{F}$$

Figure 9-5. Integration amplifier (integrator) for wide frequency-range applications.

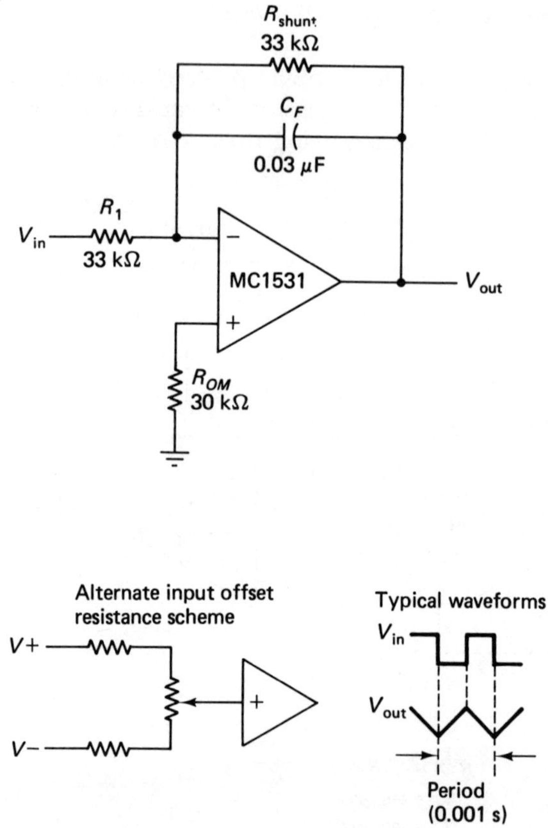

$$V_{out} \approx \frac{1}{R_1 C_F} \int V_{in}\ \Delta \text{ time}$$

$R_1 \times C_F \approx$ period of signal to be integrated

$$C_F \approx \frac{\text{Period}}{R_1} \qquad R_{shunt} \approx 10 \times R_1$$

$$R_{OM} \approx \frac{R_1 R_{shunt}}{R_1 + R_{shunt}}$$

V_{out} shifted by $+90°$ from V_{in}

Figure 9-6. Integration amplifier (integrator) for fixed-frequency operation.

from the amplifier is inversely proportional to the time constant of the feedback network and directly proportional to the integral of the input voltage.

The circuit of Fig. 9-5 is best suited to applications in which the integrator must be used over a wide range of frequencies. Keep in mind that the output amplitude depends on frequency once the values are selected. The circuit of Fig. 9-6 is best suited where only one frequency is involved. All the design considerations for the basic IC op-amp described in Sec. 9-2 apply to the integration amplifier, except as follows.

The value of the $R_1 C_F$ time constant should be approximately equal to the period of the signal to be integrated. The value of the $R_{shunt} C_F$ time constant should be substantially longer than the period of the signal to be integrated (approximately 10 times longer). Therefore, R_{shunt} should be 10 times R_1. Keep in mind that R_{shunt} and C_F form an impedance that is frequency-sensitive (most noticed at low frequencies).

The circuit values shown in Fig. 9-5 are based on the following conditions. Frequency A is 0.016 Hz, frequency B is 0.16 Hz, and frequency C is 5 kHz. Resistor R_1 is made 10 kΩ for simplicity. Thus, resistors R_2A and R_2B are 100 kΩ, as shown by the equations. The voltage gain of the circuit shown in Fig. 9-5 is dependent upon frequency. Figure 9-7 shows a gain versus frequency-response curve for a Motorola MC1531 op-amp using the component values specified in the circuit of Fig. 9-5.

If the input offset voltage of the op-amp is low, the noninverting input can be connected to ground through a fixed resistance R_3. As a first trial value, use the same resistance value as R_1. If greater precision is required or if the op-amp offset is large, use the potentiometer network shown in Fig. 9-5. With this arrangement, potentiometer R is adjusted so that there is no offset voltage under no-signal conditions.

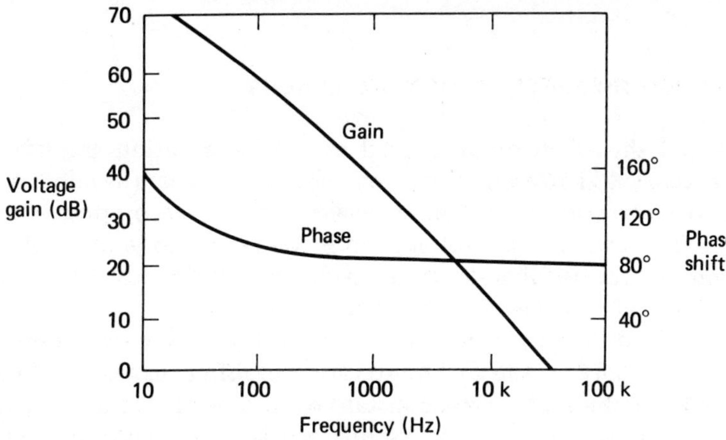

Figure 9-7. Gain versus frequency response for MC1531 op-amp used as an integration amplifier.

In the circuit of Fig. 9-6, the value of R_1 is chosen on the basis of input-bias current and voltage drop. Assuming a 5000-nA input-bias current, the 33-kΩ value for R_1 shown produces an approximate 165-mV drop, which is 10 percent of 1.65 V. Thus, the 33-kΩ R_1 is suitable for input signals 1.65 V or greater.

The value of the $R_1 C_F$ time constant must be approximately equal to the period of the signal to be integrated. The value of the $R_{shunt} C_F$ time constant must be substantially larger than the period (approximately 10 times longer). Thus, R_{shunt} is 10 times R_1. Note that R_{shunt} and C_F form an impedance that is frequency-sensitive (that is, the impedance is most noticed at low frequencies).

Assume that the circuit of Fig. 9-6 is to be used as an integrator for 1-kHz square waves. This requires a period of approximately 1 ms, or 0.001 s. Any combination of C_F and R_1 can be used, provided that the value of C_F times R_1 is approximately 0.001. Using the value of 33 kΩ for R_1, the value of C_F is 0.03 μF. The value of R_{shunt} must then be at least 330 kΩ. Note that the purpose of R_{shunt} is to provide feedback. This feedback is necessary so that an offset voltage cannot continuously charge C_F (which can result in amplifier limiting). If the offset voltage is small or can be minimized by means of R_{OM}, it is possible to eliminate R_{shunt}. Resistor R_{shunt} may have the effect of limiting gain at very low frequencies. However, above about 15 Hz, the effect of R_{shunt} is negligible (because of C_F in parallel). If greater precision is required, particularly at low frequencies, the input offset resistance R_{OM} can be replaced by the potentiometer network shown in Fig. 9-6. With this arrangement, potentiometer R is adjusted so that there is no offset voltage under no-signal conditions.

9-5 OP-AMP DIFFERENTIATION AMPLIFIERS

Figure 9-8 shows an op-amp used as a differentiation amplifier (or differentiator). Differentiation of various signals (usually square waves or sawtooth and sloping waves) can be accomplished using this circuit. The output voltage from the amplifier is inversely proportional to the feedback time constant and is directly proportional to the time rate of change of the input voltage.

All the design considerations for the basic IC op-amp described in Sec. 9-2 apply to the differentiation amplifier, except as follows. The value of the $R_F C_1$ time constant should be approximately equal to the period of the signal to be differentiated. In practical applica-

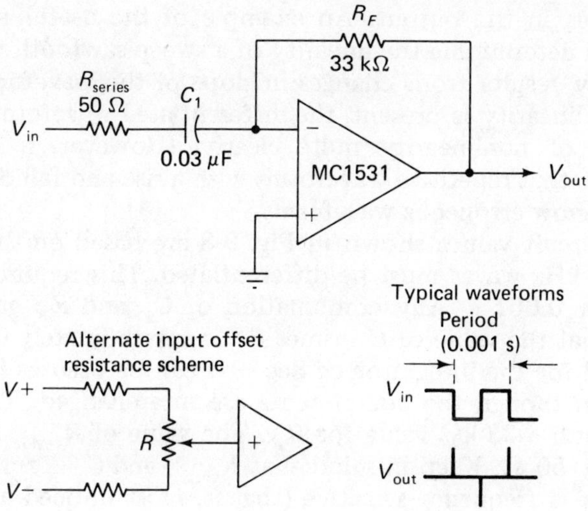

$$V_{out} \approx -R_F C_1 \frac{\Delta V_{in}}{\Delta \text{ time}}$$

$$R_F \times C_1 \approx \text{period of signal to be differentiated}$$

$$C_1 \approx \frac{\text{Period}}{R_F} \qquad R_{series} \approx 50\ \Omega$$

V_{out} shifted by $-90°$ from V_{in}

Figure 9-8. Differentiation amplifier (differentiator) circuit.

tions, the time constant is chosen on a trial-and-error basis to obtain a reasonable output level.

The main problem in the design of differentiating amplifiers is that the gain increases with frequency. Therefore, differentiators are very susceptible to high-frequency noise. The classic remedy for this effect is to connect a small resistor (on the order of 50 Ω) in series with the input capacitor so that the high-frequency gain is decreased. The addition of the resistor results in a more realistic model of the differentiator function because a resistance is always added in series with the input capacitor by the signal-source impedance.

In some applications, a differentiator may be advantageously used to detect the presence of distortion or high-frequency noise in the signal. A differentiator can often detect hidden information that could not be detected in the original signal. This is because differentiation permits slight changes in input slope to produce very significant changes in the output. An example of the usefulness of this feature is in determining the linearity of a sweep sawtooth waveform. Nonlinearity results from changes in slope of the waveform. Therefore, if nonlinearity is present, the differentiated waveform indicates the points of nonlinearity quite clearly. (However, it should be pointed out that repetitive waveforms with a rise and fall of differing slopes can show erroneous waveforms.

The circuit values shown in Fig. 9-8 are based on the assumption that 1-kHz waves must be differentiated. This requires a period of 1 ms, or 0.001 s. Any combination of C_1 and R_F can be used provided that the value of C_1 times R_F is approximately 0.001. The values used for the integrator of Sec. 9-4 can be used as initial trial values, even though the components are interchanged. Thus, C_1 is 0.03 μF, with a 33-kΩ value for R_F. The value of R_{series} (if used) is an arbitrary 50 Ω. Keep in mind that R_{series} and C_1 form an impedance which is frequency-sensitive (that is, most noticed at high frequencies).

9-6 OP-AMP NARROW-BANDPASS AMPLIFIERS

Figure 9-9 shows an op-amp used as a narrow-bandpass amplifier (also called a tuned peaking amplifier). All the design considerations for the basic IC op-amp described in Sec. 9-2 apply to the narrow-bandpass amplifier, except as follows. Circuit gain is determined by the ratio of R_1 and R_F in the usual manner. However, the frequency at which maximum gain occurs (or the narrowband peak) is the resonant frequency of the L_1C_1 circuit. Capacitor C_1 and inductance L_1 form a parallel-resonant circuit that rejects the resonant frequency. Thus, there is minimum feedback (and maximum gain) at the resonant frequency.

The circuit values shown in Fig. 9-9 are based on the assumption that the gain must be 20 dB at a peak frequency of 100 kHz. The value of R_1 is chosen on the basis of input-bias current and voltage drop in the usual manner. Assume an arbitrary value of 3.3 kΩ for R_1. The value of R_{OM} is then the same, or slightly less. With a value of 3.3 kΩ for R_1, the value of R_F must be 33 kΩ (or R_1 times 10) for a 20-dB gain.

Any combination of L_1 and C_1 can be used, provided that the resonant frequency is 100 kHz. For frequencies below 1 MHz, the value of C_1 should be between 0.001 and 0.01 μF. Assume an arbitrary 0.0015 μF for C_1. Using the equations of Fig. 9-9, the value of L_1 is approximately 1700 μH.

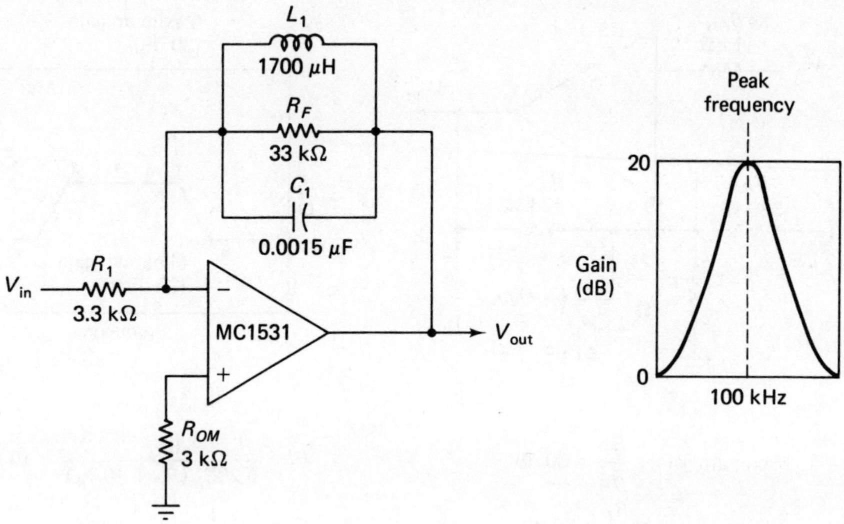

$$\text{Voltage gain} = \frac{V_{out}}{V_{in}} = \frac{R_F}{R_1} \qquad R_{OM} = \frac{R_1 R_F}{R_1 + R_F}$$

$$\text{Peak frequency (kHz)} = \frac{160}{\sqrt{L_1\,(\mu\text{H}) \times C_1\,(\mu\text{F})}} = 100 \text{ kHz}$$

$$C_1\,(\mu\text{F}) = \frac{2.54 \times 10^4}{F(\text{kHz})^2 \times L_1\,(\mu\text{H})} = 0.0015\ \mu\text{F}$$

$$L_1\,(\mu\text{H}) = \frac{2.54 \times 10^4}{F\,(\text{kHz})^2 \times C_1\,(\mu\text{F})} = 1700\ \mu\text{H}$$

Figure 9-9. Narrow-bandpass amplifier (tuned peaking).

9-7 OP-AMP WIDE-BANDPASS AMPLIFIERS

Figure 9-10 shows an op-amp used as a wide-bandpass amplifier. All the design considerations for the basic IC op-amp described in Sec. 9-2 apply to the wide-bandpass amplifier, except as follows. Maximum circuit gain is determined by the ratio of R_R and R_F. That is, the gain of the passband or flat portion of the response curve is set by R_F/R_R. Minimum circuit gain is determined by the ratio of R_1 and R_F, in the usual manner. The frequencies at which rolloff starts and ends (at both high- and low-frequency limits) are determined by impedances of the various circuit combinations, as shown by the equations of Fig. 9-10.

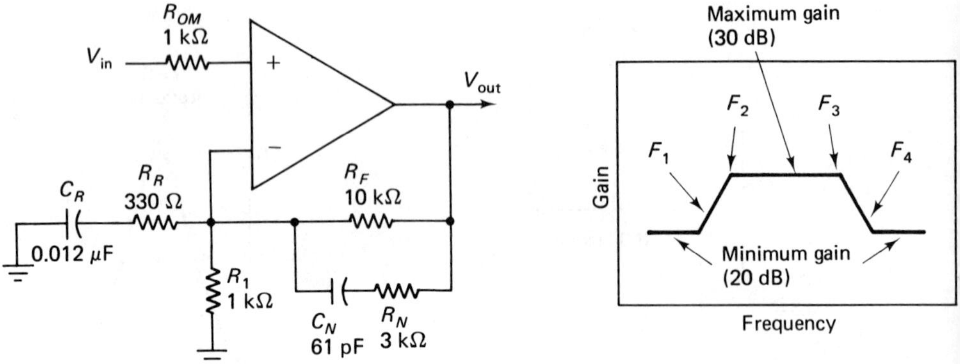

$$\text{Maximum gain} = \frac{R_F}{R_R} = 30 \text{ dB}$$

$$\text{Minimum gain} = \frac{R_F}{R_1} = 20 \text{ dB}$$

$$R_N \approx 30\% \text{ of } R_F = 3 \text{ k}\Omega$$

$$R_{OM} \approx R_1 = 1 \text{ k}\Omega$$

$$C_R = \frac{1}{6.28 F_2 R_R} = 0.012\ \mu\text{F}$$

$$C_N = \frac{1}{6.28 F_3 (R_N + R_F)} = 61 \text{ pF}$$

$$F_1 = \frac{10}{6.28 C_R (R_F + 10 R_R)} = 10 \text{ kHz}$$

$$F_2 = \frac{1}{6.28 C_R R_R} = 40 \text{ kHz}$$

$$F_3 = \frac{1}{6.28 C_N (R_N + R_F)} = 200 \text{ kHz}$$

$$F_4 = \frac{40}{6.28 C_N (40 R_N + R_F)} = 800 \text{ kHz}$$

$$R_F = R_1 \times \text{minimum gain} = 10 \text{ k}\Omega$$

$$R_R = R_F/\text{maximum gain} = 330$$

Figure 9-10. Wide-bandpass amplifier.

The values shown in Fig. 9-10 are based on the assumption that the circuit provides approximately 20-dB minimum gain at all frequencies, and approximately 30-dB gain at the passband. Gain starts increasing at frequencies above 10 kHz and rises to an approximate 30-dB passband at 40 kHz. The passband must extend to 200 kHz and then drop back to 20 dB at 800 kHz and above. Note that if phase compensation is required for the basic IC, the compensation values should be based on the minimum gain of 20 dB, not on the passband gain of 30 dB.

The value of R_1 is chosen on the basis of input-bias current and voltage drop, in the usual manner. With a value of 1 kΩ for R_1, a value of 10 kΩ is required for R_F and a value of 330 Ω for R_R. These relationships produce gains of 20 dB and 30 dB, respectively. In practice, it may be necessary to reduce both of these trial values to get the desired gain relationship.

9-8 OP-AMP UNITY-GAIN AMPLIFIERS

Figure 9-11 shows an op-amp used as a unity-gain amplifier (also known as a voltage follower). All the design considerations for the basic IC op-amp described in Sec. 9-2 apply to the unity-gain amplifier, except as follows. Feedback and input resistances are completely eliminated from the circuit. With this arrangement, the output voltage equals the input voltage (or may be slightly less than the input voltage). However, the input impedance is very high, while the output impedance is very low (as shown by the equations in Fig. 9-11).

In effect, the input impedance of the op-amp is multiplied by the open-loop gain, whereas the output impedance is divided by the open-loop gain. For example, if the IC op-amp has an open-loop gain of 1000 (60 dB), an output impedance of 200 Ω and an input impedance of 15 kΩ, the circuit of Fig. 9-11 produces an output impedance of 0.2 Ω (200/1000), and an input impedance of 15 MΩ (1000 × 15,000). Keep in mind that open-loop gain varies with frequency. Thus, the closed-loop impedances are frequency-dependent.

Another consideration that is sometimes overlooked in this application is the necessity of supplying the input bias current to the op-amp. In a conventional circuit, bias is supplied through the input resistances. In the circuit of Fig. 9-11, the input bias must be supplied through the signal source. This may alter input impedance.

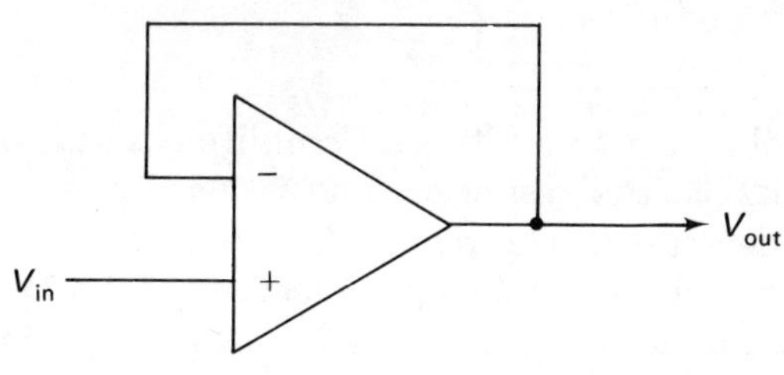

$$V_{out} \approx V_{in}$$

$$\underset{\text{(closed loop)}}{Z_{in}} \approx \text{open-loop gain} \times \text{open loop } Z_{in} \text{ of IC}$$

$$\underset{\text{(closed loop)}}{Z_{out}} \approx \frac{\text{open-loop } Z_{out} \text{ of IC}}{\text{open-loop gain}}$$

Figure 9-11. Basic IC op-amp unity-gain amplifier (voltage follower).

One more problem with the circuit of Fig. 9-11 is that the total input voltage is a common-mode voltage. That is, a voltage equal to the total input voltage appears across the two inputs. Should the input signal consist of a large d-c value, plus a large-signal variation, the common-mode input-voltage range may be exceeded. One solution for this problem is to capacitively couple the input signal to the noninverting input. This eliminates any d-c voltage, and only the signal appears at the inputs. This solves the common-mode problem but requires the use of a resistor from the noninverting input to ground. The resistance provides a path for the input-bias current. However, the resistance also sets the input impedance of the op-amp. If the resistance is large, the input-bias current produces a large offset voltage (both input and output) across the resistance.

Some manufacturers recommend that resistances (of equal value) be used in the feedback loop (from output to inverting input) and in the noninverting input. This still results in unity gain, and the circuit performs as a voltage or source follower. However, the input and output impedances are then set (primarily) by the resistance values rather than by the op-amp characteristics, as is the case with the circuit of Fig. 9-11.

9-9 OP-AMP UNITY-GAIN AMPLIFIERS WITH FAST RESPONSE

One of the problems of a unity-gain amplifier is that the slew rate is very poor. That is, the response time is very slow and the power bandwidth is decreased. The reason for poor bandwidth with unity gain is that most op-amp data sheets recommend a large-value compensating capacitor for unity gain.

As an example, assume that the op-amp has the characteristics shown in Fig. 9-3 and that the desired operating frequency is 200 kHz (the unity-gain op-amp must have a full-power bandwidth up to 200 kHz). The recommended compensation for 60-dB gain is about 0.001 μF, whereas the unity-gain compensation is 1 μF. Now assume that the op-amp also has the characteristics of Fig. 9-12 and that the load is 100 Ω. With a compensation of 0.001 μF, full power can be delivered at frequencies well above 200 kHz. However, if the compensating capacitor is 1 μF, the maximum frequency at which full power can be delivered is below 4 kHz.

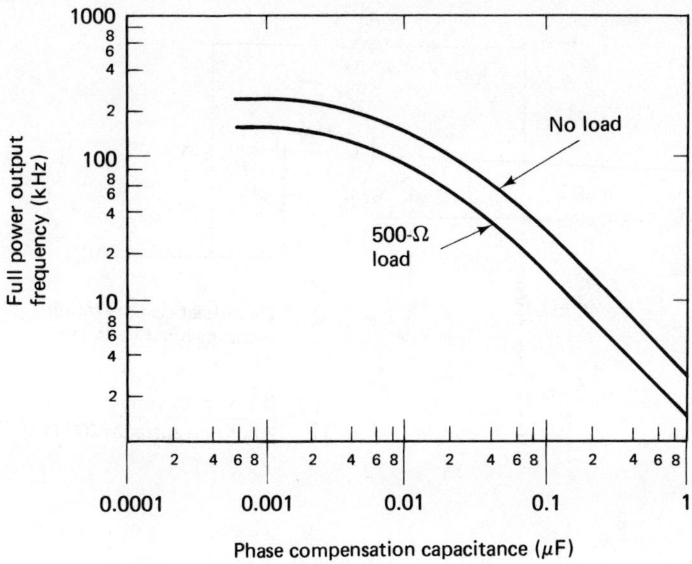

Figure 9-12. Frequency for full-power output as a function of phase-compensating capacitance.

Several methods are used to provide fast response time (high slew rate) and a power bandwidth with unity gain. One such method is described next.

9-9-1 Using Op-Amp Data-Sheet Phase Compensation

The circuit of Fig. 9-13 shows a method of connecting an op-amp for unity gain but with high slew rate (fast response and good power bandwidth). With this circuit, the phase compensation recommended on the data sheet is used, but with modification. Instead of using the unity-gain compensation, use the data-sheet phase compensation recommended for a gain of 100. Then select values of R_1 and R_3 to provide unity gain ($R_1 = R_3$). As shown by the equations, the values of R_1 and R_3 must be approximately 100 times the value of R_2. Thus, R_1 and R_3 must be fairly high values for practical design.

The values shown in Fig. 9-13 are based on the assumption that the circuit provides unity gain but with a slew rate approximately equal to that which results when a gain of 100 is used. The values also provide for an input-output signal of 8 V and that the input-bias current is 200 nA.

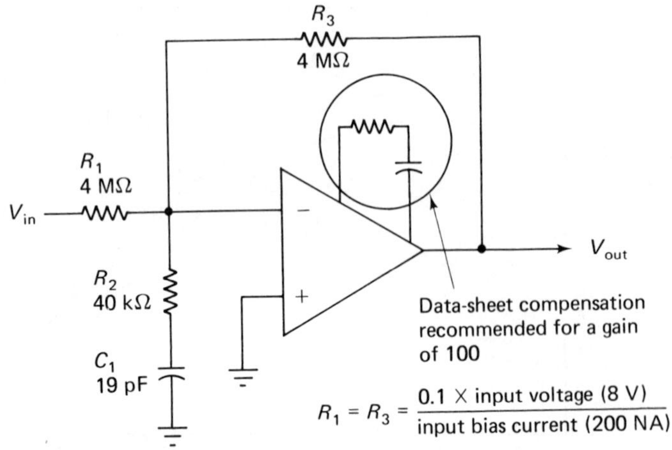

$$R_2 = \frac{R_1}{100} \qquad C = \frac{1}{6.28 \times R_2 \times F}$$

Slew rate ≈ slew rate for gain of 100

Figure 9-13. Unity-gain op-amp with fast response (good slew rate) using datasheet phase compensation.

9-10 OP-AMP HIGH-INPUT-IMPEDANCE AMPLIFIERS

Figure 9-14 shows an op-amp used as a high-input-impedance amplifier. All the design considerations for the basic IC op-amp described in Sec. 9-2 apply to the high-input-impedance amplifier, except as follows. The high-input-impedance feature of the unity-gain amplifier (Sec. 9-8) is combined with gain, as shown by the equations of Fig. 9-14. Note that the circuit of Fig. 9-14 is similar to that of the basic op-amp, except that there is no input offset compensating resistance (in series with the noninverting input) for the high-input-impedance circuit. This results in a trade-off of higher input impedance, with some increase in output offset voltage.

In the basic op-amp, an offset compensating resistance is used to nullify the input offset voltage of the op-amp. This (theoretically) results in no offset at the output. The output of the basic op-amp is at zero volts in spite of the tremendous gain. In the unity-gain amplifier (Sec. 9-8), there is no offset compensating resistance, but since there is no gain, the output is at the same offset as the input. In a typical IC op-amp, the input offset is less than 10 mV. This figure

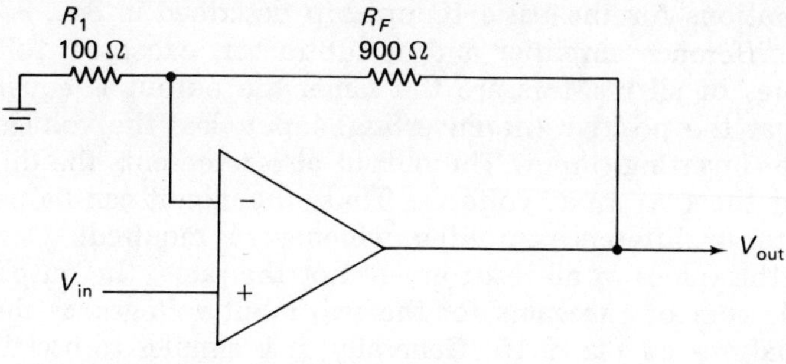

$$V_{out} = \frac{R_1 + R_F}{R_1} \times V_{in} = 10 \times V_{in}$$

$$\underset{\text{(closed loop)}}{Z_{out}} \approx \frac{\text{open-loop } Z_{out} \text{ of IC}}{1 + \text{open-loop gain}\left(\dfrac{R_1}{R_1 + R_F}\right)} \approx 2\ \Omega$$

$$R_F \approx (\text{gain} - 1) \times R_1 = 900$$

$$\underset{\text{(closed loop)}}{Z_{in}} \approx \text{open-loop } Z_{in} \text{ of IC} \times \text{closed-loop gain} \approx 150\ \text{k}\Omega$$

Figure 9-14. High input impedance amplifier.

should not be critical for the output of a typical unity-gain-amplifier application.

In the circuit of Fig. 9-14, the offset compensation resistance is omitted. The output is offset by an amount equal to the input offset voltage of the op-amp, multiplied by the closed-loop gain. However, since the circuit of Fig. 9-14 is to be used for modest gains, modest output offset results.

The values shown in Fig. 9-14 are based on the assumption that the circuit is to provide a gain of 10, with high (150-kΩ) input impedance and low (2-Ω) output impedance. The values also provide for an IC with an open-loop output impedance of 200 Ω, open-loop input impedance of 15 kΩ, and a gain of 1000 (60 dB). The value of R_1 is usually chosen on the basis of input-bias current and voltage drop and is arbitrarily set for 100 Ω in the Fig. 9-14 circuit.

9-11 OP-AMP DIFFERENCE AMPLIFIERS

Figure 9-15 shows an op-amp used as a difference amplifier and/or subtractor. One signal voltage is subtracted from another through simultaneous applications of signals to both inputs. All the design considerations for the basic IC op-amp described in Sec. 9-2 apply to the difference amplifier and/or subtractor, except as follows. If the values of all resistors are the same, the output is equal to the voltage at the positive (noninverting) input, less the voltage at the negative (inverting) input. The output also represents the difference between the two input voltages. Thus, the circuit can be used as a subtractor or difference amplifier, whichever is required.

If the values of all resistors are not the same, the output is the algebraic sum of the gains for the two input voltages, as shown by the equations of Fig. 9-15. Generally, it is simpler to use the same values for all resistors.

The values shown in Fig. 9-15 are based on the assumption that the circuit is to be used as a difference amplifier for two input voltages. Each of the voltage inputs varies from 2 to 50 mV rms. The output must vary between 20 and 500 mV rms. Since this example requires a gain of 10, all resistor values cannot be the same. To provide a gain of 10 for the negative inputs, R_2 must be 10 times R_1.

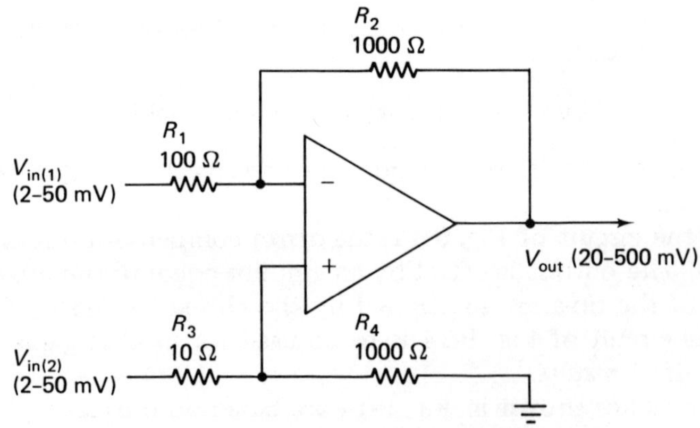

$$V_{out} = -\left(\frac{R_2}{R_1} \times V_{in(1)}\right) + \left[\left(\frac{R_4}{R_3 + R_4}\right) \left(\frac{R_1 + R_2}{R_1}\right) V_{in(2)}\right]$$

When $R_1 = R_2 = R_3 = R_4$

then $V_{out} = V_{in(2)} - V_{in(1)}$

$$R_2 = R_1 \times \text{gain}$$

$$R_3 = \frac{R_1}{\text{gain}}$$

$$R_4 = R_1 \times \text{gain}$$

Figure 9-15. Difference amplifier and/or subtractor.

The value of R_1 should be selected so that the voltage drop (with nominal input-bias current) is comparable with the minimum input signal, in the usual manner. Assume an arbitrary value of 100 Ω for R_1. With a 100-Ω value for R_1, the value of R_2 must be 1000 Ω to provide a negative input gain of 10. With a 100-Ω value for R_1, the values of R_3 and R_4 should be 10 and 1000 Ω, respectively.

9-12 OP-AMP VOLTAGE-TO-CURRENT CONVERTERS

Figure 9-16 shows an op-amp used as a voltage-to-current converter (also known as a transadmittance amplifier). This circuit can be used to supply a current to a load that is proportional to the voltage applied to the input of the amplifier. The current supplied to the load is relatively independent of the load characteristics. This circuit is essentially a current feedback amplifier.

All the design considerations for the basic IC op-amp described in Sec. 9-2 apply to the voltage-to-current converter, except as follows. Current sampling resistor R is used to provide the feedback to

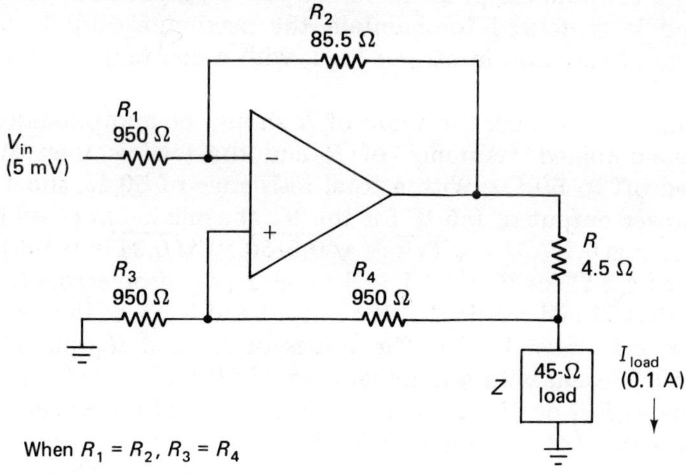

When $R_1 = R_2$, $R_3 = R_4$

and $R_3 + R_4 >> Z$

then $I_{load} \approx \frac{V_{in}}{R}$

$R \approx Z \times 0.1$

$R_3 > 20 \times Z$

$R_4 > 20 \times Z$

$R_1 = R_3$

$R_2 = R_1 \times \text{gain}$

Figure 9-16. Voltage-to-current converter.

the positive input. When R_1, R_2, R_3, and R_4 are all the same value, the feedback maintains the voltage across R at the same value as the input voltage. If a constant input voltage is applied to the amplifier, the voltage across R also remains constant (within very close tolerance), regardless of the load. If the voltage across R remains constant, the current through R must also remain constant. With R_3 and R_4 normally much higher than the load impedance, the current through the load must remain nearly constant, regardless of changes in impedance.

The values of R_1 through R_4 should normally be the same. The current sampling resistor R is then selected for the desired load current. The value of R should be selected to limit the output power, $I^2 \times (R + \text{load})$, to a value within the capability of the IC. For example, if the IC is rated at 600 mW total dissipation, with 100-mW dissipation for the basic IC, the total output power must be limited to 500 mW. As a guideline, the value of R should be approximately 1/10 of the load (Z).

The values shown in Fig. 9-16 are based on the assumption that the circuit is to be used as a voltage-to-current converter, the output load is 45 Ω (nominal), the maximum power output of the IC is 500 mW, and it is desired to maintain the maximum output current, regardless of variation in output load, with a constant input voltage of 5 mV.

With a 45-Ω load, the value of R should be approximately 4.5 Ω. The combined resistance of R and the load is then 49.5 Ω (rounded off to 50 Ω). With a total resistance of 50 Ω, and a maximum power output of 0.5 W for the IC, the maximum possible output current is 0.1 A ($I = \sqrt{P/R} = \sqrt{0.5/50} = \sqrt{0.01} = 0.1$ A). With a value of 4.5 Ω for R and 0.1 A through R, the drop across R is 450 mV. With 450 mV required at the output and 5 mV at the input, the amplifier gain must be 90. The values of R_3 and R_4 should be at least 950 Ω each, with a nominal load of 50 Ω. The value of R_1 is the same as R_3, or 950 Ω. With a value of 950 Ω for R_1, and a gain of 90, the value of R_2 should be 85.5 kΩ.

9-13 OP-AMP VOLTAGE-TO-VOLTAGE CONVERTERS

Figure 9-17 shows an op-amp used as a voltage-to-voltage converter (also known as a voltage-gain amplifier). This circuit is similar to the voltage-to-current (Sec. 9-12) except that the load and the current-sensing resistor are transposed. The voltage across the load is relatively independent of the load characteristics.

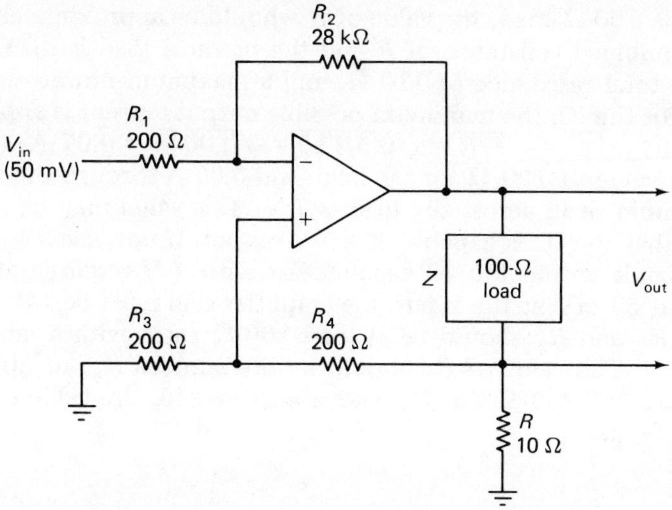

When $R_1 = R_2$, $R_3 = R_4$

then $V_{out} = V_{in}$

$R \approx Z \times 0.1$ $\quad R_1 = R_3$

$R_3 > 20 \times R$ $\quad R_2 = R_1 \times \text{gain}$

$R_4 > 20 \times R$

Figure 9-17. Voltage-to-voltage converter or amplifier.

All the design considerations for the basic IC op-amp described in Sec. 9-2 apply to the voltage-to-voltage converter, except as follows. The values of R_1 through R_4 should normally be the same. The current sampling resistor R is then selected for the desired load current. The value of R is selected to limit the output power, $I^2 \times (R + \text{load})$, to a value within the capability of the IC. For example, if the IC is rated at 600-mW total dissipation, within 100-mW dissipation for the basic IC, the total output power is limited to 500 mW. As a guideline, the value of R should be approximately 1/10 of the load (Z).

The values shown in Fig. 9-17 are based on the assumption that the circuit is to be used as a voltage-to-voltage converter, the output load is 100 Ω (nominal), the maximum power output of the IC is 500 mW, and it is desired to maintain the maximum output voltage across the load, regardless of variation in load (within the current capabilities of the IC), with a constant input voltage of 50 mW.

With a 100-Ω load, the value of R whould be approximately 10 Ω. The combined resistance of R and the nominal load is then 110 Ω. With a total resistance of 110 Ω, and a maximum power output of 0.5 W for the IC, the maximum possible output current is approximately 0.07 A ($I = \sqrt{P/R} = \sqrt{0.5/110} = \sqrt{0.0045} = 0.07$ A). With a nominal value of 100 Ω for the load, and 0.07 A through the load, the maximum drop across the load is 7 V. This value may be used, provided that the IC is capable of a 7-V output. If not, use a voltage output that is within the IC capabilities. With 7 V required at the output and 50 mV at the input, the amplifier gain must be 140. The values of R_3 and R_4 should be at least 200 Ω each, with a value of 10 Ω for R. The value of R_1 should be the same as R_3, or 200 Ω. With a value of 200 Ω for R_1 and a gain of 140, the value of R_2 should be 28 kΩ.

9-14 OP-AMP LOW-FREQUENCY SINE-WAVE GENERATORS

Figure 9-18 shows an op-amp used as a low-frequency sine-wave generator. This circuit is a parallel-T oscillator. Feedback to the negative input becomes positive at the frequency indicated in the equation, while positive feedback is provided at all times. The amount of positive feedback (set by the ratio of R_1 to R_2) is sufficient to cause the IC amplifier to oscillate. In combination with the feedback to the negative input, feedback to the positive input can be used to stabilize the amplitude of oscillations.

All the design considerations for the basic IC op-amp described in Sec. 9-2 apply to the generator, except as follows. The value of R_1 should be approximately 10 times the value of R_2. The ratio of R_1 and R_2, as set by the adjustment of R_2, controls the amount of positive feedback. Thus, the setting of R_2 determines the stability of oscillation. The amplitude of oscillation is determined by the peak-to-peak output capability of the IC and the values of the zener diodes CR_1 and CR_2. As shown by the equations, the zener voltage should be approximately 1.5 times the desired peak-to-peak output voltage. The nonlinear resistance of the back-to-back zener diodes is used to limit the output amplitude and maintain good linearity. The frequency of oscillation is set by the values of C and R. The upper-frequency limit is approximately equal to the bandwidth of the basic IC. That is, if the open-loop gain drops 3 dB at 100 kHz, the circuit should provide full-voltage output up to 100 kHz.

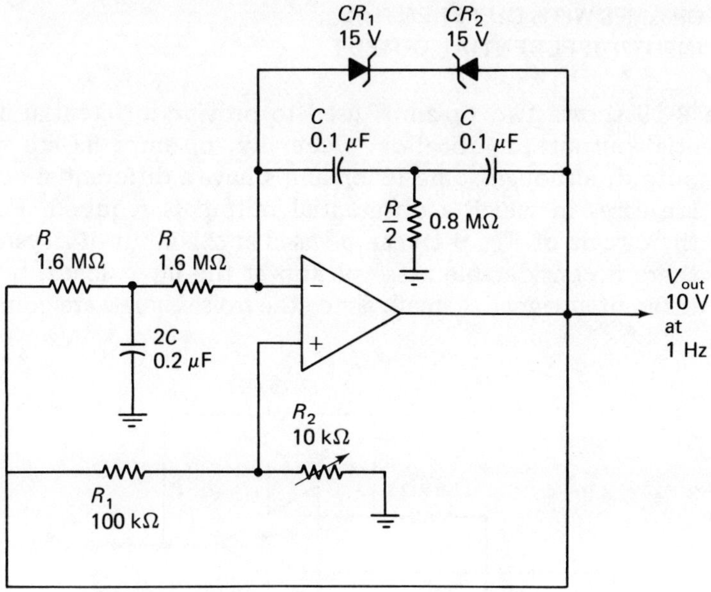

$$\text{Frequency} \approx \frac{1}{6.28RC} \qquad C \approx \frac{1}{6.28RF}$$

$$R_1 \approx R_2 \times 10 \qquad R \approx \frac{1}{6.28FC}$$

Zener point of CR_1 and $CR_2 \approx 1.5 \times V_{out}$ (peak to peak)

$R < 2\ \text{M}\Omega$

Figure 9-18. Low-frequency sine-wave generator using an IC op-amp.

The values shown in Fig. 9-18 are based on the assumption that the circuit is to provide 10-V sine-wave signals at 1 Hz. The value of R_2 should be chosen on the basis of bias current, in the usual manner. Assume a value of 10 kΩ for R_2. With a value of 10 kΩ for R_2, the value of R_1 should be 100 kΩ. With a required 10-V peak-to-peak output, the values (zener voltage) of CR_1 and CR_2 should be 15 V. It is assumed that the basic IC is capable of 10 V of peak-to-peak output.

The values of R and C are related to the desired frequency of 1 Hz. Any combination of R and C could be used provided that the combination worked out to a frequency of 1 Hz. However, for practical design, the value of R should not exceed about 2 MΩ (for a typical IC). Using a value of 1.6 MΩ for R, C should be 0.1 μF.

9-15 OP-AMPS WITH DIFFERENTIAL INPUT/DIFFERENTIAL OUTPUT

Figure 9-19 shows two op-amps used to provide differential input/differential output amplification. Generally, op-amps have a single-ended output, although some IC op-amps have a differential output. There are cases in which a differential output is required. For example the circuit of Fig. 9-19 can be used at the input of a system in which there is considerable noise pickup at the input loads, but the input differential signal is small. Since the noise signals are common-

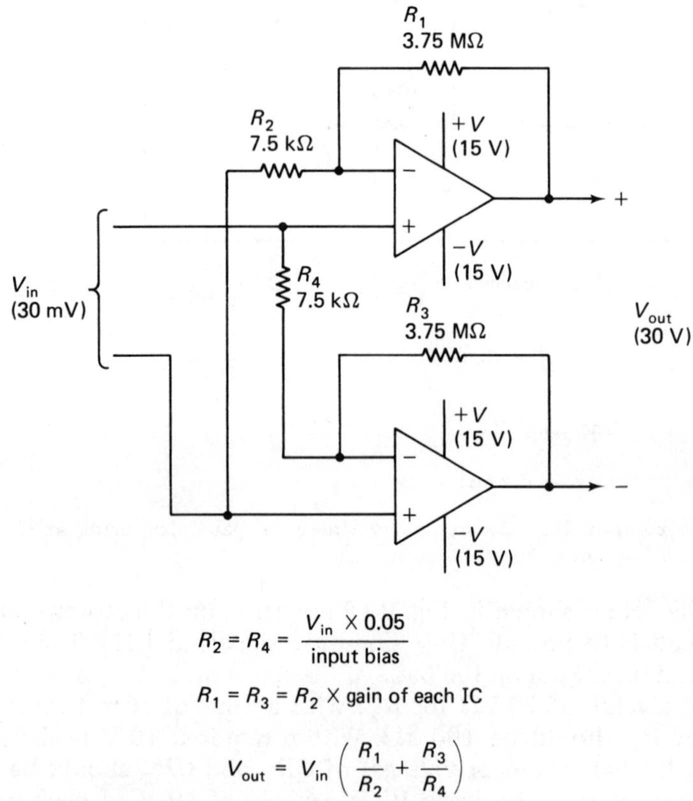

$$R_2 = R_4 = \frac{V_{in} \times 0.05}{\text{input bias}}$$

$$R_1 = R_3 = R_2 \times \text{gain of each IC}$$

$$V_{out} = V_{in}\left(\frac{R_1}{R_2} + \frac{R_3}{R_4}\right)$$

$$\frac{V_{out}}{V_{in}} = \frac{R_1}{R_2} + \frac{R_3}{R_4}$$

Figure 9-19. Differential input/differential output op-amp circuit.

mode (appearing on both input leads simultaneously at the same amplitude and polarity), they are not amplified. However, the differential input is amplified and appears as an amplified differential output voltage.

The circuit of Fig. 9-19 can be formed with two identical op-amps or, preferably, a dual-channel IC op-amp. The dual-channel IC is preferable since both channels have identical characteristics (gain, bias, etc.). This is because both channels are fabricated on the same semiconductor chip. However, the circuit will work satisfactorily if the two op-amps are carefully matched as to characteristics (particularly with regard to input bias and input offset voltage).

As shown by the equations, each op-amp (or each channel) provides one-half of the total differential output gain. Thus, if each op-amp provides a gain of 10, the differential output is 20 times the differential input. The output voltage swing is double that of the individual op-amps. Generally, the maximum output swing of a single-ended op-amp is slightly less than the IC supply voltage. Using the circuit of Fig. 9-19, the maximum differential output voltage swing is twice that of the supply voltage. For example, if the V_+ and V_- supply voltages are 15 V, the maximum differential output is slightly less than 30 V.

All the design considerations for the basic IC op-amp described in Sec. 9-2 apply to the circuit, except as follows. The values of R_2 and R_4 are chosen on the basis of input-bias current and voltage drop, in the usual manner. The values of R_1 and R_3 are chosen to provide the desired gain. Resistors R_3 and R_4 should be equal to R_1 and R_2, respectively.

The values shown in Fig. 9-19 are based on the assumption that the circuit is to provide a differential output with a swing of approximately 30 V, the available differential input is 30 mV, and the input bias current is 200 nA. With an input of 30 mV and a desired output of 30 V, the circuit must provide a total gain of 1000 (30 V/0.03 V). Each op-amp must provide a gain of 500.

The voltage drop across R_2, due to input-bias current, should be no greater than 5 percent of the input voltage. With a drop of 1.5 mV across R_2 and a 200 nA input-bias current, the value of R_2 is 1.5 mV/200 nA = 7500 Ω maximum. R_4 should also be 7500 Ω. With a value of 7500 Ω for R_2 and a required gain of 500, the value R_1 is 3.75 MΩ. R_3 should also be 3.75 MΩ. With a required differential output voltage swing of 30 V, the V_+ and V_- supply voltage of the IC op-amps must be about 15 V.

9-16 OP-AMP TEMPERATURE SENSORS

Figure 9-20 shows an op-amp used as the major active element in a temperature-sensor circuit. Temperature sensing with thermistors is popular because thermistors are inexpensive and easy to use. However, thermistors are nonlinear and can supply only very small output signals (generally a few microwatts).

The circuit of Fig. 9-20 overcomes these limitations. The output voltage is relatively linear over the temperature range of the op-amp. The output is exactly linear near the temperature at which the thermistor resistance equals the resistance of R_1. Thus, R_1 should be equal to R_T (the thermistor resistance) at the center of the desired temperature range.

The reference voltage V_{ref} is obtained from the zener diode CR_1 and the voltage divider. The upper limit is a value determined by the power rating (P_T) of the thermistor, as shown by the equations. With the values shown, R_2 is adjusted so that V_{ref} is -0.067 V. Under these conditions, the maximum power dissipated by the thermistor is about 2.5 μW, well under the 5-μW rating.

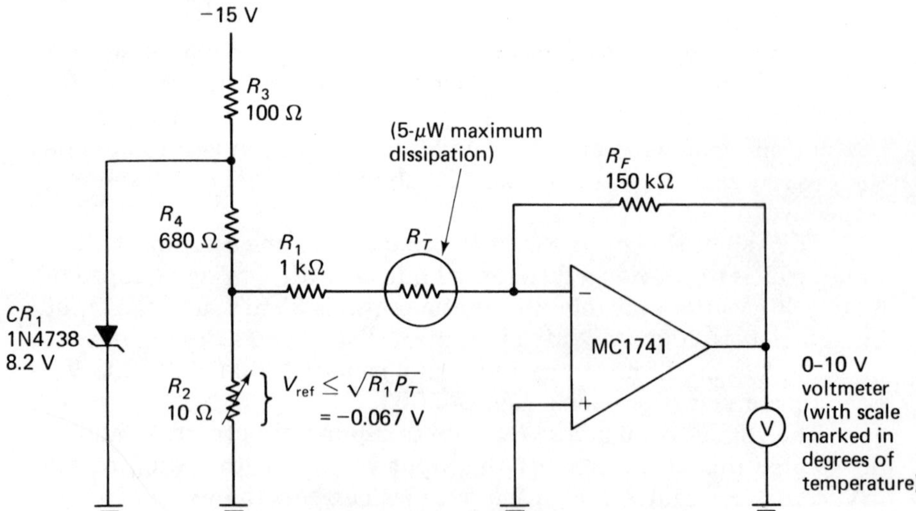

R_T = thermistor resistance

P_T = power rating of termistor

R_1 = R_T at center of temperature range

Figure 9-20. Temperature sensor using an IC op-amp.

9-17 OP-AMP ANGLE GENERATORS

Figure 9-21 shows two op-amps connected as an angle generator. As shown by the equations, the output of op-amp 1 is proportional to the sine of the input phase angle, and the output of op-amp 2 proportional to the cosine of the input phase angle. The op-amps are connected as a Scott-T transformer into a three-wire synchro line.

When all resistor values are the same, the output of the op-amp 2 is equal to twice the input voltage multiplied by the cosine of the phase angle. For example, if the phase angle is 33° and the input voltage (line-to-line, three-phase input) is 1 V, the op-amp 1 output is 2 × 1 = 2; 2 × 0.8387 (the cosine of 33°) = 1.6774 V.

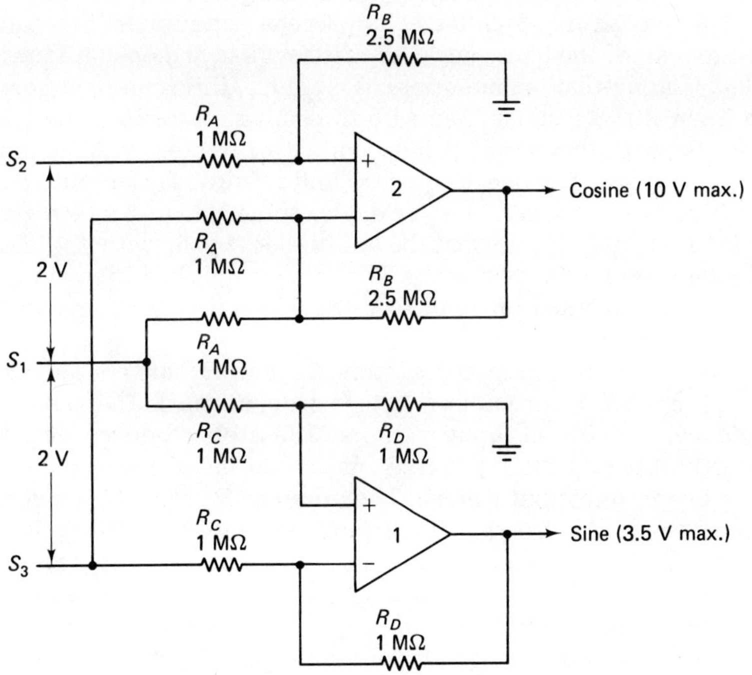

$$\text{Op amp 2} = 2\,\frac{R_B}{R_A}\,E_{max}\ \text{cosine of phase angle}$$

$$\text{Op amp 1} = 1.732\,\frac{R_D}{R_C}\,E_{max}\ \text{sine of phase angle}$$

$$E_{max} = \text{maximum line-to-line output}$$

Figure 9-21. Angle generator using an IC op-amp.

With the same conditions (phase angle 33°, input 1 V) the op-amp 1 output is 1.732 × 0.5446 (the sine of 33°) = approximately 0.94 V.

The accuracy of the circuit in Fig. 9-21 is dependent upon matching of the op-amps, as well as matching of the resistors. A dual-channel IC op-amp is ideal for the circuit because both channels are fabricated on the same chip. However, two separate op-amps with closely matched characteristics will produce satisfactory results. The resistors should have a tolerance of 1 percent (or better).

The circuit of Fig. 9-21 is most effective when the three-phase voltages are on the order of 1 V, or a fraction of 1 V (such as some analog computer servo systems), and it is desired to have output readings in the 5- to 10-V range (typical of digital logic systems).

The circuit of Fig. 9-21 has several advantages over direct measurement of the phase angle by meters. First, the op-amps present far less loading than a meter does. (Of course, this is more important with low-voltage systems than with three-phase systems in the 120-V range). Second, the circuit is independent of frequency. Most phase-angle meters are for one frequency only. Third, the output can be "weighted" or "scaled." For example, some design problems may require that the sine output be multiplied by 5, with the cosine multiplied by 10, or vice versa. This is accomplished by setting the gain of the individual op-amps to different levels (by different ratios of R_B/R_A and R_D/R_C.

The values of the input resistors R_A and R_D are chosen on the basis of input-bias current and voltage drop, as usual. The same value should be used for all input resistors. The drop produced by input-bias current across the input resistors should be 10 percent (or less) of the line-to-line input voltage. The values of R_B and R_D are chosen on the basis of desired gain for the individual outputs. The values of both R_B resistors should be the same, as should both R_D resistors. However, R_B need not equal R_D.

The values shown in Fig. 9-21 are based on the assumption that the circuit is to provide outputs for a three-phase system (S_1, S_2, S_3) with a maximum line-to-line voltage of 2 V, the op-amp has an input-bias current of 200 nA, the sine output should not exceed 3.5 V, and the cosine output should be 10 V or less.

With a 2-V input from the three-phase line and a 200-nA input-bias current, the value of each R_A and R_C resistor should be 1 MΩ (2 V × 0.1 = 0.2 V; 0.2 V/200 nA = 1 MΩ). Since the sine input is not to exceed 3.5 V, op-amp 1 need have no gain. Thus, both R_D resistors can be 1 MΩ. With no gain, the output from op-amp 1 goes from 0 V, when the phase angle is 0°, to about 3.464 V, when the phase angle is 90°.

Since the cosine output is not to exceed 10 V, find the maximum output from op-amp 2 with no gain (4 V, or twice the input of 2 V, in this case), then divide the maximum output limit by the no-gain output, or 10/4 = 2.5. Thus, a gain of 2.5 is required for op-amp 2, and the values of both R_B resistors should be 2.5 MΩ (1 MΩ × 2.5 = 2.5 MΩ). To provide a 10-V cosine output, the $V+$ and $V-$ supply for op-amp 2 should be at least 10 V (11 or 12 V, in practical use). A lower value can be used for op-amp 2. However, a better match, and a more practical circuit, is obtained if both op-amps have the same $V+$ and $V-$ supply voltage.

9-18 OP-AMP PEAK DETECTORS

Figure 9-22 shows an op-amp used as the major active element in a peak detector system. Where accuracy is required for peak detection, the conventional diode-capacitor detector is often inadequate because of changes in the diode's forward voltage drop (due to variations in the charging current and temperature). Ideally, if the forward drop of the diode can be made negligible, the peak value is the absolute

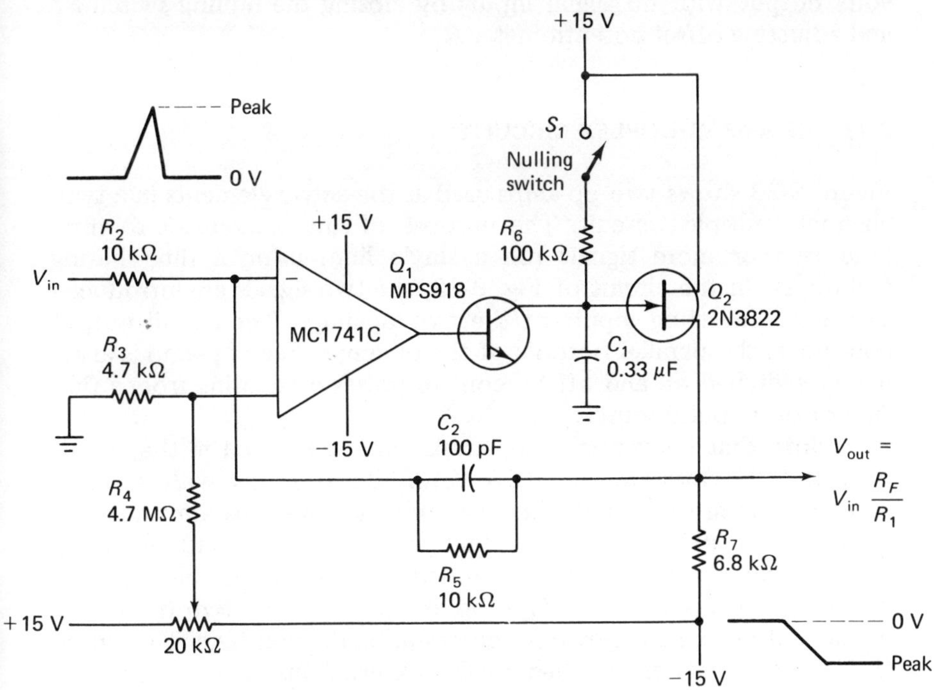

Figure 9-22. Peak detector using IC op-amp.

value of the peak input and is not diminished by the diode-voltage drop (typically, 0.5 to 0.7 V for a silicon diode). This can be accomplished by means of an op-amp.

The circuit of Fig. 9-22 uses the base-collector junction of a transistor as the detecting diode. The transistor (acting as a diode) is contained within the feedback loop (between the op-amp output and the circuit output). This reduces the effective forward voltage drop of the diode by an amount equal to the loop gain (or ratio of open-loop gain to closed-loop gain). For example, assume that the transistor has a base-emitter drop of 0.5 V, that the open-loop gain of the op-amp is 1000, and the closed-loop gain is 1 (unity). Under these conditions, the loop gain is 1000, and the effective forward voltage drop is 0.0005 V. Any transistor can be used provided that the leakage is low (preferably 10 nA or less, with +15 V at the collector).

The storage time of the circuit in Fig. 9-22 is dependent upon leakage of the diode (transistor) and the FET, as well as the value of C_1. A larger value of C_1 increases storage time. Note that the values shown in Fig. 9-22 apply to a typical op-amp that requires a $V+$ and $V-$ supply voltage of 15 V. These values can be used as a starting point for design. In use, the output is adjusted to zero offset (zero volts output with no signal input) by closing the nulling switch S_1 and adjusting offset potentiometer R_1.

9-19 OP-AMP MULTIPLEX CIRCUITS

Figure 9-23 shows two op-amps used as the active elements in a two-channel multiplex circuit. The purpose of such a circuit is to combine two or more signals (on a single line) using a time-sharing technique. In the circuit of Fig. 9-23, the two signals are introduced as the noninverting inputs of the two op-amps. The circuit output appears at the parallel outputs of the op-amps. Each op-amp is alternately switched on and off by control pulses originating from a flip-flop or other pulse source.

Note that conventional op-amps cannot be used in the circuit of Fig. 9-23. Instead, the RCA CA3078T op-amps with external variable bias are required. Such variable-bias op-amps are similar to conventional op-amps, except that standby power and operating point of the op-amp can be controlled by the external bias voltage. The pulse signals from the flip-flop are amplified by two transistors connected as grounded-base (or common-base) amplifiers. The amplified clock signals are applied to the external bias connection (terminal 5) of the op-amps. When one pulse is fully positive, the other pulse is fully negative. Thus, the op-amps are turned fully on and

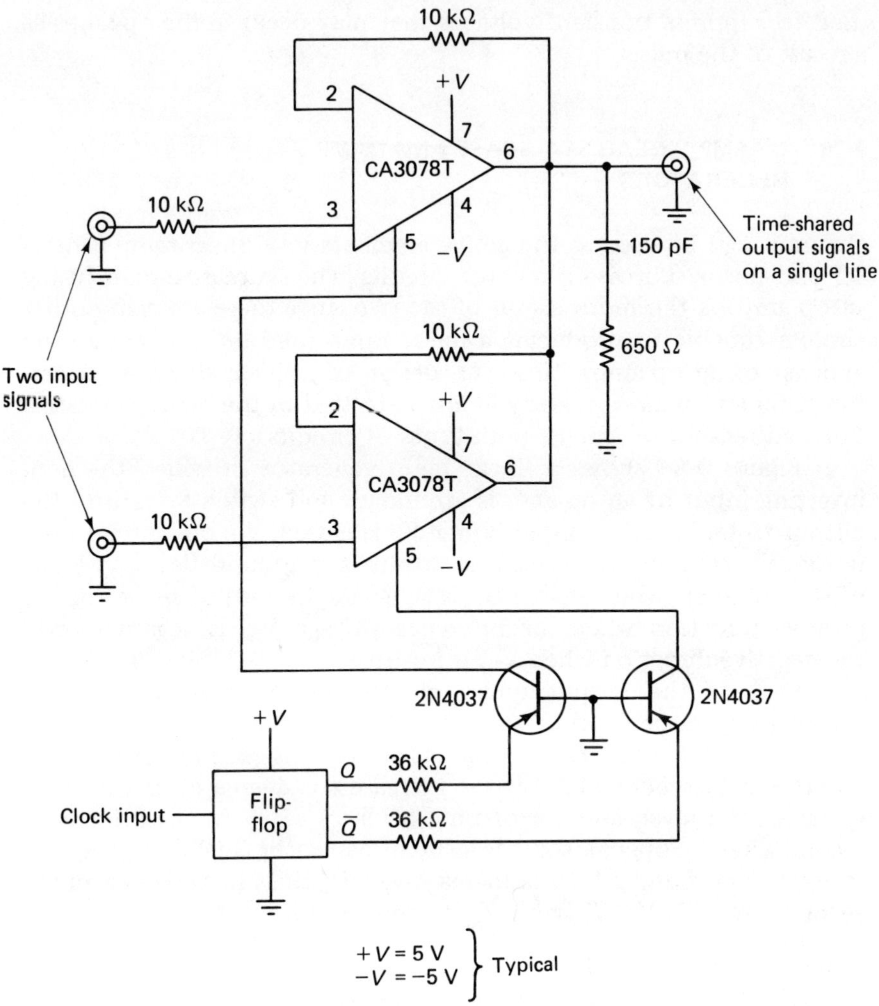

Figure 9-23. Two op-amps used as active elements in a two-channel multiplex circuit.

fully off on alternate cycles of the pulses. That is, op-amp 1 is on, and delivers the input signal to the output, when op-amp 2 is off, and vice versa.

Note that the op-amps are connected as unity-gain amplifiers. Thus, the output is at essentially the same level as the inputs. If necessary, gain can be introduced by making the values of the feedback resistors larger than the input resistors, in the usual manner. The maximum circuit output voltage swing is limited by the values of $V+$ and $V-$. The capacitor and resistor at the circuit output are used to suppress transient voltages that may occur in the op-amps as a result of the pulses.

9-20 OP-AMP LINEAR STAIRCASE AND RAMP GENERATORS

Op-amps can be used as the active elements in a linear ramp generator and linear staircase generator circuits. The staircase circuit (using an op-amp) is the more useful of the two since there are many other circuits capable of producing a linear ramp (and do not require the expense of an op-amp). However, design of a linear staircase generator using an op-amp is nearly identical to that of the ramp generator. For that reason, we discuss both types of circuits.

Figure 9-24 shows a linear ramp generator in which the non-inverting input of an op-amp is grounded, and switch S_1 returns the output to the inverting input. When S_1 is closed, the op-amp is in the unity-gain configuration and the output is at ground (less any input offset voltage). When switch S_1 is opened, the output moves in the positive direction when the reference voltage V_{ref} is negative, or in the negative direction when V_{ref} is positive.

Because the output under closed-loop conditions (feedback through capacitor C) tries to maintain the input terminals at zero volts, the charging current to the capacitor C is constant at a rate of $dV/dt = I/C$, where dV/dt is the increase (or decrease) of the ramp voltage for a given amount of time (such as a 1-V increase for each second after S_1 opens), C is the value of capacitor C (if C is expressed in microfarads and I is in amperes, then dV/dt is in volts per microsecond), and I is V_{ref}/R (with V_{ref} in volts and R in ohms).

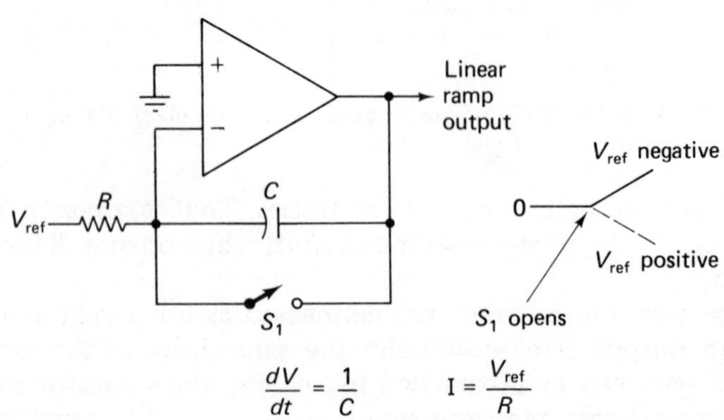

Figure 9-24. Basic linear ramp generator using an IC op-amp.

As an example, assume that V_{ref} is +10 V, R is 100 Ω, and C is 1 μF. Under these conditions, I = 0.1 A (+10 V/100 Ω) and the ramp decreases at a rate of 1 V/0.1 μs (0.1 A/μF). (The ramp decreases since V_{ref} is given as a positive voltage.)

Note that this equation for the linear ramp generator is accurate only if the charging current is much greater than the input-bias current of the op-amp. As a guideline, make the charging current at least 100 times that of the input bias. If the bias is not small in relation to the charging current, the bias will offset the charging current (either add to or subtract from). This will tend to make the ramp nonlinear, particularly at the start or end of the ramp. In any event, the dV/dt equation is no longer accurate.

The maximum possible height or amplitude of the linear ramp depends upon the supply voltages for the op-amp. That is, the maximum possible positive ramp is set by V_+, whereas V_- sets the maximum negative ramp. Assuming a given dV/dt (set by V_{ref}, R, and C), the actual height of the ramp is set by the amount of time switch S_1 is open. For example, if dV/dt is 1 V/μs, and S_1 is open for 10 μs, the ramp can increase from zero to 10 V, provided that the supply voltage is 10 V or greater. In practical applications, the maximum ramp height is slightly less than the supply voltage.

Figure 9-25 shows an op-amp used as the major active element in a linear staircase generator circuit. Note that the circuit is similar to the ramp generator except for the input. In the Fig. 9-25 circuit, charging resistance R is replaced by two diodes and capacitor C_1. Similarly, the fixed V_{ref} is replaced by pulses. In the circuit of Fig. 9-25, a pulse of amplitude E couples a charge Q to the op-amp input. The charge Q is equal to C_1 $(E - 2V_{AK})$, where $2V_{AK}$ is the forward voltage drop across the two diodes. Typically, $2V_{AK}$ is equal to about 1 V, since the drop across each diode is about 0.5 V.

When switch S_1 is open, capacitor C_2 is charged (by each pulse) in steps equal to $(E - 2V_{AK})$ (C_1/C_2). For example, assume that the pulses are 11 V, C_1 is 500 pF and C_2 is 1 μF. Under these conditions, each step is about 5 mV. This is shown as follows: (11 - 1) (0.0005/1) = 0.005 V = 5 mV.

Note that the equations for the linear staircase generator are accurate only if the charging current is much greater than the input-bias current of the op-amp. Thus, op-amps with minimum input-bias are recommended for staircase generator circuits. Of more importance, the pulse amplitude must be much greater than the voltage/temperature variations of the diodes. If the diode voltage variations are not small in relation to the charging pulses, the voltage/temperature variations add to (or subtract from) the pulse amplitudes and produce uneven staircase steps.

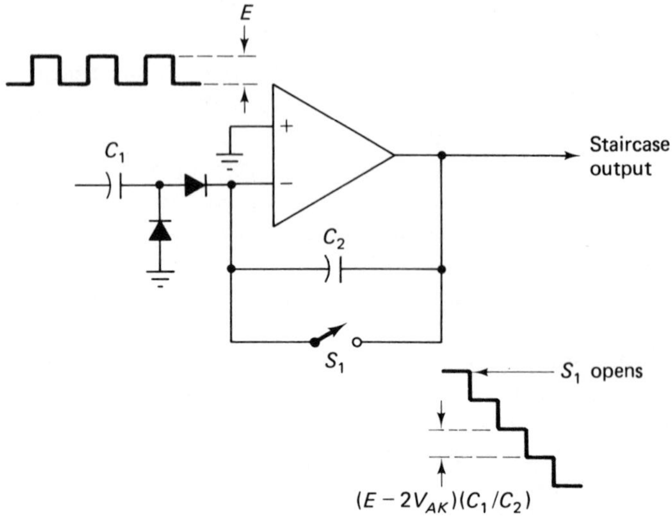

Figure 9-25. Basic linear staircase generator using an IC op-amp.

The maximum possible height of the staircase is set by the op-amp supply voltages. The actual height of the staircase is set by the amount of time switch S_1 is open (assuming an infinite number of pulses and steps).

In practical ramp and staircase circuits, switch S_1 is replaced with an electronic switch. Figure 9-26 shows a staircase generator circuit controlled by a FET switch. This circuit is used as part of a digital voltmeter. Clock pulses are applied to the FET gates to control operation of the staircase circuit. When the clock pulses are on, the drain-source resistance of the FET is reduced to zero, and capacitor C_3 is shorted, or closed. When the clock pulses are off, the drain-source resistance rises to several hundred megohms, removing the "short" from C_3. With C_3 in the feedback circuit, the staircase generator produces steps in response to each signal pulse applied through C_1 and C_2. The generator continues to produce steps until C_3 is again shorted by the FET in response to "off" clock pulses.

Note that capacitor C_1 is adjustable so that the staircase output can be set to exactly 5-mV steps. The signal pulses at the input are fixed at approximately 11 V. This is reduced to about 10 V by the drop across diodes D_1 and D_2. In theory, when C_1 is adjusted so that the parallel capacitance of C_1 and C_2 is 500 pF, the ratio of input capacitance to output capacitance is 0.0005:1, and the 10-V input produces 5-mV steps at the output. However, because the diodes do not always have the same voltage drop and the op-amp input bias is not always the same (due to temperature variations, etc.), the value of C_1-C_2 is not always precisely 500 pF. Instead, C_1 is simply adjusted until the output steps are precisely 5 mV.

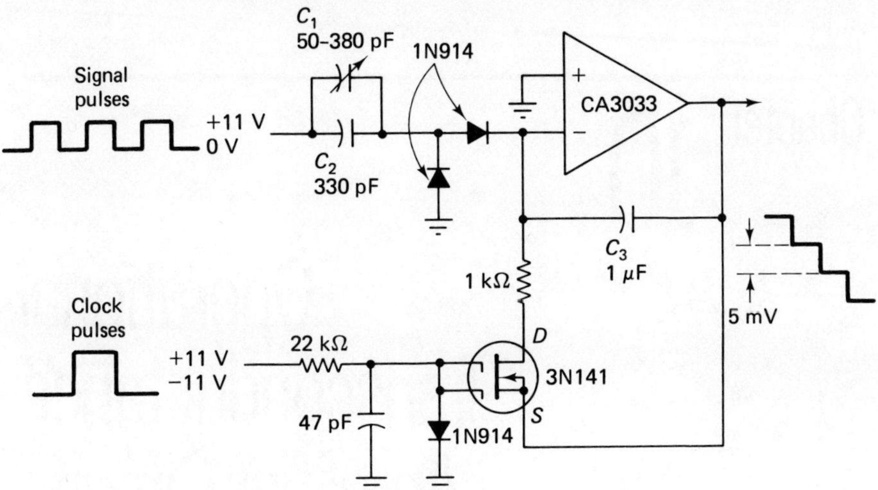

Figure 9.26. Staircase generator circuit controlled by a FET electronic switch.

Chapter 10

Operational Transconductance Amplifier Circuits

This chapter is devoted entirely to operational transconductance amplifier (OTA) circuits. An OTA is similar in form to the conventional op-amp described in Chapter 9. However, OTAs and op-amps are not always interchangeable. For that reason, an explanation of the unique characteristics found in OTAs is in order. The OTA not only includes the usual differential inputs of an op-amp, but also contains an additional control input in the form of an *amplifier bias current* (or I_{ABC}). This control input increases the OTA's flexibility for use in a wide range of applications.

The characteristics of an ideal OTA are similar to those of an ideal op-amp except that the OTA has an extremely high output impedance. Because of this difference, the output signal of an OTA is best described in terms of current that is proportional to the difference between the voltages of the two inputs (inverting and noninverting). Thus, the transfer characteristic (or input/output relationship) is best defined in terms of *transconductance* rather than voltage gain. Transconductance, or g_m, is the ratio between the difference in current output (I_{out}) for a given difference in voltage input (E_{in}). Except for the high output impedance, and the definition of input/

output relationships, the characteristics of a typical OTA are similar to those of a typical op-amp.

This chapter describes operation of the OTA and features various circuits using the OTA. As is discussed, the OTA provides the user with a wider variety of circuit arrangements than does the conventional op-amp. This is because the user can select the optimum circuit conditions for a specific application simply by varying the bias (I_{ABC}) conditions of the OTA. For example, if low power consumption, low bias, and low offset current, or high input impedance are desired, then low I_{ABC} is selected. On the other hand, if operation into a moderate load impedance is the main consideration, then higher levels of I_{ABC} are used.

10-1 BASIC OTA CIRCUITS

Figure 10-1 shows a simplified circuit diagram of an OTA. The output signal is a "current" that is proportional to the transconductance of the OTA established by the I_{ABC} and the differential input voltage.

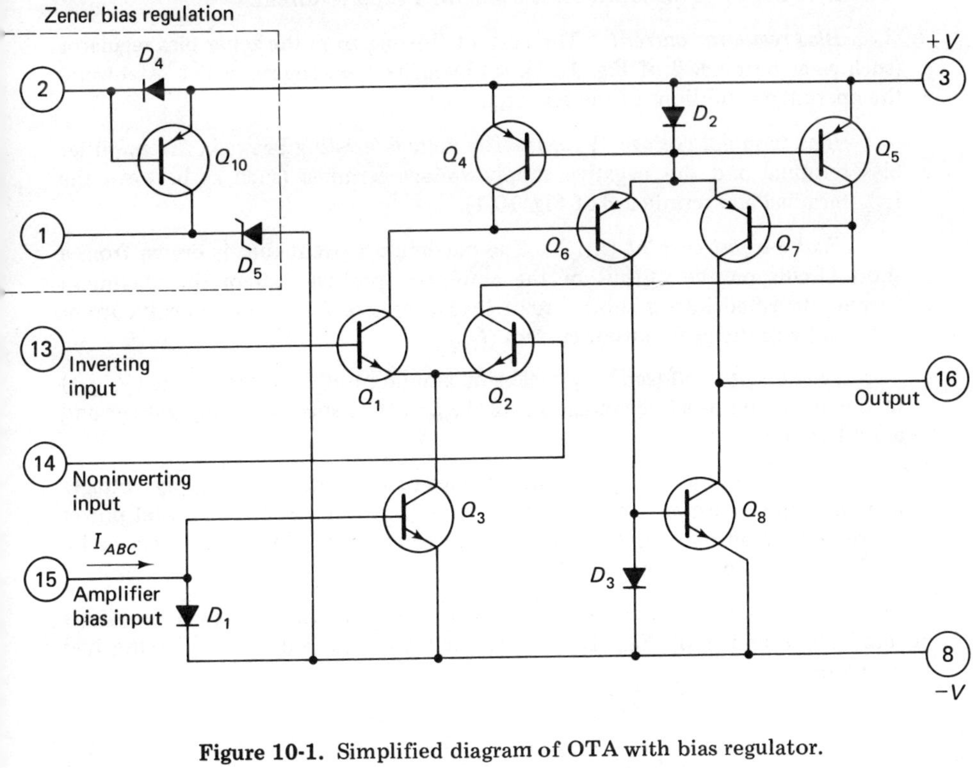

Figure 10-1. Simplified diagram of OTA with bias regulator.

The OTA can either source or sink current at the output, depending on the polarity of the input signal. The circuit shown in Fig. 10-1 is for an RCA OTA (designated in CA3060) that is available in IC form. All of the OTAs in the CA3060 incorporate a unique zener diode regulator system (D_4, D_5, Q_{10}) that permits current regulation below supply voltages normally associated with such systems.

10-1-1 Definition of OTA Terms

The following terms apply to all types of OTA circuits. However, the terms were first applied to IC OTA devices developed by RCA.

Amplifier bias current (I_{ABC}): The current supplied to the amplifier bias terminal to establish the operating point (such as the I_{ABC} current at the base of Q_3 in Fig. 10-1).

Amplifier supply current (I_A): The current drawn by the amplifier from the positive supply source. The total supply current (which includes the sum of the amplifier supply current, the amplifier bias current, and the regulator bias current) is not to be mistaken for the amplifier supply current.

Bias regulator current: The current flowing from the zener bias regulator (such as at terminal 2 of Fig. 10-1), set by an external source, which establishes the operating conditions of the bias regulator.

Bias terminal voltage (V_{ABC}): The voltage existing between the amplifier bias terminal and the negative supply voltage terminal (such as between the I_{ABC} terminal and terminal 8 of Fig. 10-1).

Peak output current (I_{OM}): The maximum current that is drawn from a short circuit on the output of the amplifier (positive I_O) or the maximum current delivered into a short circuit load (negative I_O). Peak-to-peak current swing is twice the peak output current (I_{OM}).

Peak output voltage (V_{OM}): The maximum positive voltage swing ($V_{OM}+$) or the maximum negative voltage swing ($V_{OM}-$) for a specific supply voltage and amplifier bias.

Power consumption (P): The product of the sum of the supply voltages and the supply current, or ($V+$ plus $V-$) $\times$ (I_A). This is not the total power consumed by an operating circuit. The power in the regulator must also be included for total power consumed.

Zener regulator voltage (V_Z): The regulator voltage (such as across terminals 1 and 8 of Fig. 10-1), measured with current flowing in the bias regulator.

10-1-2 Effects of Control Bias on OTA Characteristics

Unlike conventional op-amps, the characteristics of OTAs can be altered by adjustment of amplifier bias current, I_{ABC}. In effect, many of the OTA characteristics can be programmed to meet specific design problems. The following is a summary of the effects of bias on typical OTA units. Note that the characteristics listed here for OTAs are the same as for conventional op-amps described in Chapter 9.

Input offset current (I_{IO}) is directly affected by I_{ABC}. I_{IO} increases almost in direct proportion with increases in I_{ABC}. The same is essentially true for *input bias current* (I_{IB}), *amplifier supply current* (I_A), *device dissipation* (P_D), *transconductance* (g_m), and *peak output current* (I_{OM}).

Input offset voltage (V_{IO}) is not drastically affected by variations in I_{ABC}. A possible exception is when the OTA is operated at high temperatures. The same is essentially true of *peak output voltage* (V_{OM}), which is set (primarily) by supply voltage, as is the case with a conventional op-amp.

Input and output capacitance (C_I and C_O), as well as *amplifier bias voltage* (V_{ABC}) all increase with I_{ABC}. However, these characteristics do not increase in direct proportion to I_{ABC}. That is, a large increase in I_{ABC} produces a small increase in C_I, C_O, and V_{ABC}.

Input and output resistance (R_I and R_O) both decrease with increases in I_{ABC}.

10-2 BASIC OTA SYSTEM CIRCUITS

Figure 10-2 is the working schematic of a basic OTA system circuit, complete with external circuit components. The following paragraphs provide a specific design example for the circuit. Note that the OTA used in Fig. 10-2 is the RCA CA3060 shown in Fig. 10-1.

10-2-1 Design Example for Basic OTA

Assume that the circuit is to provide a closed-loop voltage gain of 10 (20 dB), the input offset voltage must be adjustable to zero, current drain must be as low as possible, the supply voltage is ±6 V, maximum input voltage is ±50 mV, input resistance is 20 kΩ, and load resistance is 20 kΩ.

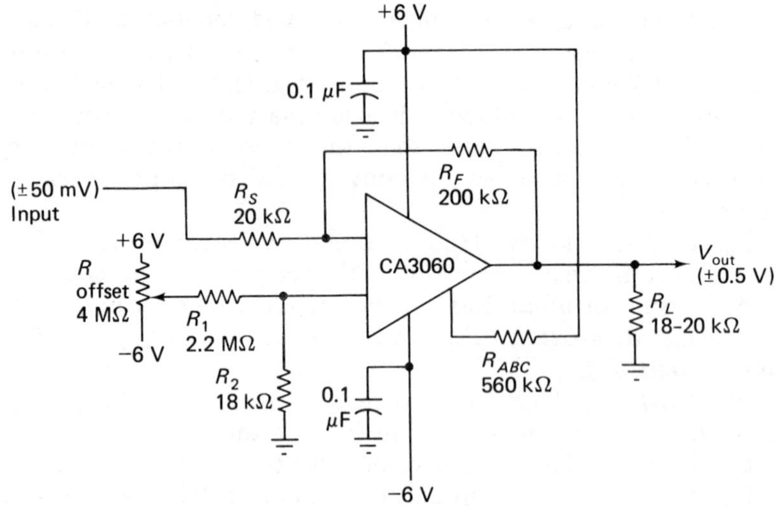

Figure 10-2. Basic OTA system circuit.

Transconductance. As in the case of a conventional op-amp, closed-loop gain is set by the ratio of feedback resistance R_F to input resistance R_S. With R_S specified as 20 kΩ, and a desired closed-loop gain of 10, the value of R_F is 10 × 20, or 200 kΩ. The next step is to calculate the required transconductance g_m (also shown as g_{21} on some OTA data sheets) to produce a suitable open-loop gain. Assume that the open-loop gain A_{OL} must be at least 10 times the closed-loop gain. With a closed loop gain of 10, the open-loop gain must be 10 × 10, or 100.

Open-loop gain A_{OL} is related directly to load resistance R_L and transconductance g_m. However, the actual load resistance is the parallel combination of R_L and R_F, or approximately 18 kΩ. With an A_{OL} of 100 and an actual load of 18 kΩ, the g_m should be 100/18,000, or approximately 5.5 millimhos (mmhos).

The transconductance g_m is set by I_{ABC}. With a data-sheet curve similar to that of Fig. 10-3, select an I_{ABC} from the minimum-value curve to assure that the OTA provides sufficient gain. As shown in Fig. 10-3, for a g_m of 5.5 mmhos, the required I_{ABC} is approximately 20 μA.

Output swing. Check that the calculated I_{ABC} of 20 μA will produce the desired output swing capability. With an input of ±50 mV and a gain of 10, the output voltage swing is ±0.5 V and the 0.5 V appears across the output load (approximately 18 kΩ). With a 0.5-V swing and an approximate load of 18 kΩ, the total amplifier

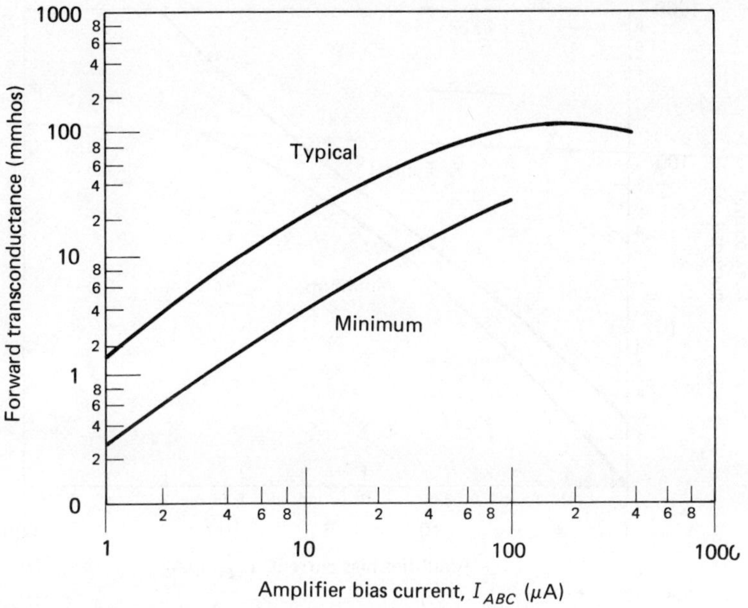

Figure 10-3. Forward transconductance versus amplifier bias current for OTA.

current output is approximately 0.5 V/18 kΩ, or 27.7 μA. With a data-sheet curve similar to that of Fig. 10-4, use the minimum-value curve to check that an I_{ABC} of 20 μA produces an I_{OM} of at least 27.7 μA. As shown by Fig. 10-4, an I_{ABC} of 20 μA produces an I_{OM} of about 40 μA, well above the required 27.7.

Calculating R_{ABC}. As shown in Fig. 10-2, R_{ABC} is connected to the +6-V supply. With this arrangement, R_{ABC} and diode D_1 are then in series between the $+V$ and $-V$ supplies, and there is a total of 12 V across the series components (as shown in Fig 10-1). The drop across D_1, which is V_{ABC}, can be found by reference to a curve similar to Fig. 10-5, which shows that an I_{ABC} of 20 μA produces a V_{ABC} of approximately 0.63 V. The drop across R_{ABC} is 12 - 0.63 V, or 11.37 V. For a drop of 11.37 V and an I_{ABC} of 20 μA, the value of R_{ABC} is 11.37/20 μA, or about 568 kΩ. Use the next lowest standard resistor of 560 kΩ, as shown in Fig. 10-2, to assure that a minimum I_{ABC} of 20 μA will flow.

Input offset circuit. The final step is to calculate values for the input offset adjustment circuit R_{offset}, R_1, and R_2 shown in Fig. 10-2. To reduce the loading effect of the offset adjustment circuit on

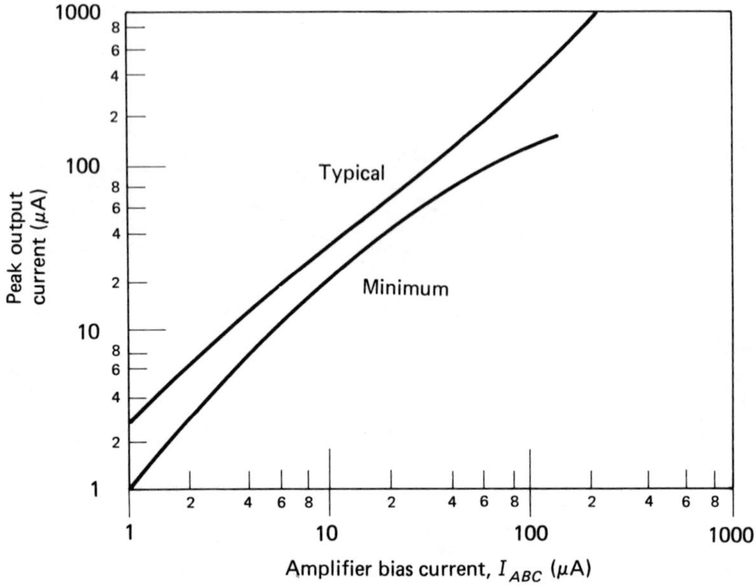

Figure 10-4. Peak output current versus amplifier bias current for OTA.

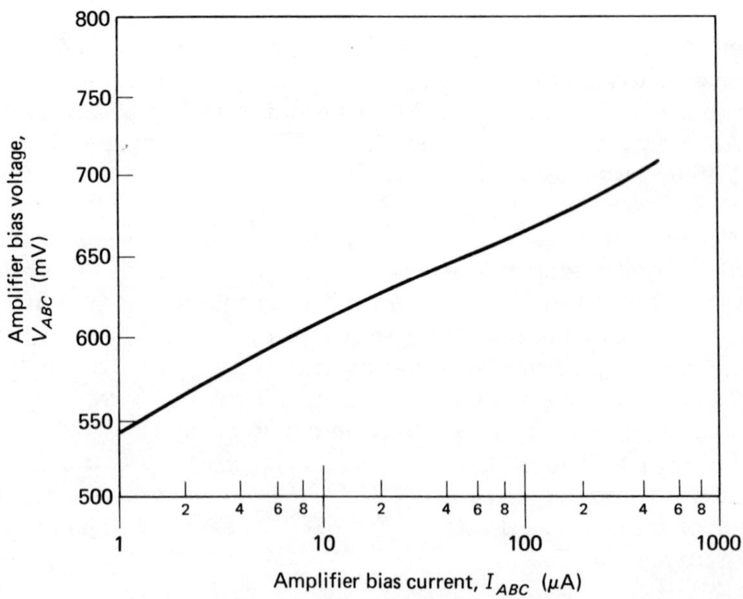

Figure 10-5. Amplifier bias voltage versus amplifier bias current for OTA.

the power supply, the values should be selected in a manner similar to that of a conventional op-amp. For example, the value of R_2 should be approximately equal to the parallel combination of R_S and R_F, or approximately 18 kΩ.

With a data-sheet curve similar to that of Fig. 10-6, find the input offset current. As shown in Fig. 10-6, for an I_{ABC} of 20 μA, the input offset current should be a maximum of 200 nA. With 200 nA flowing through R_2, the voltage across R_2 is 3.6 mV. This 3.6 mV must be added to the maximum input offset voltage possible for the OTA. The CA3060 data sheet shows a maximum input offset voltage of 5 mV. Thus, the maximum voltage required at the noninverting input is 5 mV + 3.6 mV, or 8.6 mV.

The current necessary to provide a possible offset voltage of 8.6 mV across an 18-kΩ resistance is about 0.48 μA. This current must flow through R_1. A possible ±6 V is available from R_{offset} to R_1. However, for a more stable circuit, assume that ±1 V is available to R_1. With 1 V available and a required current of 0.48 μA, the value of R_1 is approximately 2 MΩ. Use the next larger standard value of 2.2 MΩ. The value of R_{offset} is not critical. A larger value of R_{offset} draws less current from the supplies. As a guideline, the maximum

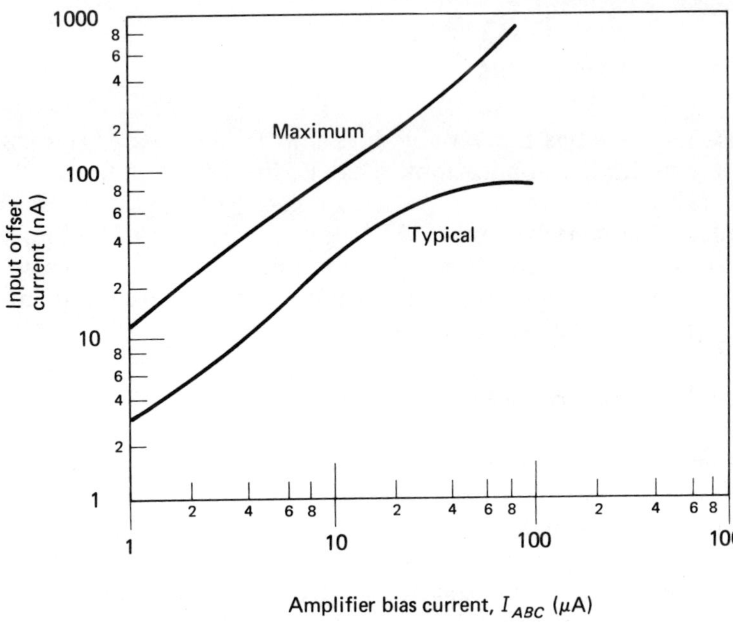

Figure 10-6. Input offset current versus amplifier bias current for OTA.

value of R_{offset} should be less than twice the value of R_1, or less than 4.4 MΩ in this case. Use a standard value of 4 MΩ.

This completes the final step in design of the circuit. Other circuit design considerations are essentially the same as for conventional op-amps. For example, the data sheet for the CA3060 recommends open-loop compensation by means of a resistor and capacitor across the differential input terminals. As always, the user should follow all data-sheet recommendations.

10-2-2 Typical OTA Applications

The remaining sections of this chapter describe how OTA units can be used in practical circuits. Keep in mind that an OTA can be used as a substitute for a conventional op-amp in any application provided that the OTA has comparable characteristics (frequency response, power output, etc.). The area in which an OTA can not be substituted for a conventional op-amp is one in which low output impedance is required. Even when OTA output impedance is reduced to the minimum by adjustment of I_{ABC}, the impedance is generally higher than for a comparable op-amp.

10-3 OTA MULTIPLEXERS

Because an OTA has the variable bias (I_{ABC}) feature, the OTA can be gated for multiplex applications. That is, the OTA can be gated full-on or full-off by means of pulses applied to the I_{ABC} input. In a multiplex circuit, two or more OTA units are connected so that their outputs are summed together, with the inputs receiving signals from separate sources. The resultant output is the combination of the input signals.

10-3-1 Two-Channel Multiplexer Circuits

Figure 10-7 shows a simple two-channel multiplex system using two RCA CA3080 OTA devices. In this example, positive and negative 5-V power supplies are used for the OTAs. The IC flip-flop is powered by the positive supply. If necessary to satisfy the logic supply requirements, the negative supply may be increased to −15 V, with the positive supply at +5 V.

Outputs from the clocked flip-flop are applied through PNP transistors to gate the OTAs (on alternate half-cycles) at the I_{ABC} input. The transistors are connected in the grounded-base configura-

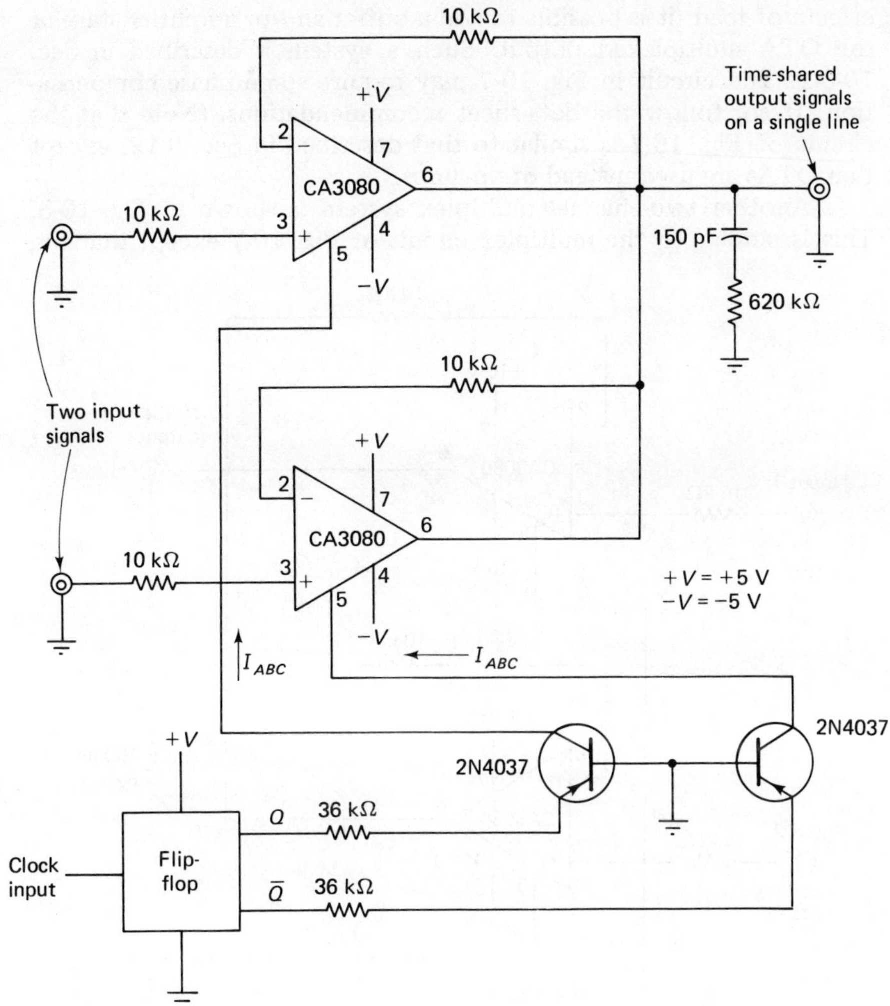

Figure 10-7. Two-channel linear time-shared multiplex circuit using OTAs.

tion to minimize capacitive feedthrough (from flip-flop to OTA) via the base-collector junction of the PNP transistors.

There is some level shift between input and output of the multiplex system. In the case of the OTA units shown, the level shift is about 2 mV for the CA3080A and 5 mV for the CA3080. Of course, the level shift depends on characteristics of the OTA used. In general, the level shift is due to input offset voltage of the OTA, rather than gain. Typically, open-loop gain of the system is about 100 dB, with normal loading at the output. To further increase gain and reduce the effects of load, it is possible to add a buffer and/or amplifier stage at the OTA multiplexed output. Such a system is described in Sec. 10-3-2. The circuit in Fig. 10-7 may require some phase compensation. If so, follow the data-sheet recommendations. (Note that the circuit of Fig. 10-7 is similar to that described in Sec. 9-19, except that OTAs are used instead of op-amps.)

Another two-channel multiplex system is shown in Fig. 10-8. This is similar to the multiplex circuit of Fig. 10-7 except that the

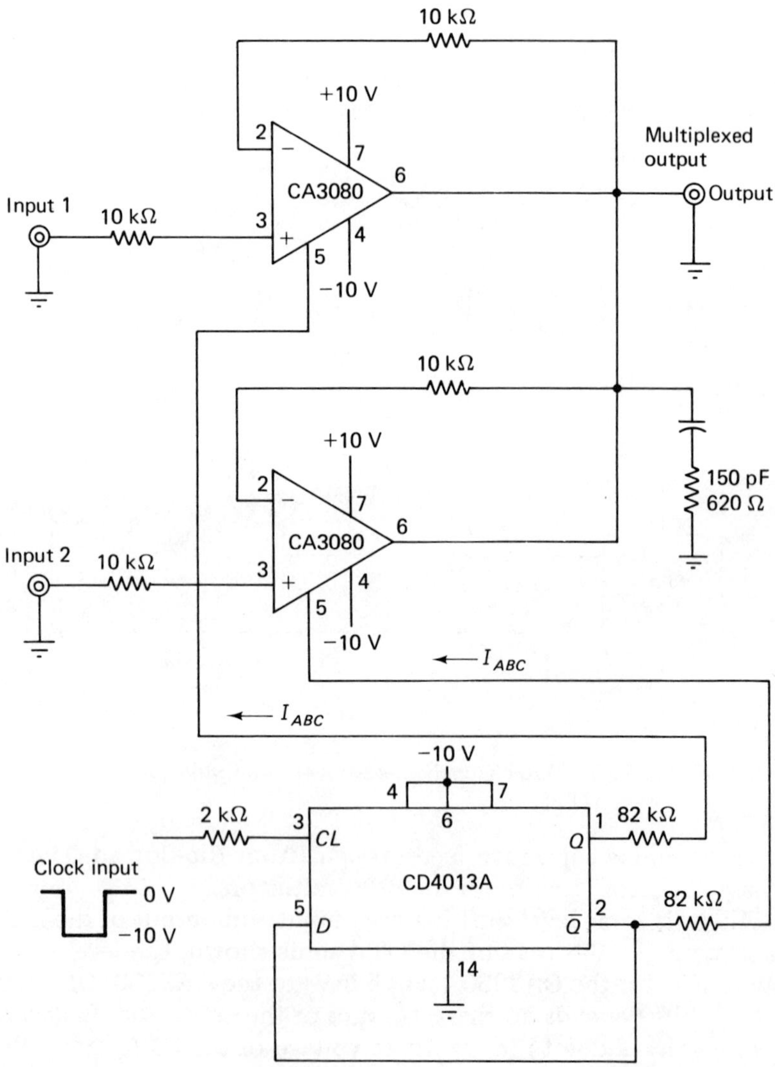

Figure 10-8. Two-channel linear multiplex system using COS/MOS flip-flop to gate two OTAs.

level-shifting transistors are omitted. These transistors are not needed in the circuit of Fig. 10-8 because of the higher output voltage from the flip-flop, which is an IC unit with outputs in the order of ±10 V.

10-3-2 Three-Channel Multiplexer Circuit

Figure 10-9 shows a three-channel multiplex system using an RCA CA3060 and a 3N138. The CA3060 is an IC package that con-

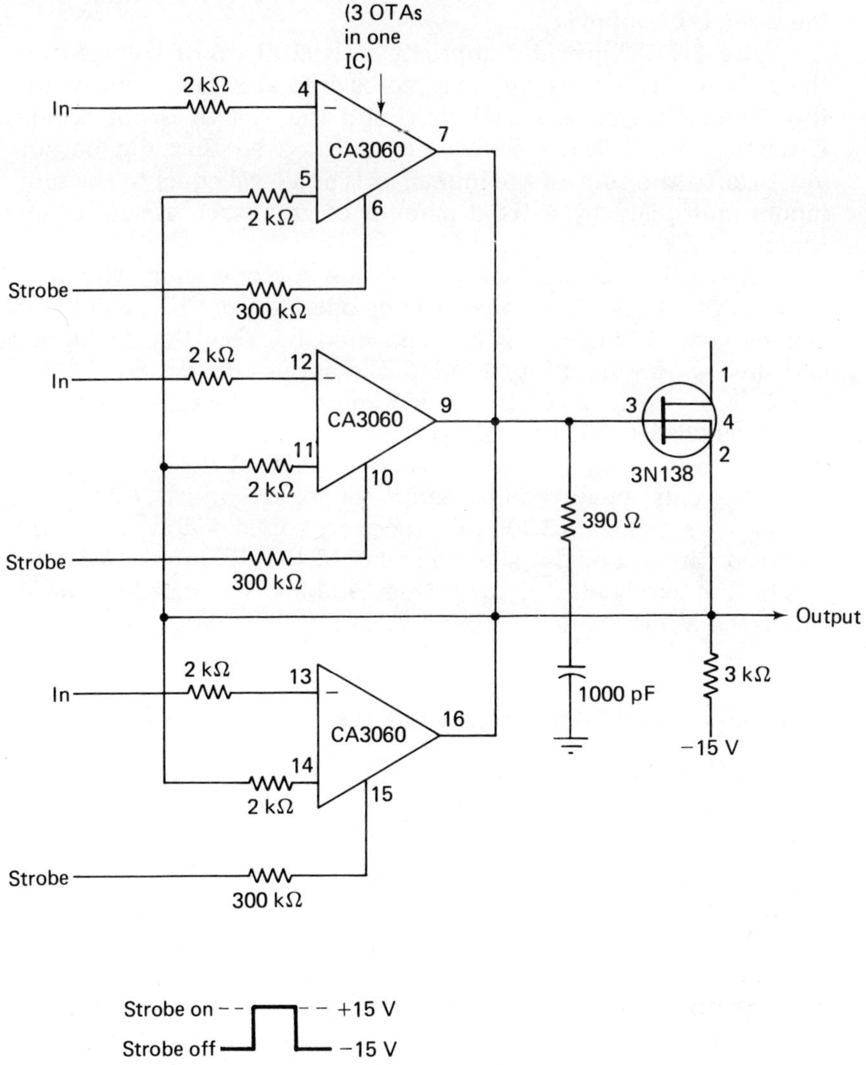

Figure 10-9. Three-channel multiplexer circuit using three OTAs in one IC.

tains three identical OTA units. The 3N138 is a MOSFET used as a buffer/amplifier.

Each OTA is connected as a high-input impedance voltage follower (similar to the op-amp voltage followers described in Chapter 9). A gating or "strobe" pulse is applied to each of the OTA units at some specific time interval. When the "strobe" pulse is applied, each OTA is activated and the OTA output swings to the approximate level of the input (at that OTA unit). All three OTAs can be strobed simultaneously, if required. The resultant output is then the sum of the three OTA inputs.

The 3N128 provides approximately 100 dB of voltage gain at the output. Part of the output is fed back to each OTA unit. Without this feedback, gain is unstabilized and the system is not accurate. Generally, multiplex systems are designed so that the output is identical to the sum of the inputs, or is at a level equal to the sum of inputs multiplied by a fixed amount of gain (such as sum of input times 100 dB).

The values shown in Fig. 10-9 are for operation with supply voltages of ±15 V. The system can be operated at ±6 V, with modifications. The 300-kΩ resistors connected between the strobe pulses and the I_{ABC} inputs must be reduced to about 100 kΩ. Similarly, the 3-kΩ drain resistance of the 3N128 must be reduced to about 1 kΩ. When reduced (±6 V) supply voltages are used, it is possible to use other MOSFET amplifiers, such as the RCA 40811.

The only phase compensation required consists of the single 390-Ω resistor and a 1000-pF capacitor, connected at the interface between the OTA outputs and the MOSFET input. With these values, the bandwidth of the system is about 1.5 MHz and the slew rate is 0.3 V/ms.

10-4 OTA SAMPLE-AND-HOLD CIRCUITS

The multiplex system described in Sec. 10-3 can be modified to produce sample-and-hold functions, using the strobe or gating characteristics of an OTA. That is, the I_{ABC} input of the OTA is pulsed to switch the OTA on and off. The output of the OTA then represents a "sample" of the OTA signal input (taken during the "on" period). The "hold" function is provided by means of a capacitor at the OTA output.

Figure 10-10 shows the basic sample-and-hold system using an RCA CA3080A as the amplifier. In this circuit, the OTA functions as a simple voltage follower, with the phase-compensation capacitor C serving the additional function of sampled-signal storage. When the

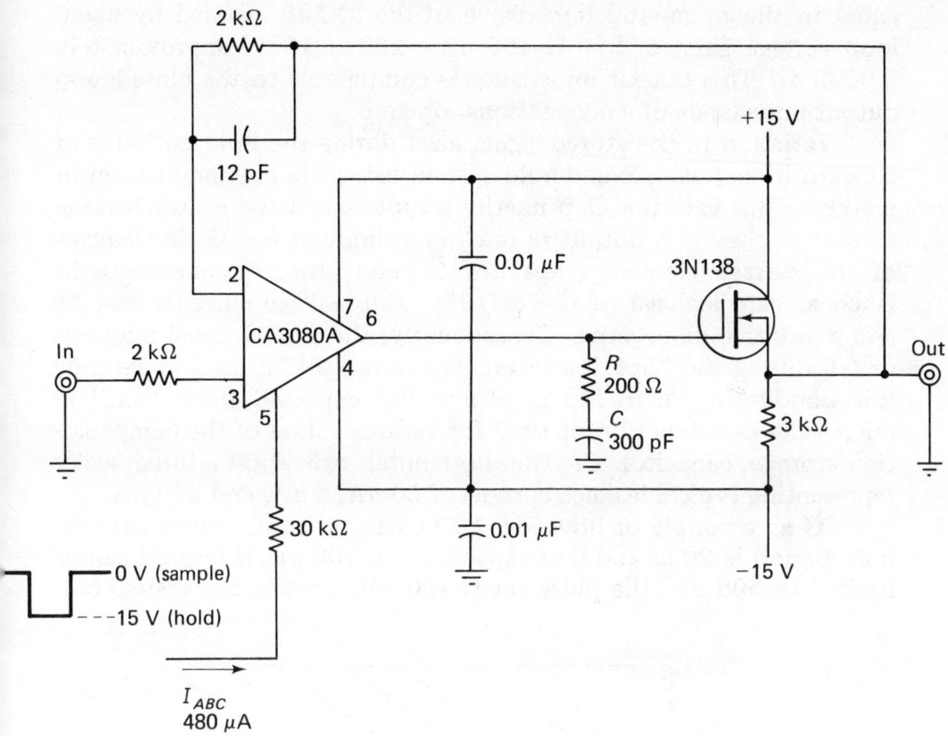

Figure 10-10. Sample-and-hold circuit using an OTA.

I_{ABC} input is at 0 V (SAMPLE), the OTA is on. Under these conditions, the OTA output is at the same level as the signal input, and the 300-pF phase-compensation capacitor C charges to the level of the OTA output. When the I_{ABC} input is at −15 V (HOLD), the OTA is off. However, capacitor C remains charged.

The main problem with any sample-and-hold system using a charging capacitor is that the capacitor may discharge through leakage during the OFF or HOLD cycle. Such leakage can occur through the amplifier output circuit, or the 3N138 input circuit. However, since an OTA has a very high output resistance, the leakage path through the OTA is practically nil. (The CA3080 has an output resistance in excess of 1000 MΩ under cutoff conditions.) Similarly, the gate leakage of the 3N138 is very low (typically 10 pA) since the transistor is a MOSFET (with insulated gate).

The open-loop voltage gain of the system is approximately 100 dB. The open-loop output impedance of the 3N138 is approximately 220 Ω (with a g_m of about 4600 μmhos at an operating current of 5 mA). The system closed-loop output impedance is approximately equal to the open-loop impedance of the 3N138, divided by open-loop voltage gain, or 220 Ω/100 dB = 220 Ω/10^5 = approximately 0.0022 Ω. This output impedance is comparable to the closed-loop output impedance of a conventional op-amp.

Variation in the stored signal level during the hold period is of concern in any sample-and-hold system using the charging capacitor method. This variation is primarily a function of the cutoff leakage current in the OTA output (a maximum limit of 5 nA), the leakage of the storage element (capacitor *C*), and other extraneous paths (such as gate leakage of the 3N138). The leakage currents may be either positive or negative. Consequently, the stored signal may rise or fall during the "hold" interval. The term "tilt" is used to describe this condition. Figure 10-11 shows the expected pulse "tilt" in microvolts as a function of time for various values of the compensation/storage capacitor *C*. The horizontal axis shows three scales representing typical leakage current of 50 nA, 5 nA, and 500 pA.

As an example of how Fig. 10-11 can be used, assume that the hold period is 20 μs and that capacitor *C* is 100 pF. If leakage can be limited to 500 pA, the pulse tilt is 100 μV. That is, the system out-

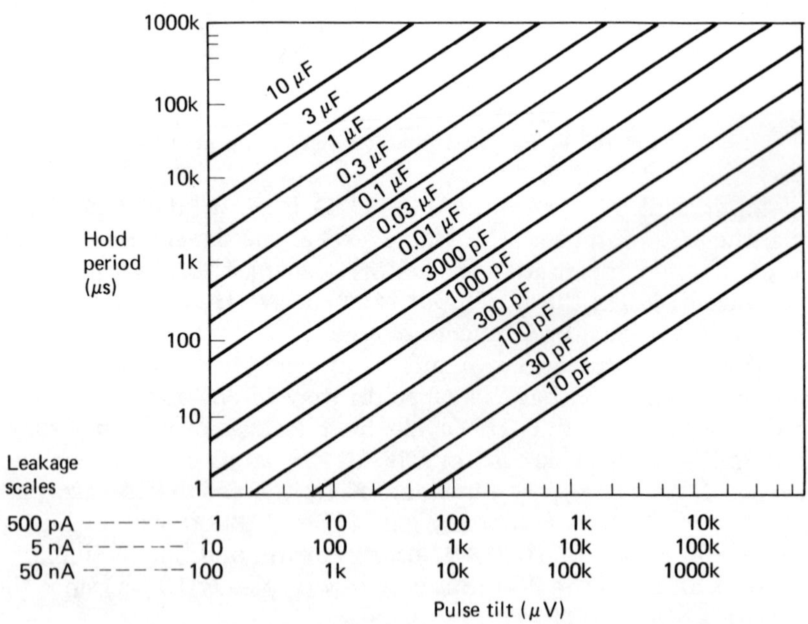

Figure 10-11. Chart showing "tilt" in sample-and-hold potentials as a function of hold time with load capacitance as a parameter.

put level can be shifted by 100 μV during the hold period. If the leakage current increases to 5 nA, the pulse tilt increases to 1000 μV. The effects of level shift (or pulse tilt) are cumulative. For example, if 10 hold periods are required for a system signal input, the total level shift is 10 times that for each hold period.

The basic circuit can be modified to operate at pulse levels found in DTL/TTL logic systems (typically 0 V and 5 V). Figure 10-12 shows the required modification. The control pulse to the I_{ABC} input is applied through a PNP transistor (RCA 2N4037) connected in the grounded-base configuration. Grounded-base is used to minimize capacitive feedthrough from the control pulse source to the I_{ABC} input. The 9.1-kΩ resistor connected to the 2N4037 emitter establishes I_{ABC} conditions similar to those used in the circuit of Fig. 10-10.

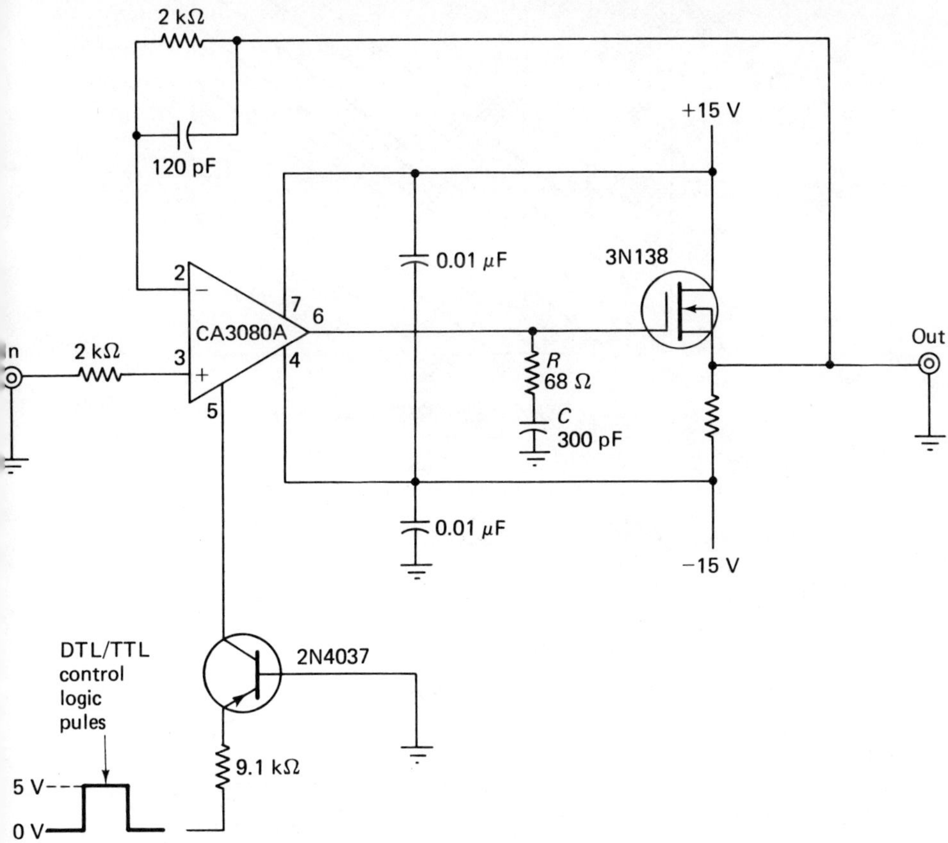

Figure 10-12. Sample-and-hold circuit for DTL/TTL control logic using an OTA.

There is a trade-off on the size of capacitor C. If the value of capacitor C is increased, larger signals can be sampled and held. However, each increase in capacitor C decreases slew rate. In general, use the largest possible capacitor C compatible with required slew rate. Figure 10-13 shows slew rate as a function of I_{ABC} and various capacitors C. The information of Fig. 10-13 applies to capacitor C in both Figs. 10-10 and 10-12. The magnitude of the current being supplied to capacitor C is equal to the I_{ABC} when the OTA is supplying maximum output current. As a guideline, let the RC network equal the lowest frequency at which OTA gain begins to drop.

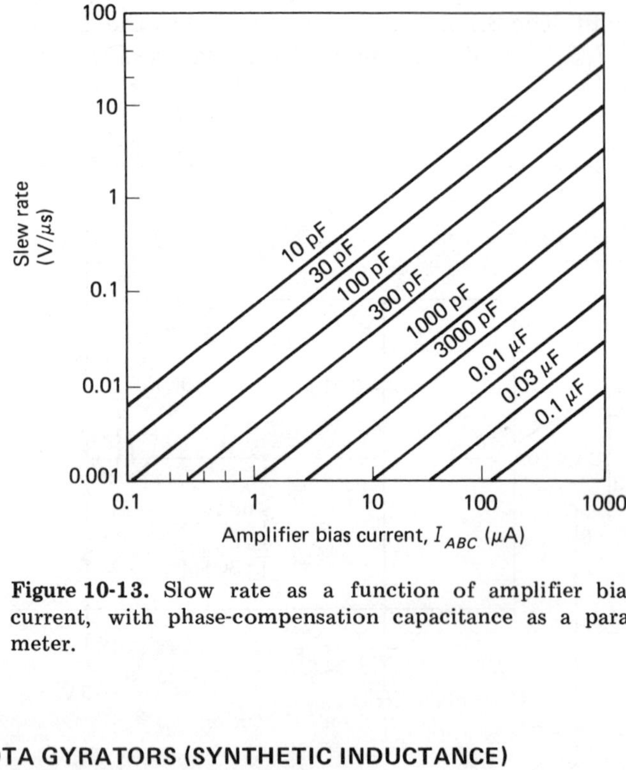

Figure 10-13. Slow rate as a function of amplifier bias current, with phase-compensation capacitance as a parameter.

10-5 OTA GYRATORS (SYNTHETIC INDUCTANCE)

One of the problems in designing filters used at very low frequencies is the need for a very large inductance. This problem can be overcome by means of active filter gyrator circuits. A gyrator is a circuit that appears as a variable inductance of high value (typically in the kilohenry range) but does not contain any inductive components.

Figure 10-14 shows a gyrator circuit composed of two OTA units. The circuit appears as an inductance of up to 10 kH across terminals A and B. The circuit can be tuned to exact inductance values

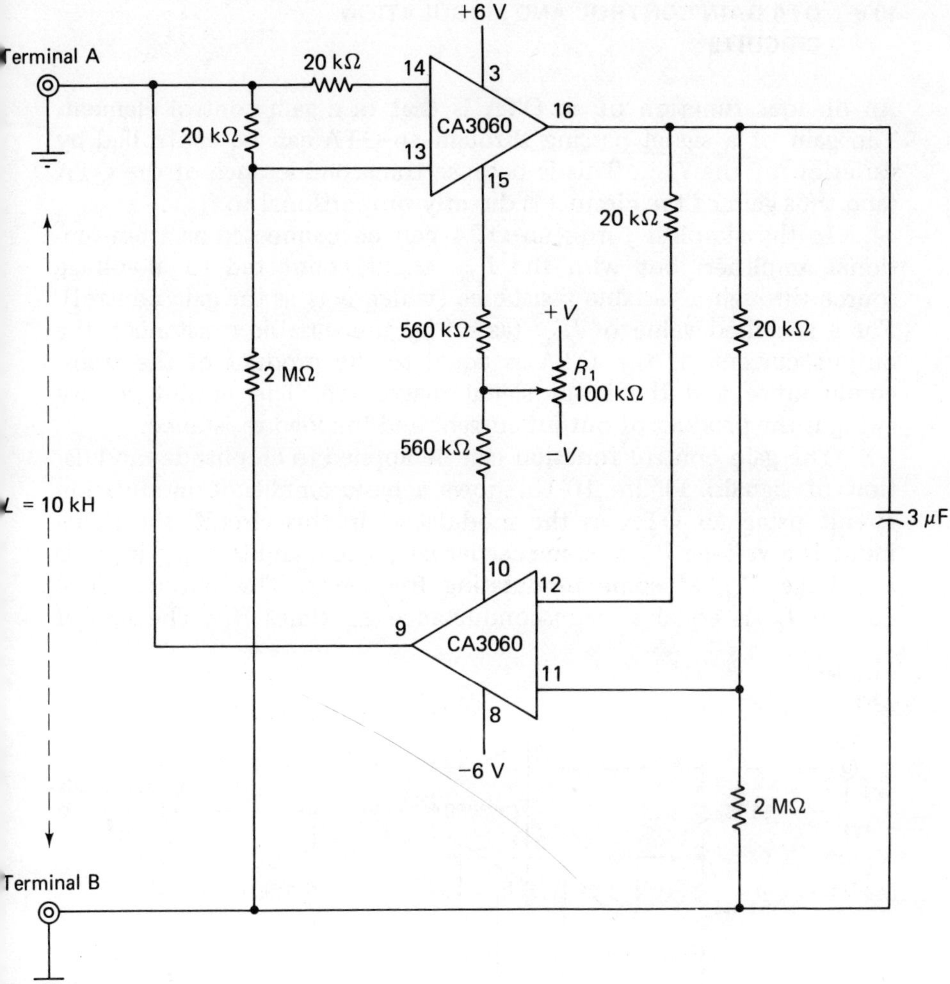

Figure 10-14. Gyrator (synthetic inductance) for an active filter circuit using two OTAs.

by means of R_1. The setting of R_1 varies the I_{ABC} current of both OTAs simultaneously. When one I_{ABC} current is increased, the opposite I_{ABC} current is decreased. This action varies the transconductance of both OTAs simultaneously and serves to "tune" the circuit by changing the resistance of the OTAs. The circuit of Fig. 10-14 makes use of the high-impedance output available in OTA units. The Q of the 10-kH variable inductance is approximately 13, using the circuit values shown. The 20-kΩ to 2-MΩ attenuators in this circuit extend the dynamic range of each OTA by a factor of 100.

10-6 OTA GAIN CONTROL AND MODULATION CIRCUITS

An obvious function of an OTA is that of a gain control element. The gain of a signal passing through an OTA can be controlled by variation of the I_{ABC}. This is because transconductance of the OTA (and thus gain of the circuit) is directly proportional to I_{ABC}.

In the simplest form, an OTA can be connected as a conventional amplifier, but with the I_{ABC} input connected to a voltage source through a variable resistance (which acts as the gain control). For a specified value of I_{ABC} (as set by the variable resistance), the output current of the OTA is equal to the product of the transconductance and the input signal magnitude. The output voltage swing is the product of output current and the load resistance.

The gain control function can be applied to amplitude modulation of signals. Figure 10-15 shows a basic amplitude modulation circuit using an OTA as the modulator. In this circuit, the signal input is a voltage V_X at some carrier frequency, and the I_{ABC} input is a voltage V_M at some modulating frequency. The output signal current I_O is equal to transconductance g_m times V_X. The sign of

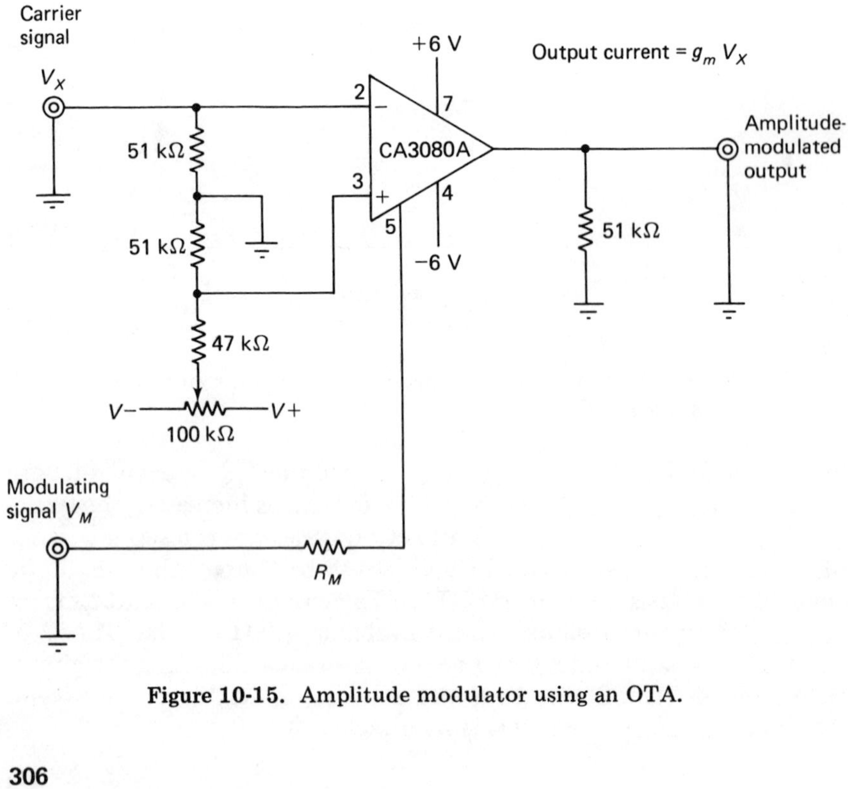

Figure 10-15. Amplitude modulator using an OTA.

the output signal is negative because the input signal is applied to the inverting input of the OTA.

The transconductance of the OTA is controlled by adjustment of I_{ABC}, as usual. However, in this circuit, the level of the unmodulated carrier output is established by a particular I_{ABC} through resistor R_M. Amplitude modulation of the carrier frequency occurs because variations of the voltage V_M force a change in the I_{ABC} supplied via resistor R_M. When V_M goes in the negative direction (toward the I_{ABC} terminal potential), the I_{ABC} decreases and reduces g_m of the OTA. When V_M goes positive, the I_{ABC} increases, resulting in an increase of the g_m.

For the particular OTA shown (an RCA CA3080A), the g_m is approximately equal to $19.2 \times I_{ABC}$, where g_m is mmhos and I_{ABC} is in mA. In this case, I_{ABC} is approximately equal to

$$\frac{V_M - V^-}{R_M} = I_{ABC}$$

$$I_O = -g_m V_X$$

$$g_m V_X = (19.2)(I_{ABC})(V_X)$$

$$I_O = \frac{-19.2(V_M - V^-)V_X}{R_M}$$

$$I_O = \frac{19.2(V_X)(V^-)}{R_M} - \frac{19.2(V_X)(V_M)}{R_M} \quad \text{(modulation equation)}$$

Note that there are two terms in the modulation equation. The first term represents the fixed carrier input, independent of V_M. The second term represents the modulation, which either adds to or subtracts from the first term. When V_M is equal to the V^- term, the output is reduced to zero.

In the preceding modulation equations, the term

$$(19.2)(V_X)\frac{V_{ABC}}{R_M}$$

involving the amplifier bias current terminal voltage V_{ABC} was neglected. This term was assumed to be small because V_{ABC} is small compared with V^- in the equation. If the amplifier bias current terminal is driven by a current-source (such as from the collector of a PNP transistor), the effect of V_{ABC} variation is eliminated. Instead, any variation is dependent upon the PNP transistor base-emitter junction characteristics. Figure 10-16 shows a method of driving the I_{ABC} input using a PNP transistor.

If an NPN transistor is added to the circuit of Fig. 10-16, as an emitter follower to drive the PNP transistor, variations resulting from base-emitter characteristics of the PNP are considerably reduced because of the complementary nature of the NPN and PNP base-emitter junctions. Also, the temperature coefficients of the two base-emitter junctions tend to cancel one another. Figure 10-17 shows a configuration using one transistor (in an RCA CA3018A NPN transistor array) as an emitter follower (with the three remaining transistors of the transistor array connected as a current source for the emitter followers).

Note that the circuits of Figs. 10-15 and 10-16 use ±6-V supplies, whereas the circuit of Fig. 10-17 requires a ±15-V supply. The 100-kΩ potentiometer shown in all three circuits is used to null the effects of amplifier input offset voltage (V_{IO}). This potentiometer is

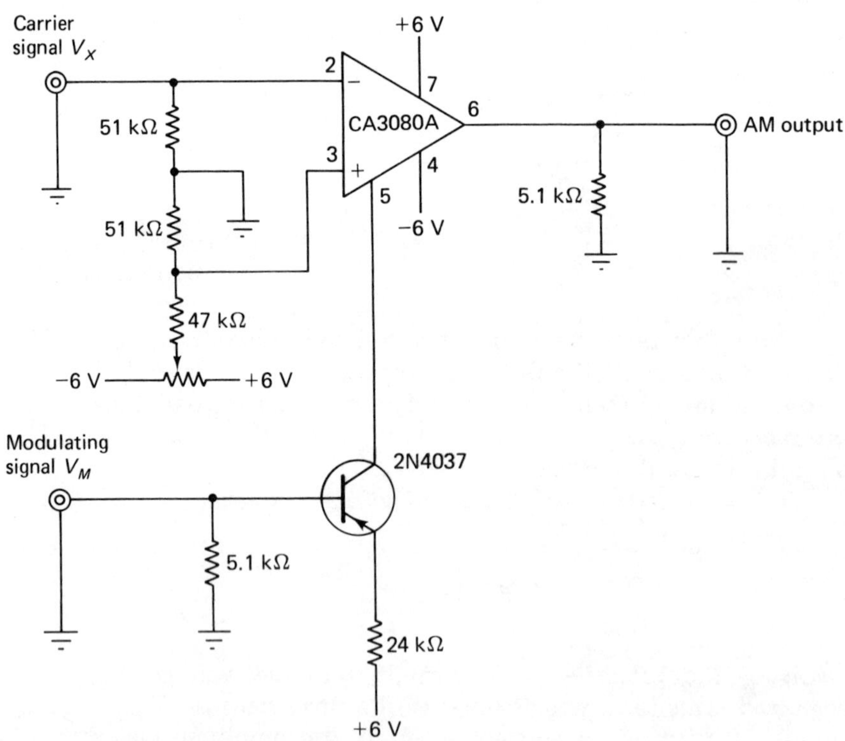

Figure 10-16. Amplitude modulator using an OTA controlled by a PNP transistor.

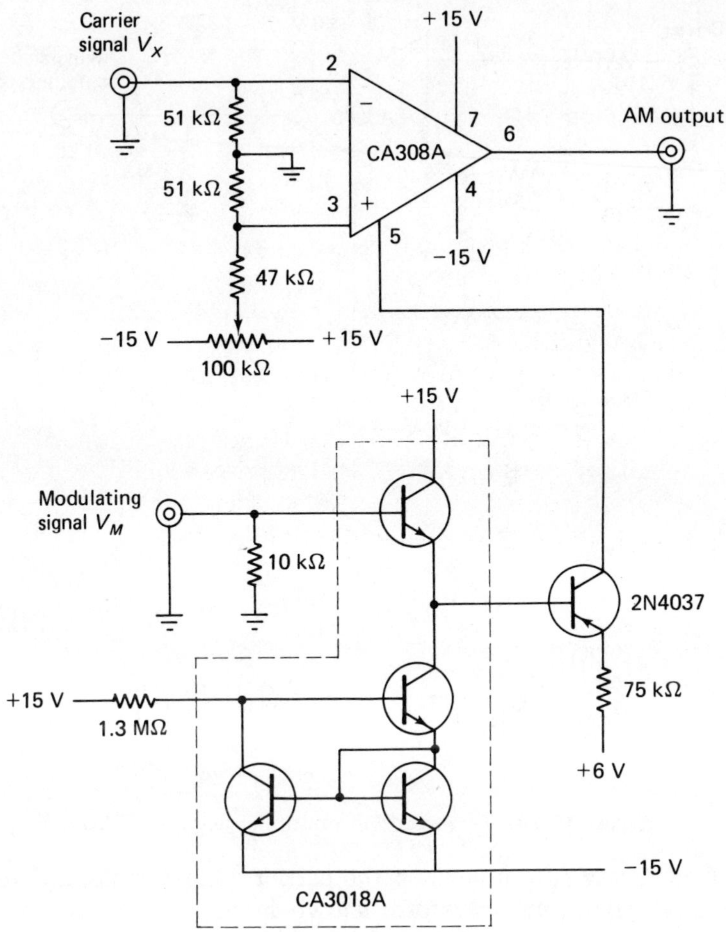

Figure 10-17. Amplitude modulator using an OTA controlled by PNP and NPN transistors.

used to set the output voltage symmetrically about zero. The OTA modulation circuit described here permits a range exceeding 1000:1 in gain, and thus provides modulation of the carrier signal input in excess of 99 percent.

10-7 OTA TWO-QUADRANT MULTIPLIERS

Figure 10-18 shows an OTA used as a two-quadrant multiplier. Note that the circuit of Fig. 10-18a is essentially the same as for the modulator circuits described in Sec. 10-6. That is, when modulation

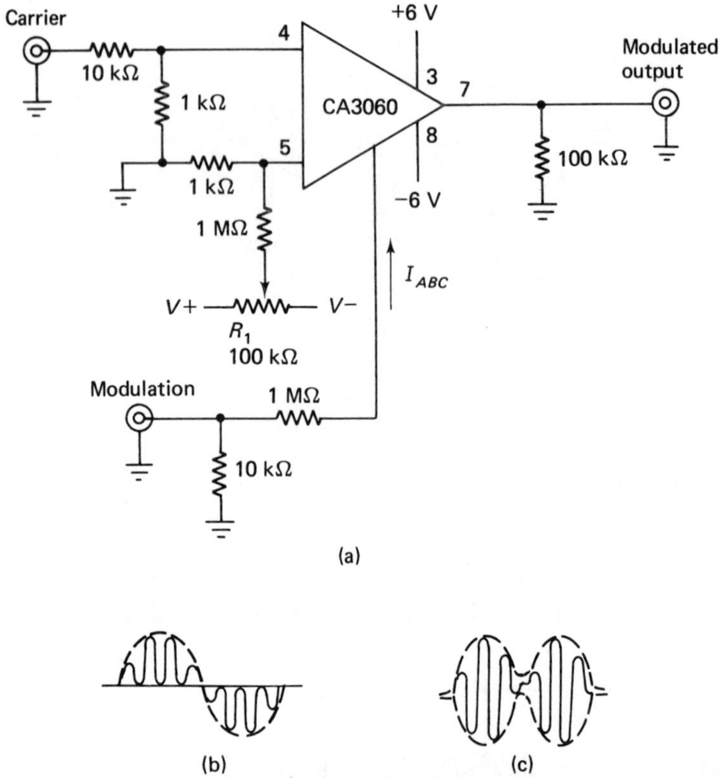

Figure 10-18. Two-quadrant multiplier using an OTA.

is applied to the I_{ABC} input and the carrier voltage is applied to the differential input, the waveform shown in Fig. 10-18b is obtained. However, in the circuit of Fig. 10-18, the input offset control R_1 is adjusted to balance the circuit so that no modulation can occur at the output without a carrier input. The linearity of the modulator is indicated by the solid trace of the superimposed modulating frequency, as shown in Fig. 10-18b. The maximum depth of modulation (or percentage of modulation) is determined by the ratio of the peak input modulating voltage to V^-.

The two-quadrant multiplier characteristic of the circuit is seen if modulation and carrier are reversed (modulation to differential input, carrier to I_{ABC}), as shown in Fig. 10-18c. The polarity of the output must follow that of the differential input. Thus, the output is positive only during the first (or positive) half-cycle of the modulation and is negative only during the second half-cycle. Note that both input and output signals are reference to ground. The output signal is zero when either the differential input or I_{ABC} is zero.

10-8 OTA FOUR-QUADRANT MULTIPLIERS

OTA units can be used as four-quadrant multipliers. Two circuit configurations are possible. One circuit uses three identical OTA devices. The second circuit uses a single OTA.

10-8-1 Four-Quadrant Multiplier with Three OTA Units

Figure 10-19 shows a complete schematic for the four-quadrant multiplier, including all the adjustment controls associated with differential input and an adjustment for equalizing the gains of amplifiers 1 and 2. Adjustment of the circuit is quite simple. With both the X and Y voltages at zero, connect terminal 10 to terminal 8. This procedure disables amplifier 2. Adjust the offset voltage of amplifier 1 to zero by means of R_1. Remove the short between terminals 8 and 10. Connect terminal 15 to terminal 8. This disables amplifier 1. Adjust the offset voltage of amplifier 2 to zero by means of R_2.

With a-c signals on both the X and Y inputs, adjust R_3 and R_{11} for symmetrical output signals. Figure 10-20 shows the output waveform with the circuit adjusted. Figure 10-20a shows suppressed carrier modulation of a 14-kHz carrier with a triangular wave. Figures 10-20b and c, respectively, show the squaring of a triangular wave and a sine wave. Notice that in both cases the outputs are always positive and return to zero after each cycle.

10-8-2 Four-Quadrant Multipliers Using a Single OTA

A single OTA can be used in many low-frequency, low-power, four-quadrant multiplier applications. The basic multiplier circuit shown in Fig. 10-21, using an RCA CA3080A, is particularly useful in waveform generation, double-balanced modulation, and other signal-processing applications, in portable equipment where low-power consumption is essential and accuracy requirements are moderate. The multiplier circuit of Fig. 10-21 is basically an extension of the gain-control function discussed in Sec. 10-6. Note that the single OTA version of the four-quadrant multiplier does not provide the accuracy of the system using three OTA units.

To obtain a four-quadrant multiplier, the first term of the modulation equation (which represents the fixed carrier) must be reduced to zero. In the circuit of Fig. 10-21, this is done by placing a feedback resistor R between the output and the inverting input terminal of the OTA. The value of the feedback resistor R is equal to $1/g_m$.

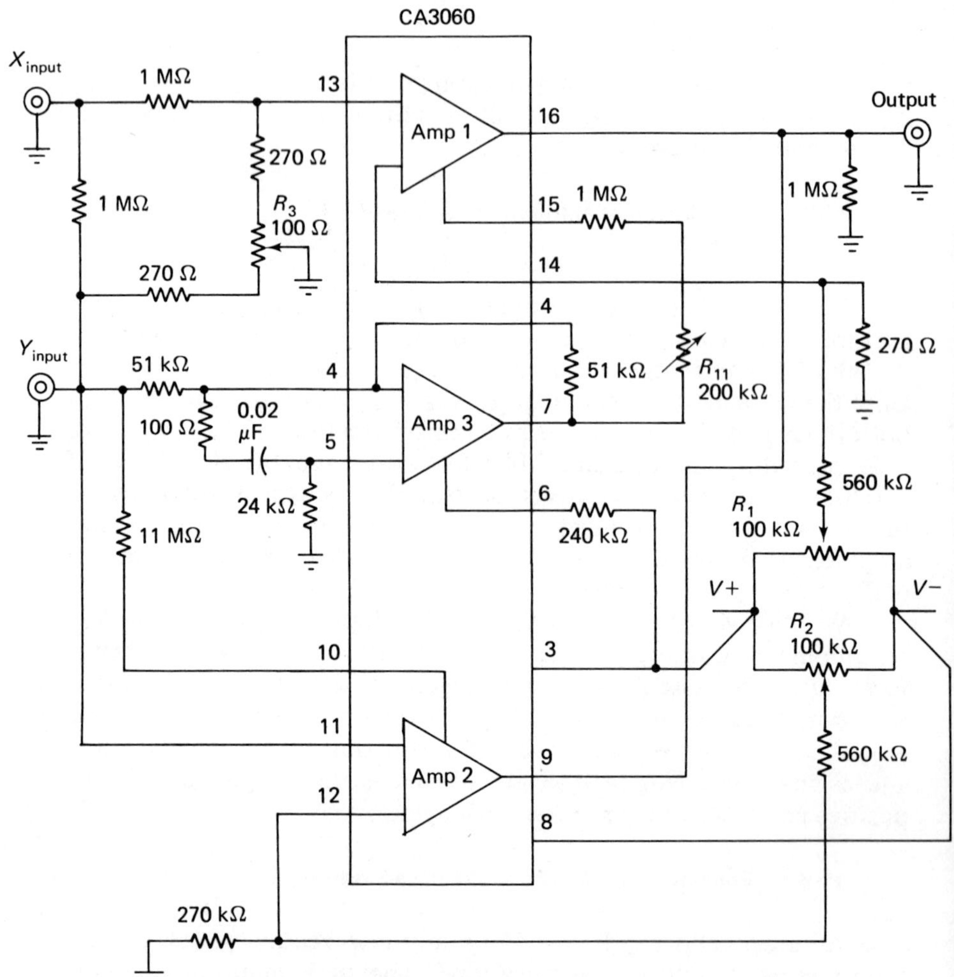

Figure 10-19. Typical four-quadrant multiplier circuit using CA3060 OTA.

The output current I_O is equal to g_m $(-V_X)$ since the input is applied to the inverting terminal of the OTA. The output current due to the resistor R is equal to V_X/R. Thus, the two signals (V_X and V_Y) cancel when R equals $1/g_m$.

For the particular OTA shown, the current is

$$I_O = \frac{-19.2 V_X V_Y}{R_M}$$

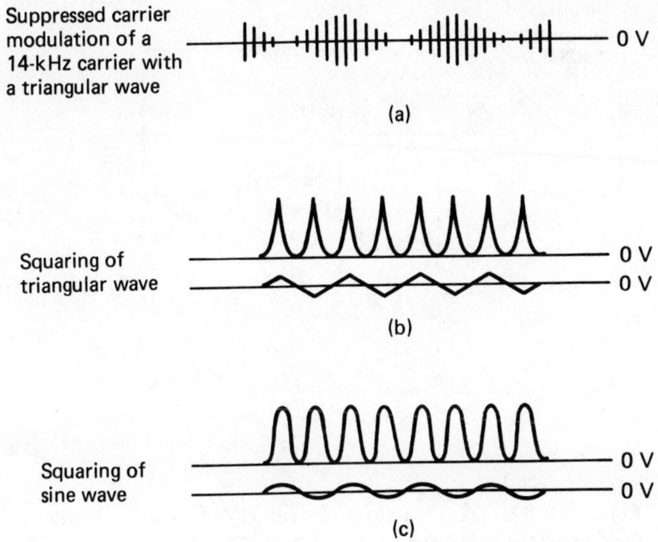

Figure 10-20. Voltage waveforms of four-quadrant multiplier circuit.

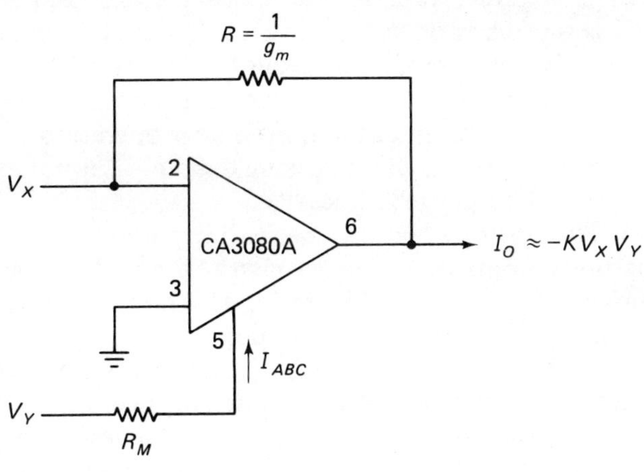

Figure 10-21. Basic four-quadrant analog multiplier using an OTA.

The output signal for the single OTA configuration is a "current" that is best terminated by a short circuit. This condition can be satisfied by making the load resistance for the multiplier output very small. An alternative method for lowering the load resistance is to feed the output of the multiplier into a current-to-voltage converter, as shown in Fig. 10-22.

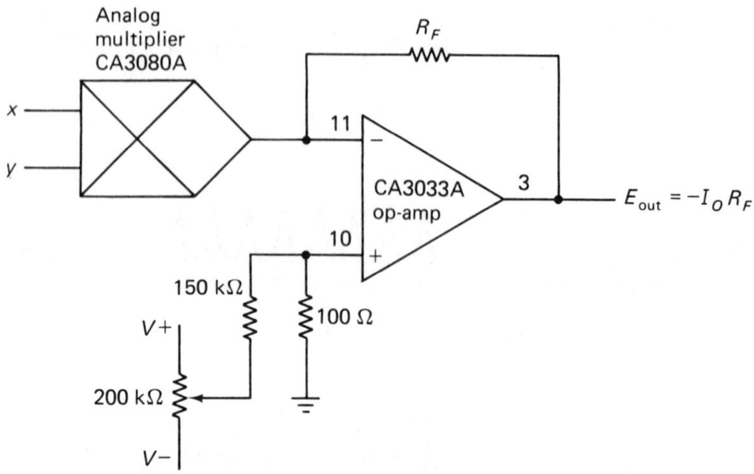

Figure 10-22. OTA analog multiplier driving an op-amp that operates as a current-to-voltage converter.

The current "cancellation" in the feedback resistance (R of Fig. 10-21) is a direct function of the OTA differential amplifier linearity. If the OTA is operated on a nonlinear portion of its range, "cancellation" is not complete and the first term of the modulation may not be reduced to zero. To prevent this condition, operate the OTA at low signal levels. In general, nonlinear operation occurs when the OTA is overdriven. In the following examples, signal excursion is limited to ±10 mV to preserve linearity.

Figure 10-23 shows the schematic diagram of a basic multiplier, with adjustment controls added to give the circuit an accuracy of approximately ±7 percent of full scale. There are only three adjustment controls: R_1 at the output compensates for slight variations in the current-transfer ratio of the OTA (which would otherwise result in a symmetrical output about some current level other than zero); R_2 in the V_Y input establishes the g_m of the system equal to the value of the fixed resistor shunting the system when the Y-input is zero; and R_3 compensates for an error that may arise from input offset voltage.

The following procedure is used to adjust the circuit of Fig. 10-23:

1. Set R_1 to the center of its range.
2. Ground the X- and Y-inputs.
3. Adjust R_3 until a zero volt reading is obtained at the output.

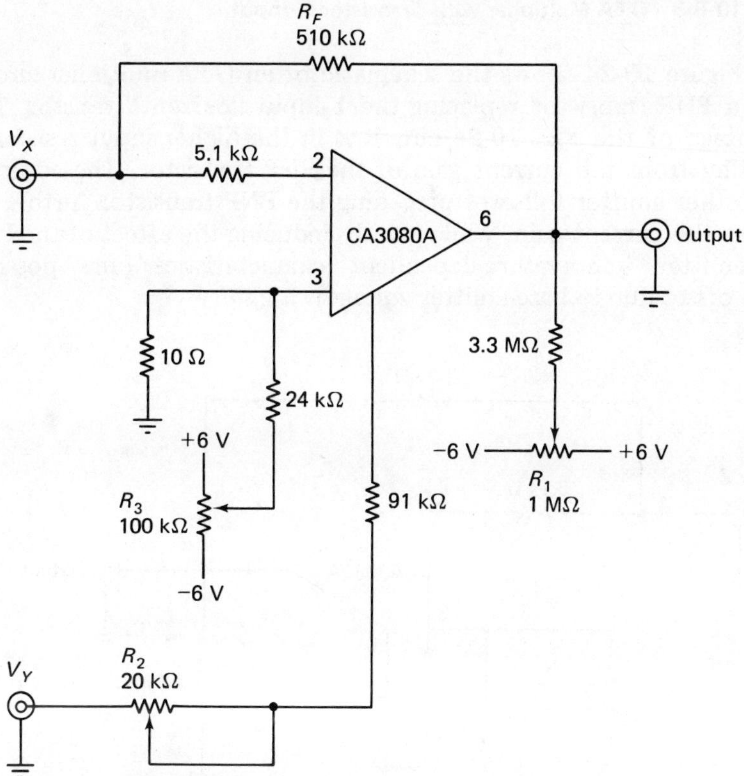

Figure 10-23. Analog multiplier using an OTA.

4. Ground the Y-input and apply a signal to the X-input through a low source impedance generator. (It is essential that a low-impedance source be used. This minimizes any change in the g_m balance for zero point due to the Y-input bias current, which is an I_{ABC} of about 50 μA.)
5. Adjust R_2 until a reading of zero volts is obtained at the output. This adjustment establishes the g_m of the OTA at the proper level to cancel the output signal. The output current is diverted through the feedback resistance R_F.
6. Ground the X-input and apply a signal to the Y-input through a low-source-impedance generator.
7. Adjust R_3 for an output of zero volts.

There will be some interaction among the adjustments and the procedure should be repeated for best circuit performance.

10-8-3 OTA Multiplier with Transistor *Y*-Input

Figure 10-24 shows the schematic of an OTA multiplier circuit with a PNP transistor replacing the *Y*-input "current" resistor. The advantage of the Fig. 10-24 circuit is in the higher input resistance resulting from the current gain of the PNP transistor. The addition of another emitter follower preceding the PNP transistor further increases the current gain, while greatly reducing the effect of the PNP base-emitter temperature-dependent characteristics (and possible input offset due to base-emitter voltage).

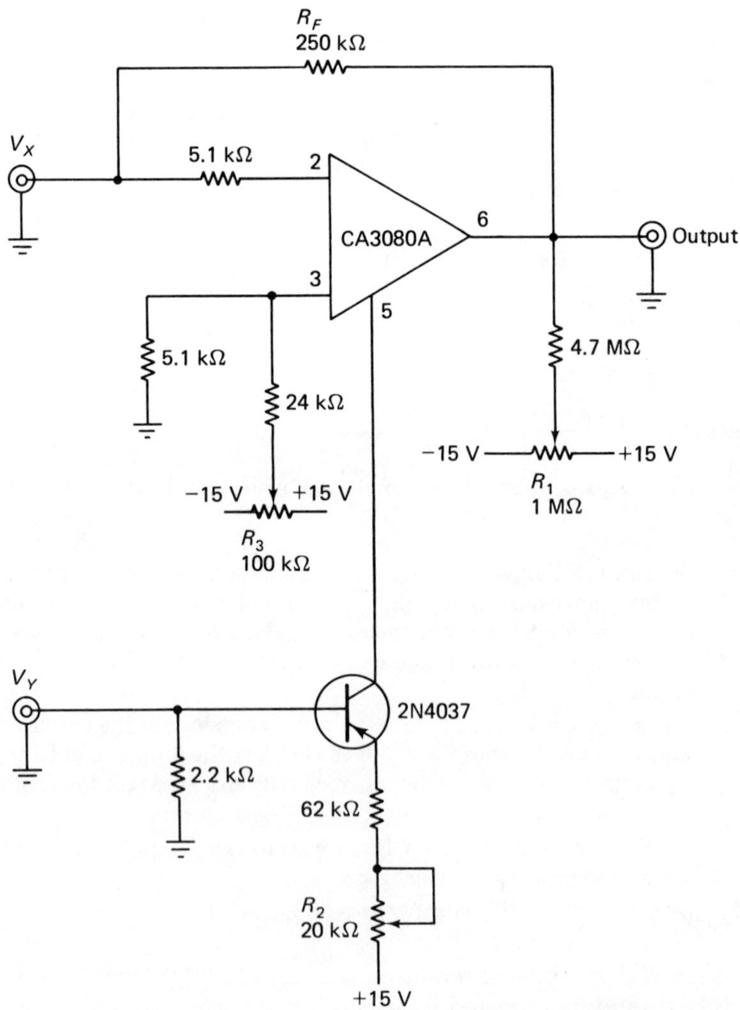

Figure 10-24. OTA multiplier with transistor Y-input.

Figure 10-25 shows output signals of the Fig. 10-24 circuit. Figure 10-25a and b shows outputs when the circuit is used as a suppressed-carrier generator. Figure 10-25b and c shows outputs

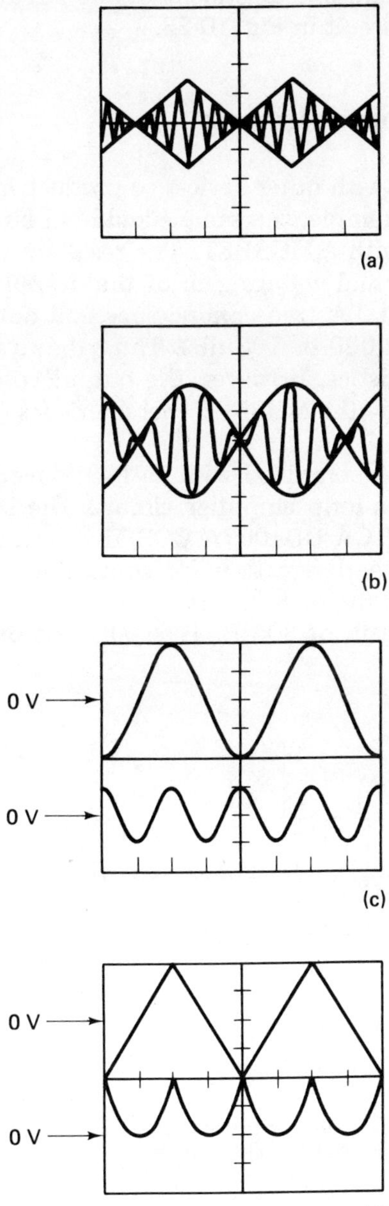

(a) 500 mV/div.
200 μs/div.
Triangular input 700 Hz to V_Y, 5 V p-p
Carrier input 21 kHz to V_X, 13.5 V p-p

(b) 500 mV/div.
200 μs/div.
Modulating frequency 700 Hz to V_Y, 5 V p-p
Carrier input 21 kHz to V_X, 13.5 V p-p

(c) Top trace: input to X and Y, 2 V/div., 1 ms/div., 200 Hz

Bottom trace: output, 500 mV/div., 1 ms/div., 400 Hz

(d) Same as (c)

Figure 10-25. Waveforms of analog multiplier.

when the circuit is used in signal squaring (that is, in squaring sine-wave and triangular-wave inputs).

If ±15-V power supplies are used (as shown in Fig. 10-24), both inputs can accept ±10-V input signals. Adjustment of the circuit in Fig. 10-24 is the same as for the circuit in Fig. 10-23.

10-9 OTA WITH HIGH-CURRENT-OUTPUT STAGES

It is possible to combine an OTA with other devices to produce high-gain, high-current circuits. For example, the sample-and-hold circuit of Sec. 10-4 combines an OTA with a MOSFET. The resultant voltage gain is about 100 dB. The actual voltage gain of the overall circuit is equal to the product of OTA transconductance and output resistance (which is typically 142,000 or 103 dB). Thus, the overall gain is set by the OTA characteristics. However, the output voltage swing and current swing are set by the MOSFET characteristics (and the source-terminal load).

Figure 10-26 shows an OTA combined with an MOS inverter/amplifier to form a simple, open-loop amplifier circuit. The MOS device shown is one-third of an RCA CD4007A COS/MOS inverter. (The term COS/MOS is the RCA designation for complementary-symmetry MOS devices.) Each of the three inverter/amplifiers in the CD4007A has a typical voltage gain of 30 dB. This gain, combined

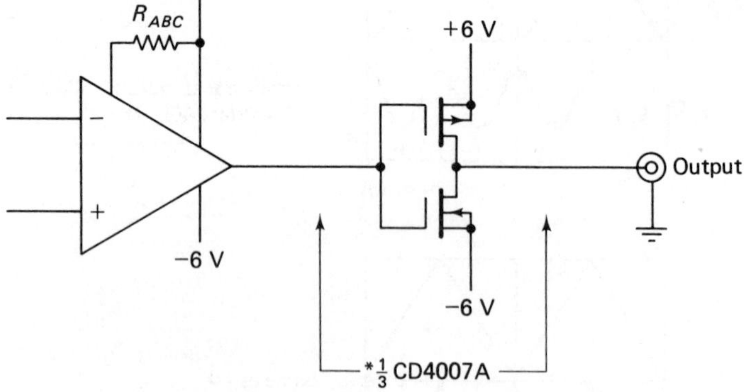

*Additional current output can be obtained when remaining two amplifiers of the CD4007A are connected in parallel with the single stage shown.

Figure 10-26. OTA combined with MOS to form open-loop amplifier circuit with high current output.

with the typical 100-dB gain of the OTA, results in a total voltage gain of about 130 dB. (Note that OTA gain is affected by I_{ABC}, which, in turn, is set by R_{ABC}.)

There are several circuit configurations available when an OTA is combined with MOS inverter/amplifiers. Some of these configurations provide additional gain, while others provide additional current capacity.

The circuit shown in Fig. 10-27 provides unity gain (with the same resistance values for R_F and R_I), and increased current capacity. The MOS inverter/amplifier (one-third of a CD4007A) can source or sink a current of about 6 mA.

The circuit shown in Fig. 10-28 provides a voltage gain of about 160 dB and a source or sink current capacity of about 12 mA. The voltage gain results from open-loop operation. The OTA provides about 100 dB, MOS inverter/amplifier *A* provides about 30 dB, and the parallel-connected MOS inverter/amplifiers *B* and *C* provide the

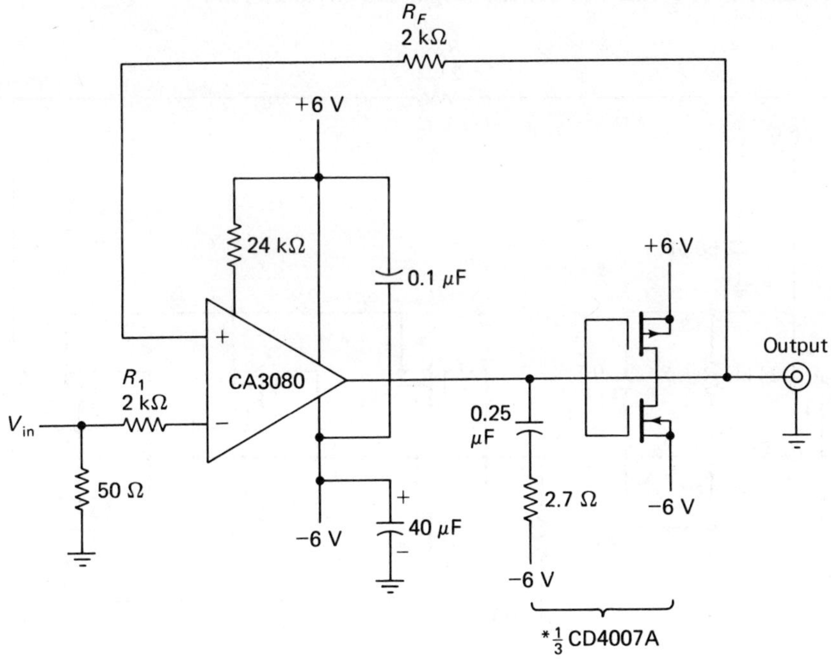

*Additional current output can be obtained when remaining two amplifiers of the CD4007A are connected in parallel with the single stage shown.

Figure 10-27. OTA combined with MOS to form unity gain amplifier with increased current capacity.

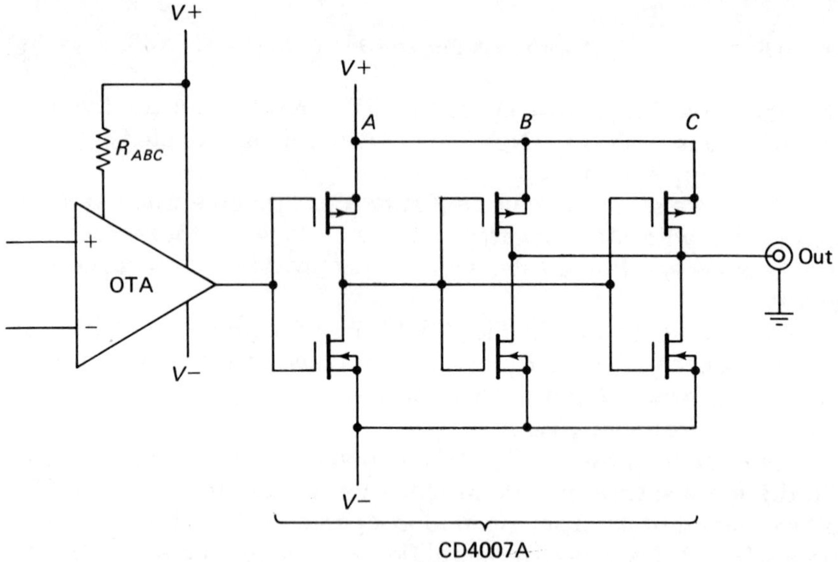

Figure 10-28. OTA combined with three MOS stages to form open-loop amplifier circuit with high current output and increased gain.

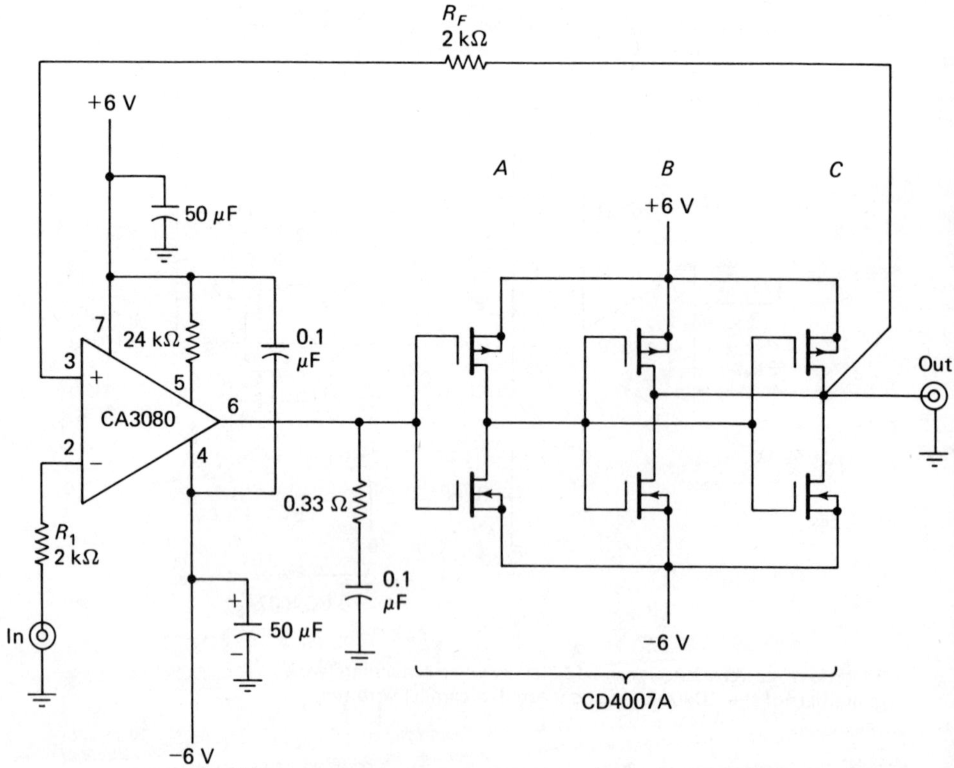

Figure 10-29. OTA combined with three MOS stages to form unity gain amplifier with increased current capacity.

remaining 30 dB. Since the outputs of B and C are connected in parallel, the normal source/sink capacity of 6 mA is doubled to 12 mA.

The circuit of Fig. 10-29 provides unity gain (with the same resistance values for R_F and R_I) but with increased current capacity. The parallel-connected B and C outputs provide approximately 12 mA of sink or source current capacity.

10-10 OTA MULTISTABLE CIRCUITS

An OTA can be set to draw very little standby current by proper adjustment of I_{ABC}. Similarly, MOS inverter/amplifiers draw very little current when used in switching applications. A MOS inverter/amplifier switch draws current when changing states but very little current when the output signal voltage swings either positive or negative. The low standby power capabilities of the OTA, when combined with the characteristics of the MOS inverter/amplifier, are ideally suited for use in precision multistable circuits (similar to the multivibrator circuits discussed in Chapter 4).

10-10-1 OTA Astable (Free-Running) Multivibrator

Figure 10-30 shows an OTA combined with a MOS inverter/amplifier to form an astable (free-running) multivibrator. As shown by the equations, the frequency of operation is set by feedback resistance R and capacitor C, as well as resistors R_1 and R_2. Since resistors R_1 and R_2 also set output impedance of the circuit, a trade-off may be necessary in selecting values of R and C. The value of R_{ABC} is set by the I_{ABC} requirements as usual (Sec. 10-2). For a multistable circuit, I_{ABC} is usually set so that the OTA draws minimum power, but enough that the MOS inverter/amplifier is properly driven to provide the desired output voltage swing. Using the values shown, the output frequency is about 7.7 kHz, with an output voltage swing equal to $V+$ and $V-$ (about 12 V).

10-10-2 OTA Monostable (One-Shot) Multivibrator

Figure 10-31 shows an OTA combined with a MOS inverter/amplifier to form a monostable (one-shot) multivibrator. Generally, monostable multivibrators are used to produce output pulses of some specific time duration (T), regardless of the trigger input pulse duration and frequency. As shown by the equations, the time duration T of output pulses is set by feedback capacitance C and resistance R, as well as resistors R_1 and R_2.

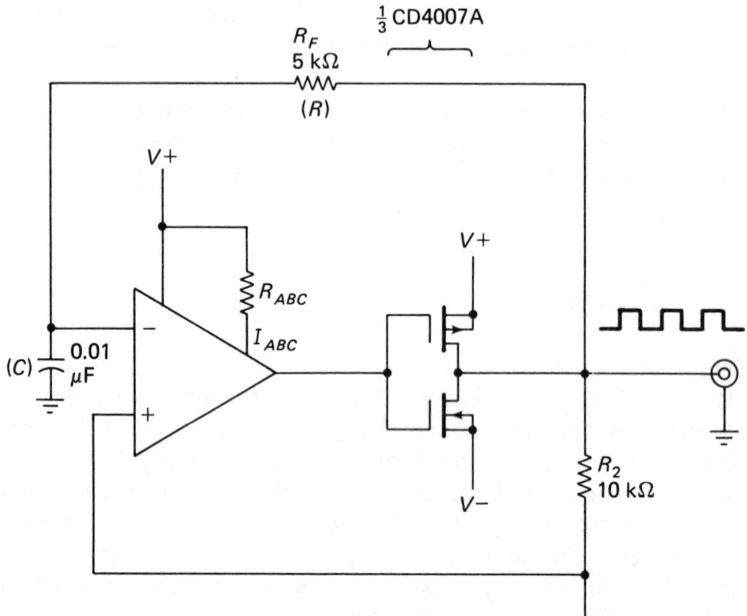

$$F \approx \frac{1}{2RC \times 1\left(\frac{2R_1}{R_2} + 1\right)}$$

Figure 10-30. OTA combined with a MOS to form an astable (free-running) multivibrator.

The relationship of $V+$, $V-$, and V_D also affects output pulse duration T. Note that V_D, the voltage across the input diode, is typically 0.5 V. Again, R_{ABC} is adjusted so that the OTA provides just enough output to drive the MOS inverter/amplifier for the desired output voltage swing. Also, it may be necessary to trade off the values of R_1, R_2, R, and C, since R_1 and R_2 set the output impedance of the circuit.

10-10-3 OTA Threshold Detector

Figure 10-32 shows an OTA combined with a MOS inverter/amplifier to form a *threshold detector.* The threshold point is set by R_1 and R_2, as well as the supply voltage. Standby power consumption is set by I_{ABC}, which, in turn, is set by R_{ABC} in the usual manner (Sec. 10-2).

The standby power consumption of the circuits shown in Figs. 10-30 through 10-32 is typically 6 mW. However, the standby power can be made to operate in the micropower region by changes in the

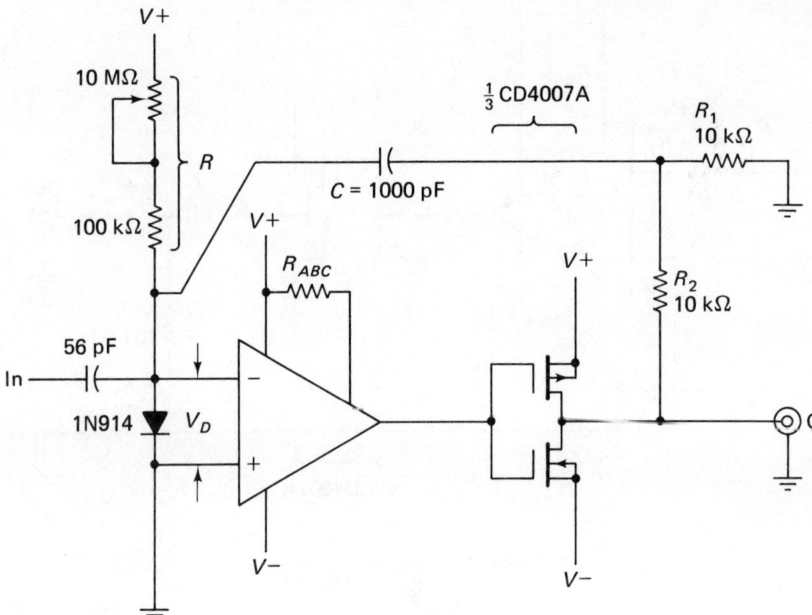

Time duration of output pulse = T

$$T = RC \times 1\left[\frac{\frac{R_1}{R_1 + R_2}(V^+ - V^-) + (V^+ - V_D)}{V+}\right]$$

Figure 10-31. OTA combined with MOS to form a monostable (one-shot) multivibrator.

value of R_{ABC}. Also, for greater current output from any of the circuits shown in Figs. 10-30 through 10-32, the remaining MOS inverter/amplifiers in the RCA CD4007A can be connected in parallel with the single stage. Each of the three elements in the CD4007A will sink or source about 6 mA. Thus, with all three elements in parallel, the circuit should be able to sink or source about 18 mA.

10-11 OTA MICROPOWER COMPARATORS

Figure 10-33 shows an OTA combined with two MOS inverter/amplifiers to form a micropower comparator. Circuit output is proportional to the differential signal at the OTA inputs. If both inputs are at the same level, there is no output. If desired, either of the inputs (inverting or noninverting) can be adjusted to some reference level by means of the voltage-divider network shown in Fig. 10-33.

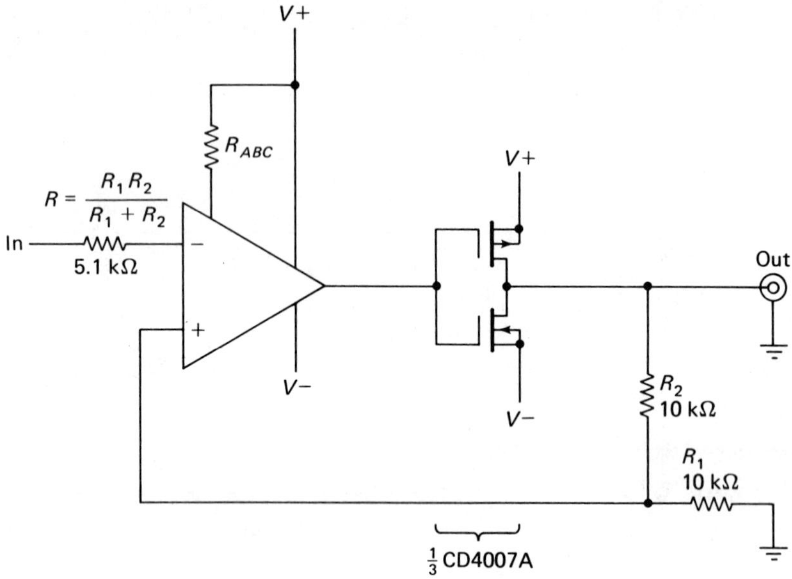

$$\pm \text{threshold} = \pm \text{supply} \left(\frac{R_1}{R_1 + R_2}\right)$$

Figure 10-32. OTA combined with MOS to form a threshold detector.

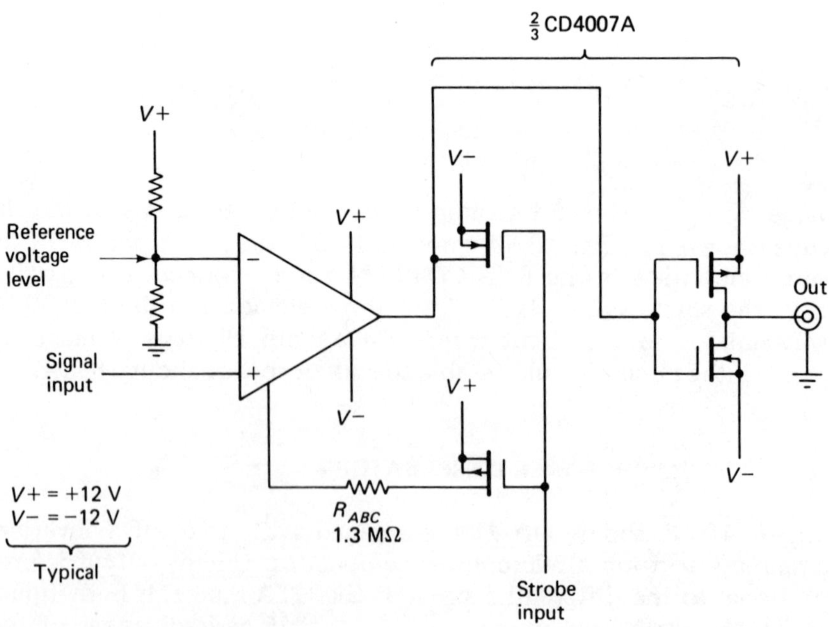

Figure 10-33. OTA combined with two MOS stages to form a micropower comparator.

Under these conditions, the output is proportional to the difference between the signal input level and the reference level.

The circuit is "on" only when a strobe or gate pulse is applied. The standby power consumption of the circuit is about 10 μW. When the circuit is strobes, the OTA consumes about 420 μW. Under these conditions, the circuit responds to a differential input voltage signal in about 6 μs. By decreasing the value of R_{ABC}, the circuit response time can be decreased to about 150 ns. However, the standby power consumption rises to about 20 μW.

The differential input common-mode range of the circuit is approximately −1 V to +10 V. The voltage gain of the circuit is about 130 dB.

Index

A

B

C

D